Optical Rheometry of Complex Fluids

TOPICS IN CHEMICAL ENGINEERING
A Series of textbooks and monographs

SERIES EDITOR

KEITH E. GUBBINS
Cornell University

ASSOCIATE EDITORS

MARK A. BARTEAU
University of Delaware

EDWARD L. CUSSLER
University of Minnesota

KLAVS F. JENSEN
Massachusetts Institute of Technology

DOUGLAS A. LAUFFENBURGER
University of Illinois

MANFRED MORARI
California Institute of Technology

W. HARMON RAY
University of Wisconsin

WILLIAM B. RUSSEL
Princeton University

SERIES TITLES

Receptors: Models for Binding, Trafficking, and Signalling
D. Lauffenburger and J. Linderman

Process Dynamics, Modeling, and Control
B. Ogunnaike and W. H. Ray

Optical Rheometry of Complex Fluids
G. Fuller

Nonlinear and Mixed Integer Optimization: Fundamentals and Applications
C. Floudas

Optical Rheometry of Complex Fluids

GERALD G. FULLER
Professor
Department of Chemical Engineering
Stanford University

New York Oxford
OXFORD UNIVERSITY PRESS
1995

Oxford University Press

Oxford New York
Athens Auckland Bangkok Bombay
Calcutta Cape Town Dar es Salaam Delhi
Florence Hong Kong Istanbul Karachi
Kuala Lumpur Madras Madrid Melbourne
Mexico City Nairobi Paris Singapore
Taipei Tokyo Toronto

and associated companies in
Berlin Ibadan

Published by Oxford University Press, Inc.,
200 Madison Avenue, New York, New York 10016

Library of Congress Cataloging-in-Publication Data
Fuller, Gerald G.
Optical Rheometry of Complex Fluids/Gerald G. Fuller.
p. cm. (Topics in chemical engineering)
Includes bibliographical references and index.
ISBN 0-19-509718-1
1. Fluid dynamic measurements. 2. Rheology. I. Title.
II. Series: Topics in chemical engineering (Oxford University Press)
TA357.5.M43F85 1995 620.1'064'0287—dc20 94-44678

9 8 7 6 5 4 3 2 1

Printed in the United States of America
on acid-free paper

This book is dedicated to my family:
Darci, Adam, Leslie and Chloe.

Preface

The use of optical methods to study the dynamics and structure of complex polymeric and colloidal liquids subject to external fields has a long history. The choice of an optical technique is normally motivated by the microstructural information it provides, its sensitivity, and dynamic range. A successful application of an optical measurement, however, will depend on many factors. First, the type of interaction of light with matter must be correctly chosen so that the desired microstructural information of a sample can be extracted. Once selected, the arrangement of optical elements required to perform the required measurement must be designed. This involves not only the selection of the elements themselves, but also their alignment. Finally, a proper interpretation of the observables will depend on one's ability to connect the measurement to the sample's microstructure.

The title of the book, "*Optical Rheometry of Complex Fluids,*" refers to the strong connection of the experimental methods that are presented to the field of rheology. Rheology refers to the study of deformation and orientation as a result of fluid flow, and one principal aim of this discipline is the development of constitutive equations that relate the macroscopic stress and velocity gradient tensors. A successful constitutive equation, however, will recognize the particular microstructure of a complex fluid, and it is here that optical methods have proven to be very important. The emphasis in this book is on the use of *in situ* measurements where the dynamics and structure are measured in the presence of an external field. In this manner, the connection between the microstructural response and macroscopic observables, such as stress and fluid motion can be effectively established. Although many of the examples used in the book involve the application of flow, the use of these techniques is appropriate whenever an external field is applied. For that reason, examples are also included for the case of electric and magnetic fields.

This book has been written for the practitioner, as well as researchers seeking to either predict the optical response of complex liquids or to interpret optical data in terms of microstructural attributes. For these purposes, the book is meant to be self contained, beginning with sections on the fundamental Maxwell field equations describing the interaction of electromagnetic waves with anisotropic media. These interactions include transmission, reflection, and scattering and are covered in the first four chapters of the book. Spectroscopic interactions, such as absorption, Raman scattering, and fluorescence are discussed in Chapter 5. Since complex fluids subject to external fields possess anisotropic optical properties, a great deal of attention is devoted to the effects of light polarization. Although the majority of the book is devoted to techniques for the measurement of dynamics and structure in the presence of external fields, methods of measurement of flow field kinematics are also presented. These are discussed in Chapter 6 and include dynamic light scattering and laser Doppler velocimetry.

The connection between the observables extracted from optical measurements, and the microstructure of polymeric and colloidal liquids is presented in Chapter 6. This is developed in terms of current models of molecular and particulate dynamics. The study of the dynamics and structure of complex liquids is interdisciplinary, involving physicists, chemists, and chemical engineers. Recognition of this wide audience is reflected in the applications that are included, where examples are drawn from each segment of the community.

The design of an optical instrument must include an analysis of the detected signal. For many problems, this can be conveniently accomplished using Mueller calculus. The methodology of this procedure is explained, and matrices are developed for most optical elements encountered in the laboratory. This treatment is offered at a sufficiently general level so that the reader is supplied with the methods to generate Jones or Mueller matrices for complex, composite optical elements. The particular optical arrangement that is chosen must meet many requirements, and these are presented in Chapter 8. Most important, sufficient information must be extracted to isolate the desired optical properties of the sample. In addition, the timescale of the measurement must be commensurate with the response time of the sample if a time-dependent field is applied. For this purpose, polarization modulation and wavelength modulation techniques are discussed. These considerations are used to guide the design of polarimeters for the measurement of birefringence and dichroism, scattering experiments for measurement of the structure factor, and spectroscopic schemes capable of extracting the dynamics of individual components in complex mixtures.

The success of an optical measurement will be controlled by the quality of the optical components that make up the instrument, and the accuracy of their alignment. For this reason, Chapter 9 is devoted to the selection of specific components to accomplish a required task. Since such a decision is influenced by the construction of an element, the underlying physics and design criteria used in the manufacture of optical components are presented. This discussion is combined with alignment protocols that the author and his students have found useful in their own laboratory.

The final chapter on applications of optical rheometric methods brings together examples of their use to solve a wide variety of physical problems. A partial list includes the use of birefringence to measure spatially resolved stress fields in non-Newtonian flows, the isolation of component dynamics in polymer/polymer blends using spectroscopic methods, the measurement of the structure factor in systems subject to field-induced phase separation, the measurement of structure in dense colloidal dispersions, and the dynamics of liquid crystals under flow.

Acknowledgments

It is a pleasure to thank the fine graduate students with whom I have had the pleasure of working at Stanford University. I am particularly grateful to Jeff Zawada, who granted permission to use Figure 10.1; Jan Van Egmond and Ellee Meyer, who provided Figure 10.4; Lynden Archer, for putting together the figures used in Case Study 3; and Kelly Huang for providing the data used in Case Study 4.

I have also benefitted from many collaborations that have influenced the content of this book. In this respect, I must acknowledge the contributions of the late Dale Pearson who not only participated in work using infrared dichroism, but who also engendered many other important collaborative arrangements. I would especially like to thank Reimund Stadler of the University of Mainz for his involvement in many joint projects, and to his students Volker Abetz and Ulf Seidel, for their contributions. In particular, Ulf Seidel kindly provided the data used in Case Study 2. I am likewise appreciative of the collaborations I have enjoyed with Jan Mewis and Paula Moldenaers of the Katholieke Universiteit Leuven, and Ralph Colby of Kodak. The data in Case Study 5 was generously provided by Hans Baaijens of the Technical University of Eindhoven working in the research group of Han Meijer and Frank Baaijens. Figure 10.6 was contributed by Tsutomu Takahashi of Nagaoka University of Technology. I would also like to extend a special thanks to Joe Starita, the founder of Rheometrics, for his encouragement and support of developments of rheo-optical methods. Finally, I would like to thank Patrick Navard for his kind hospitality during the sabbatical leave at the Ecole des Mines de Paris, Sophia Antipolis, France,where the final parts of this book were completed.

CONTENTS

Optical Rheometry of Complex Fluids

1 Propagation of Electromagnetic Waves

The experimental methods presented in this monograph concern the interaction of light with complex materials for the purpose of elucidating aspects of structure and dynamics. The interactions range from simple transmission and reflection, to scattering and nonlinear responses. Measurement of changes in the properties of the light (polarization, intensity, or frequency) is used to infer the microstructural characteristics of the sample and for this reason, the basic nature of light propagation in macroscopic media must be understood. This chapter presents the basic field equations governing the propagation of electromagnetic waves and the boundary conditions by which they are constrained.

1.1 The Maxwell Equations

Electromagnetic waves combine the propagation of two vector fields, **E** and **B**. These are the electric and magnetic induction fields, respectively, and in a vacuum are governed by the Maxwell equations 1,2,3]:

$$
\begin{aligned}
\varepsilon_0 \nabla \cdot \mathbf{E} &= \rho, \\
\nabla \times \mathbf{B} - \mu_0 \varepsilon_0 \frac{\partial \mathbf{E}}{\partial t} &= \mu_0 \boldsymbol{J}, \\
\nabla \times \mathbf{E} + \frac{\partial \mathbf{B}}{\partial t} &= 0, \\
\nabla \cdot \mathbf{B} &= 0,
\end{aligned} \tag{1.1}
$$

where ε_0 and μ_0 are the permittivity and permeability of free space. The density of charges in the system, $\rho(\mathbf{x})$, and the current density, $\boldsymbol{J}(\mathbf{x})$, obey the following equation of continuity:

$$\frac{\partial}{\partial t}\rho + \nabla \cdot \boldsymbol{J} = 0 . \tag{1.2}$$

Equation (1.2) is not independent and is obtained by combining the first two equations in (1.1).

Light propagating within a medium will interact with the constituent molecules or particles and the induction fields **E** and **B** will be altered. In principle, the equations in (1.1) can be applied to any situation as long as the charge and current densities are specified. In a condensed media, however, this approach is not practical and modified electric and magnetic fields are introduced to represent the interaction of the electric and magnetic induction fields with the material. These are the electric displacement, **D**, and the magnetic field, **H**. These are inserted into the Maxwell equations involving $\rho(\mathbf{x})$ and $\boldsymbol{J}(\mathbf{x})$ to yield

$$\begin{aligned} \nabla \cdot \mathbf{D} &= \rho, \\ \nabla \times \mathbf{H} - \frac{\partial \mathbf{D}}{\partial t} &= \boldsymbol{J}, \\ \nabla \times \mathbf{E} + \frac{\partial \mathbf{B}}{\partial t} &= 0, \\ \nabla \cdot \mathbf{B} &= 0. \end{aligned} \tag{1.3}$$

Constitutive relations connecting **D**, **H** and **J** to **E** and **B** are required to close the full set of equations. When only linear interactions are important, the following empirical equations are used:

$$\begin{aligned} D_i &= \varepsilon_{ij}E_j + \zeta_{ij}H_j, \\ B_i &= \zeta'_{ij}E_j + \mu_{ij}H_j, \\ J_i &= \sigma_{ij}E_j, \end{aligned} \tag{1.4}$$

where the Einstein notation has been used and repeated indices imply summation over dyadic components in (1.4). The material property tensors in (1.4) are

ε_{ij}: dielectric tensor,

μ_{ij}: magnetic permeability tensor,

ζ_{ij}, ζ'_{ij}: optical rotation tensors,

σ_{ij}: conductivity tensor.

Conditions can arise when the material response to the imposition of electric fields is nonlinear. Under such circumstances, more complex constitutive relationships must be employed and it is most common to expand the electric displacement vector, **D**, as a power series in the electric field according to

$$D_i = \varepsilon_{ij}^{(1)} E_j + \varepsilon_{ijk}^{(2)} E_j E_k + \varepsilon_{ijkl}^{(3)} E_j E_k E_l + \ldots, \tag{1.5}$$

$\varepsilon_{ij}^{(1)}$ is the dielectric tensor, and the higher order terms, $\varepsilon_{ijk}^{(2)}$ and $\varepsilon_{ijkl}^{(3)}$, are the nonlinear permittivities.

The dielectric tensor describes the linear response of a material to an electric field. In many experiments, and particularly in optical rheometry, anisotropy in $\varepsilon_{ij}^{(1)}$ is the object of measurement. This anisotropy is manifested as birefringence and dichroism, two quantities that will be discussed in detail in Chapter 2. The nonlinear terms are responsible for such effects as second harmonic generation, electro-optic activity, and frequency tripling. These phenomena occur when certain criteria are met in the material properties, and at high values of field strength.

1.2 Plane Waves in Nonconducting Media

The solutions to the Maxwell field equations that are most often used in the applications discussed in this book are referred to as "plane waves" of monochromatic light. These are derived from the Maxwell "curl" equations. In a system free of charges and currents, these are

$$\nabla\times\mathbf{H} = \frac{\partial\mathbf{D}}{\partial t}, \qquad \nabla\times\mathbf{E} = -\frac{\partial\mathbf{B}}{\partial t}. \tag{1.6}$$

For an isotropic medium without optical rotation, use of the constitutive relations (1.4) yields

$$\nabla\times\mathbf{H} = \varepsilon\frac{\partial}{\partial t}\mathbf{E}, \qquad \nabla\times\mathbf{E} = -\mu\frac{\partial\mathbf{H}}{\partial t}, \tag{1.7}$$

which can be combined to give

$$\varepsilon\mu\frac{\partial^2\mathbf{E}}{\partial t^2} + \nabla\times\nabla\times\mathbf{E} = 0. \tag{1.8}$$

Using the general vector identity, $\nabla\times\nabla\times\mathbf{E} = \nabla(\nabla\cdot\mathbf{E}) - \nabla^2\mathbf{E}$, and that $\nabla\cdot\mathbf{E} = 0$ in a space free of charge, the following wave equation results:

$$\frac{\partial^2\mathbf{E}}{\partial t^2} - \frac{1}{\varepsilon\mu}\nabla^2\mathbf{E} = 0, \tag{1.9}$$

with a similar equation for $\mathbf{H}$.

A plane wave is defined as a field propagating in a single direction (taken to be z in this example), and that is uniform in the perpendicular plane. In this case, there are no gradients in x and y, and equation (1.9) becomes,

$$\left(\frac{d^2}{dt^2} - v^2\frac{\partial^2}{\partial z^2}\right)\mathbf{E} = \left(\frac{\partial}{\partial t} - v\frac{\partial}{\partial z}\right)\left(\frac{\partial}{\partial t} + v\frac{\partial}{\partial z}\right)\mathbf{E} = 0, \tag{1.10}$$

where

$$v = \frac{1}{\sqrt{\varepsilon\mu}} \tag{1.11}$$

is the wave speed. This equation admits a solution of the general form

$$\mathbf{E} = \mathbf{f}(z \pm vt), \tag{1.12}$$

where $\mathbf{f}(x)$ is an arbitrary function. The choice of the sign in the argument is a matter of convention, and we will use the negative sign. We seek a periodic solution and are led to the form,

$$\mathbf{E} = \mathbf{E}_0 e^{i(kz-\omega t)}; \qquad \mathbf{H} = \mathbf{H}_0 e^{i(kz-\omega t)}; \tag{1.13}$$

where $\mathbf{E}_0$ and $\mathbf{H}_0$ are complex amplitudes. The wave number, k, and angular frequency, ω, are connected to the speed by

$$v = \frac{\omega}{k}. \tag{1.14}$$

The speed of light in vacuum is $c = 1/\sqrt{\varepsilon_0\mu_0}$. Defining the refractive index as

$$n = \sqrt{\frac{\varepsilon\mu}{\varepsilon_0\mu_0}}, \tag{1.15}$$

the speed of light in a material relative to its value in vacuum is the familiar result,

$$v = \frac{c}{n}. \tag{1.16}$$

If the material is conducting ($\sigma \neq 0$), the refractive index is

$$n^2 = \frac{\varepsilon\mu}{\varepsilon_0\mu_0}\left(1 + \frac{i\sigma}{\varepsilon\omega}\right), \tag{1.17}$$

indicating that the refractive index is *complex*, thereby inducing attenuation of the electric vector as well as a phase shift.

The wavelength of the light can also been introduced. For light of speed, c, and frequency, ω, it is

$$\lambda = \frac{2\pi c}{\omega}, \tag{1.18}$$

so that the wave number can be written as

$$k = \frac{2\pi n}{\lambda}. \tag{1.19}$$

Since the divergences of **E** and **H** are zero, and since these fields are uniform in the (x,y) plane, these are *transverse* waves, with no component in the propagation direction, z. If the propagation direction is taken to be parallel to an arbitrary unit vector, $\hat{\mathbf{k}}$, these results may be rewritten in the following, more general form:

$$\mathbf{E} = \mathbf{E}_0 e^{i(\mathbf{k}\cdot\mathbf{x} - \omega t)}; \qquad \mathbf{H} = \mathbf{H}_0 e^{i(\mathbf{k}\cdot\mathbf{x} - \omega t)}; \tag{1.20}$$

where $\mathbf{k} = \frac{2\pi}{\lambda} n\hat{\mathbf{k}}$ is the wave vector.

Examination of either of the curl equations leads to a useful relationship between **E** and **H**. Inserting the results of (1.20) into the curl equations leads to

$$\mathbf{H} = \frac{1}{\mu\omega}\mathbf{k} \times \mathbf{E}. \tag{1.21}$$

Evidently, the electric and magnetic fields have identical phase and are orthogonal to one another.

1.2.1 Plane Waves in Anisotropic Materials without Optical Rotation

The following example concerns the simplest case of light propagation through an optically anisotropic material. The problem to be solved is pictured in Figure 1.1 and shows light transmitted along the z axis through a material with a dielectric tensor of the form:

$$\varepsilon_{ij} = \begin{bmatrix} \varepsilon_{xx} & 0 & 0 \\ 0 & \varepsilon_{yy} & 0 \\ 0 & 0 & \varepsilon_{zz} \end{bmatrix}, \tag{1.22}$$

and $\zeta_{ij} = \zeta'_{ij} = 0$; $\mu_{ij} = \mu\delta_{ij}$.

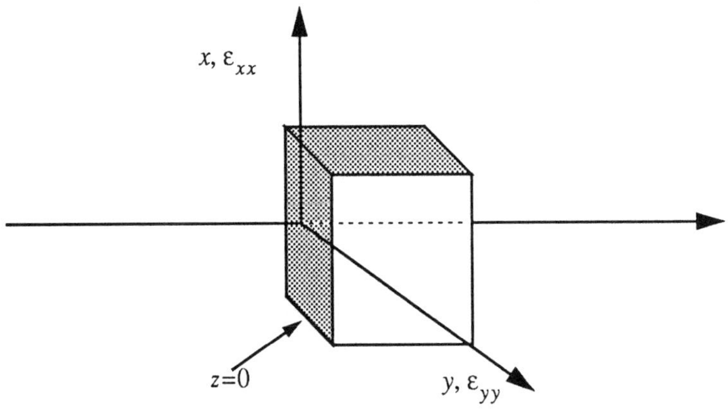

Figure 1.1 Propagation of light through a uniaxial, anisotropic material.

Since the Maxwell's equations are linear in **E** and **H**, the solution to this problem is easily solved by decomposing the fields into their x and y components. The electric field is

$$\mathbf{E} = \begin{bmatrix} E_x \\ E_y \\ 0 \end{bmatrix} = \begin{bmatrix} E_{x_0} e^{i\left(\frac{2\pi n_{xx} z}{\lambda} - \omega t\right)} \\ E_{y_0} e^{i\left(\frac{2\pi n_{yy} z}{\lambda} - \omega t\right)} \\ 0 \end{bmatrix}, \tag{1.23}$$

where the refractive indices n_{xx} and n_{yy} are

$$n_{xx} = \sqrt{\frac{\varepsilon_{xx}\mu}{\varepsilon_0\mu_0}}; \qquad n_{yy} = \sqrt{\frac{\varepsilon_{yy}\mu}{\varepsilon_0\mu_0}}. \tag{1.24}$$

As will be discussed further in Chapter 2, this is an example of a *birefringent* material, with birefringence, $\Delta n_{xy} = n_{xx} - n_{yy}$. Clearly, two other birefringences, Δn_{yz} and Δn_{xz}, can be defined for this sample and would be measured by sending the light along the x and y axes, respectively.

1.2.2 Plane Waves in Materials with Optical Rotation

In this case the material is characterized by $\varepsilon_{ij} = \varepsilon\delta_{ij}$, $\mu_{ij} = \mu\delta_{ij}$, $\zeta_{ij} = i\sqrt{\varepsilon\mu}\gamma\delta_{ij}$, and $\zeta'_{ij} = 0$. The parameter γ is the *rotary power.* This material couples the electric and magnetic fields, and the Maxwell curl equations are

$$\begin{aligned} \nabla\times\mathbf{H} - \varepsilon\frac{\partial\mathbf{E}}{\partial t} - i\sqrt{\varepsilon\mu}\gamma\frac{\partial\mathbf{H}}{\partial t} &= 0, \\ \nabla\times\mathbf{E} + \mu\frac{\partial\mathbf{H}}{\partial t} &= 0. \end{aligned} \tag{1.25}$$

These can be combined to give

$$\frac{\partial^2\mathbf{E}}{\partial t^2} - \frac{1}{\varepsilon\mu}\nabla^2\mathbf{E} - \frac{i\gamma}{\sqrt{\varepsilon\mu}}\nabla\times\frac{\partial\mathbf{E}}{\partial t} = 0. \tag{1.26}$$

The time dependence of the electric field is still $e^{-i\omega t}$, so that its spatial dependence is described by

$$\nabla^2\mathbf{E} + \gamma k\nabla\times\mathbf{E} + k^2\mathbf{E} = 0. \tag{1.27}$$

For light propagating along the z axis, this leads to the following equations for the components of the electric vector:

$$\begin{aligned} \frac{d^2E_x}{dz^2} - \gamma k\frac{dE_y}{dz} + k^2E_x &= 0, \\ \frac{d^2E_y}{dz^2} + \gamma k\frac{dE_x}{dz} + k^2E_y &= 0. \end{aligned} \tag{1.28}$$

This coupled set of equations is easily solved as

$$\begin{bmatrix} E_x \\ E_y \end{bmatrix} = e^{i\sqrt{1+\frac{\gamma^2}{4}}kz}\,\mathbf{R}\left(\frac{k\gamma}{2}z\right)\cdot\begin{bmatrix} E_x^0 \\ E_y^0 \end{bmatrix}, \tag{1.29}$$

where

$$\mathbf{R}(\theta) = \begin{bmatrix} \cos\theta & \sin\theta \\ -\sin\theta & \cos\theta \end{bmatrix} \tag{1.30}$$

is the rotation matrix and E_x^0 and E_y^0 are the components of the electric field incident on the material at the interface, $z=0$.

This result demonstrates the tendency of an optically active material to rotate the electric vector as it propagates through the sample. Materials possessing this property are normally composed of molecules having chiral symmetry. This effect leads to circular birefringence and circular dichroism, two optical properties that are frequently used in the characterization of biomaterials.

1.3 Green's Function Solutions of the Wave Equations

We seek here general solutions to Maxwell's equations and begin with equation (1.8) which resulted by combining the Maxwell curl equations,

$$(\underline{\varepsilon}\mu)\cdot\frac{\partial^2\mathbf{E}}{\partial t^2} + \nabla\times\nabla\times\mathbf{E} = 0, \tag{1.31}$$

where the dielectric properties are taken in the form of a tensor, $\underline{\varepsilon}$. Assuming the time dependence to be $e^{-i\omega t}$, equation (1.31) becomes

$$-(\underline{\varepsilon}\mu\omega^2)\cdot\mathbf{E} + \nabla\times\nabla\times\mathbf{E} = 0, \tag{1.32}$$

and since $\underline{\varepsilon}\mu\omega^2 = \left(\frac{2\pi}{\lambda}\right)^2 n^2$, we have

$$\nabla\times\nabla\times\mathbf{E} - k_0^2\mathbf{E} = k_0^2(n^2 - \mathbf{I})\cdot\mathbf{E}, \tag{1.33}$$

where $k_0 = (2\pi)/\lambda$.

Equation (1.33) has been written in a form that motivates the use of a Green's function solution. The term on the righthand side represents the action of the dielectric properties of the material and formally renders this equation inhomogeneous. The Green's function, $\mathbf{G}(\mathbf{x}, \mathbf{x}')$, is the solution to the following inhomogeneous equation,

$$\nabla\times\nabla\times\mathbf{G}(\mathbf{x}, \mathbf{x}') - k_0^2\mathbf{G}(\mathbf{x}, \mathbf{x}') = \mathbf{I}\delta(\mathbf{x}-\mathbf{x}') . \tag{1.34}$$

Once this tensor function is determined, the solution to equation (1.34) is the following convolution over the Green's function,

$$\mathbf{E}(\mathbf{x}) = \mathbf{E}_0(\mathbf{x}) + k_0^2\int d\mathbf{x}'\mathbf{G}(\mathbf{x}, \mathbf{x}') \cdot (n^2\mathbf{x}' - \mathbf{I}) \cdot \mathbf{E}(\mathbf{x}') , \tag{1.35}$$

where $\mathbf{E}_0(\mathbf{x})$ is the incident field.

A simple version of the Green's function can be found if $\nabla \cdot \mathbf{E} = 0$ everywhere in space. Under these circumstances, $\nabla \cdot \mathbf{G} = 0$, and equation (1.34) reduces to

$$(\nabla^2 + k_0^2)\mathbf{G}(\mathbf{x}, \mathbf{x}') = -\mathbf{I}\delta(\mathbf{x}-\mathbf{x}') . \tag{1.36}$$

In an unbounded medium, and with the condition that $\mathbf{G} \to 0$ as $|\mathbf{x}| \to \infty$ permits the application of Fourier transformation of the Green's function. This is defined as

$$\tilde{\mathbf{G}}(\mathbf{k}', \mathbf{x}') = \frac{1}{(\sqrt{2\pi})^3}\int d\mathbf{x} e^{i\mathbf{k}' \cdot \mathbf{x}}\mathbf{G}(\mathbf{x}, \mathbf{x}') , \tag{1.37}$$

with the inversion formula,

$$\mathbf{G}(\mathbf{x}, \mathbf{x}') = \frac{1}{(\sqrt{2\pi})^3}\int d\mathbf{k} e^{-i\mathbf{k}' \cdot \mathbf{x}}\tilde{\mathbf{G}}(\mathbf{x}, \mathbf{x}') . \tag{1.38}$$

Applying this transformation to equation (1.36) gives the following algebraic equation,

$$(-k'^2 + k_0^2)\tilde{\mathbf{G}} = -\frac{\mathbf{I}}{(\sqrt{2\pi})^3}e^{i\mathbf{k}' \cdot \mathbf{x}'} , \tag{1.39}$$

which is easily solved for $\tilde{\mathbf{G}}$. Using the inversion formula leads to the following result for the Green's function,

$$\mathbf{G}(\mathbf{x}, \mathbf{x}') = \frac{\mathbf{I}}{(2\pi)^3}\int d\mathbf{k}'\frac{e^{-i\mathbf{k}' \cdot (\mathbf{x}-\mathbf{x}')}}{k'^2 - k_0^2} . \tag{1.40}$$

This integral is evaluated by first converting to spherical coordinates so that $d\mathbf{k}' = d\phi d\theta \sin\theta dk' k'^2$ and choosing the z axis in this system so that $\mathbf{k}' \cdot (\mathbf{x}-\mathbf{x}') = k'|\mathbf{x}-\mathbf{x}'|\cos\theta$. The two angular integrals are easily carried out, leading to

$$\mathbf{G}(\mathbf{x}, \mathbf{x}') = \frac{\mathbf{I}}{8\pi^2 i|\mathbf{x}-\mathbf{x}'|} \int_{-\infty}^{\infty} dk'k' \frac{(e^{ik|\mathbf{x}-\mathbf{x}|} - e^{-ik|\mathbf{x}-\mathbf{x}|})}{k'^2 - k_0^2} . \tag{1.41}$$

To calculate the remaining integral, the poles of the integrand at $\pm k_0$ are moved into the complex plane by an infinitesimal amount, $-i\eta$, into the *lower* complex plane [4]. This choice is motivated by requiring the proper causal response for the Green's function which demands that it be analytic in the *upper* complex plane.† Equation (1.41) is rewritten as

$$\mathbf{G}(\mathbf{x}, \mathbf{x}') = \frac{\mathbf{I}}{8\pi^2 i|\mathbf{x}-\mathbf{x}'|} \lim_{\eta \to 0^+} \int_{-\infty}^{\infty} dk'k' \frac{(e^{ik'|\mathbf{x}-\mathbf{x}'|} - e^{-ik'|\mathbf{x}-\mathbf{x}'|})}{k'^2 - (k_0 + i\eta)^2} , \tag{1.42}$$

which can be computed using contour integration and the method of residues to yield,

$$\mathbf{G}(\mathbf{x}, \mathbf{x}') = \frac{e^{ik_0|\mathbf{x}-\mathbf{x}'|}}{4\pi|\mathbf{x}-\mathbf{x}'|}\mathbf{I} . \tag{1.43}$$

In general, it is not possible to set $\nabla \cdot \mathbf{G} = 0$ and equation (1.34) must be used. Applying the Fourier transformation to this equation leads to

$$\mathbf{k}' \times (\mathbf{k}' \times \tilde{\mathbf{G}}) + k_0^2 \tilde{\mathbf{G}} = -\frac{\mathbf{I}}{(\sqrt{2\pi})^3} e^{i\mathbf{k}' \cdot \mathbf{x}'} , \tag{1.44}$$

or

$$(k'^2 \mathbf{I} - \mathbf{k}'\mathbf{k}' + k_0^2 \mathbf{I}) \cdot \tilde{\mathbf{G}} = -\frac{\mathbf{I}}{(\sqrt{2\pi})^3} e^{i\mathbf{k}' \cdot \mathbf{x}'} . \tag{1.45}$$

Inverting the tensor multiplying $\tilde{\mathbf{G}}$ gives

$$\tilde{\mathbf{G}}(\mathbf{k}', \mathbf{x}') = \frac{1}{(\sqrt{2\pi})^3} \frac{k_0^2 \mathbf{I} - \mathbf{k}'\mathbf{k}'}{k_0^2 (k'^2 - k_0^2)} e^{i\mathbf{k}' \cdot \mathbf{x}'} . \tag{1.46}$$

Using the inversion formula in equation (1.38) the Green's function is

---

†The electric field is a function of time which must obey the causality condition. This requirement simply means that a function *F(t)* is only nonzero *after* the appearance of a disturbance. Therefore, such a function is zero whenever *t<0*. The Green's function discussed here is the temporal Fourier transform of the time dependent function and depends upon frequency, ω. The causality principle for the Fourier transform, $\tilde{F}(\omega)$, of $F(t)$ states the it must be an analytic function of ω throughout the *upper* complex halfplane. For a detailed discussion of this principle, see reference 4.

$$\mathbf{G}(\mathbf{x}, \mathbf{x}') = \frac{1}{(2\pi)^3} \lim_{\eta \to 0^+} \int d\mathbf{k}' \frac{k_0^2 \mathbf{I} - \mathbf{k}'\mathbf{k}'}{k_0^2 (k'^2 - (k_0 + i\eta)^2)} e^{-i\mathbf{k}' \cdot (\mathbf{x} - \mathbf{x}')} . \tag{1.47}$$

This result can be further manipulated in the following way:

$$\mathbf{G}(\mathbf{x}, \mathbf{x}') = \left(\mathbf{I} + \frac{1}{k_0^2} \nabla\nabla \right) \frac{1}{(2\pi)^3} \lim_{\eta \to 0^+} \int d\mathbf{k}' \frac{e^{-i\mathbf{k}' \cdot (\mathbf{x} - \mathbf{x}')}}{(k'^2 - (k_0 + i\eta)^2)} , \tag{1.48}$$

which, upon recognizing that the integral in equation (1.48) is the Green's function expressed in (1.43), becomes

$$\mathbf{G}(\mathbf{x}, \mathbf{x}') = (\mathbf{I} + k_0^{-2} \nabla\nabla) \frac{e^{ik_0|\mathbf{x} - \mathbf{x}'|}}{4\pi|\mathbf{x} - \mathbf{x}'|} . \tag{1.49}$$

The Green's function solutions in (1.43), (1.48), and (1.49) will form the starting point for the solution of scattering problems in Chapter 4.

1.4 Polarization: The Jones and Stokes Vectors

1.4.1 The Jones Vector

As demonstrated by the example of section 1.2.1, interaction of the electric field with anisotropic materials can cause its orthogonal components to have dissimilar phases and amplitudes. These properties of the electric vector describe the state of polarization of the electric vector. Since the electric vector lies in the plane perpendicular to the axis of propagation, a convenient description for this purpose is the two-component vector:

$$\mathbf{A} = \begin{bmatrix} \tilde{A}_x e^{i\delta_x} \\ \tilde{A}_y e^{i\delta_y} \end{bmatrix} e^{i\frac{2\pi}{\lambda}nz - i\omega t} . \tag{1.50}$$

This describes the x and y components of the electric vector of light propagating along the z axis and through an isotropic material of refractive index n. Evidently, the light has had a prior interaction with an anisotropic material and these two components have differing amplitudes and phases. For example, in section 1.2.1, a sample with a uniaxial dielectric tensor was observed to induce a phase difference, $\delta_x - \delta_y = (2\pi/\lambda)(n_1 - n_2)d$, where d is the sample thickness. Equation (1.50), with the term $\exp[i(2\pi nz/\lambda - \omega t)]$ suppressed, is the *Jones vector* representation of the electric vector[5,6]. In this simple description of the light, it is assumed that only a single wavelength is present and the light is perfectly coherent. It is also classified as perfectly polarized. Partial polarization and quasi-chromatic light with a finite distribution of wavelengths are discussed in the next section along with the introduction of the second representation of the electric vector, the *Stokes vector.*

As the electric vector propagates, it will rotate in space and time according to the difference in the phases δ_x and δ_y. To illustrate the form of the electric vector, it is convenient to map it onto the (x, y) plane using time as a parameter. Taking the real part of the vector $\mathbf{A}$, we have

$$\mathbf{A} = \begin{bmatrix} A_x \\ A_y \end{bmatrix} = \begin{bmatrix} \tilde{A}_x \cos\left(\delta_x + \frac{2\pi n z}{\lambda} - \omega t\right) \\ \tilde{A}_y \cos\left(\delta_y + \frac{2\pi n z}{\lambda} - \omega t\right) \end{bmatrix}. \tag{1.51}$$

As an example, consider the case where each component has the same amplitude $(\hat{A}_x = \tilde{A}_y = A)$, but with a phase difference, $\delta_x - \delta_y = \pi/2$. Then the components of $\mathbf{A}$ are

$$A_x = \tilde{A}\cos\left(\delta_x + \frac{2\pi n z}{\lambda} - \omega t\right),$$
$$A_y = \tilde{A}\cos\left(\delta_x - \frac{\pi}{2} + \frac{2\pi n z}{\lambda} - \omega t\right) = \tilde{A}\sin\left(\delta_x + \frac{2\pi n z}{\lambda} - \omega t\right), \tag{1.52}$$

and these equations map out a circle of radius $\tilde{A}$. As time increases, the electric vector precesses clockwise and this defines right-circularly polarized light. If the phase difference was of opposite sign, $\delta_x - \delta_y = -\pi/2$, the electric vector would rotate counter-clockwise and produce left-circularly polarized light. It is important to note, however, that these definitions are linked to the convention chosen for the time dependent term, $e^{\pm i\omega t}$. If this was chosen to be $e^{i\omega t}$, the handedness of the polarization would be reversed. These two mappings for circularly polarized light are shown in Figure 1.2.

If the x and y components have the same phase $(\delta_x - \delta_y = 0)$, the light will be linearly polarized (see Fig. 1.2). Its orientation will be along the angle $\theta = \operatorname{atan}(\tilde{A}_y/\tilde{A}_x)$ and its magnitude will be $\sqrt{\tilde{A}_x^2 + \tilde{A}_y^2}$. More general values of the ratio $\tilde{A}_x/\tilde{A}_y$ and finite differences in phase produce elliptically polarized light.

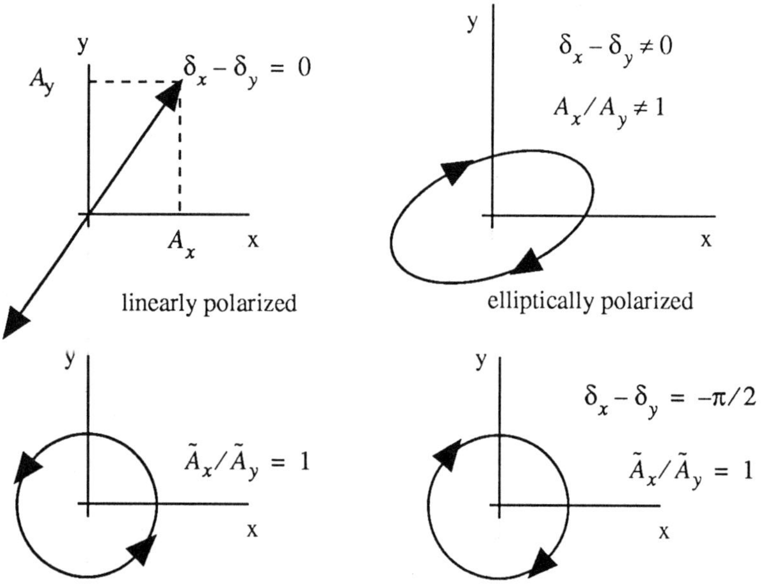

Figure 1.2 Parametric plots of various classes of polarized light. The light is propagating out of the plane of the paper.

The Jones vector is not directly observable. Most experiments use square law detectors that measure the intensity [1],

$$I = \mathbf{A} \cdot \mathbf{A}^* = \tilde{A}_x^2 + \tilde{A}_y^2 . \tag{1.53}$$

For this reason, the Stokes vector discussed in section 1.4.3 is often more convenient since each of its elements is an observable quantity.

1.4.2 Linear and Circular Polarization Basis Sets

It is convenient to describe the polarization vector in terms of an orthogonal basis set of unit vectors, $(\mathbf{a}_1, \mathbf{a}_2)$, such that $\mathbf{a}_1{}^* \cdot \mathbf{a}_2 = 0$. Two choices are commonly used, one for linearly polarized light[1]:

$$\mathbf{a}_x = \begin{bmatrix} 1 \\ 0 \end{bmatrix}; \ \mathbf{a}_y = \begin{bmatrix} 0 \\ 1 \end{bmatrix}, \tag{1.54}$$

and one for circularly polarized light:

$$\mathbf{a}_l = \frac{1}{\sqrt{2}} \begin{bmatrix} 1 \\ i \end{bmatrix}; \ \mathbf{a}_r = \frac{1}{\sqrt{2}} \begin{bmatrix} 1 \\ -i \end{bmatrix}. \tag{1.55}$$

The subscripts l and r refer to left- and right-circularly polarized light, respectively. Any electric field, **A**, can be expressed in terms of either of these two sets. The Jones vectors for the two representations are related by the transformation:

$$\mathbf{A}_{circular} = \mathbf{F} \cdot \mathbf{A}_{linear}, \tag{1.56}$$

where

$$\mathbf{F} = \frac{1}{\sqrt{2}}\begin{bmatrix} 1 & 1 \\ -i & i \end{bmatrix}, \tag{1.57}$$

and has the properties that

$$\mathbf{F}^{-1} = (\mathbf{F}^*)^T = \frac{1}{\sqrt{2}}\begin{bmatrix} 1 & i \\ 1 & -i \end{bmatrix}. \tag{1.58}$$

1.4.3 The Stokes Vector

In addition to the intensity of light, defined in equation (1.53), it is desirable to formulate observable quantities that characterize both the amplitude and phase of the electric vector projected onto orthogonal directions. This is accomplished using the Stokes vector, **S**, with components [1,5]:

$$\mathbf{S} = \begin{bmatrix} S_0 \\ S_1 \\ S_2 \\ S_3 \end{bmatrix} = \begin{bmatrix} |A_x|^2 + |A_y|^2 \\ |A_x|^2 - |A_y|^2 \\ 2Re\,(A_x{}^* A_y) \\ 2Im\,(A_x{}^* A_y) \end{bmatrix} = \begin{bmatrix} \tilde{A}_x^2 + \tilde{A}_y^2 \\ \tilde{A}_x^2 - \tilde{A}_y^2 \\ 2\tilde{A}_x\tilde{A}_y\cos(\delta_x - \delta_y) \\ 2\tilde{A}_x\tilde{A}_y\sin(\delta_x - \delta_y) \end{bmatrix}. \tag{1.59}$$

Although the Stokes vector, with its greater number of components, appears to be a more cumbersome representation of the electric vector, it is often more convenient to use than the Jones vector. This is because its components are observable quantities. For monochromatic, perfectly polarized light, the four components of the Stokes vector are not linearly independent, but related according to

$$S_0^2 = S_1^2 + S_2^2 + S_3^2. \tag{1.60}$$

Linearly polarized light is recognized as having a Stokes vector of the form $\mathbf{S}^T = (S_0, S_1, S_2, 0)$. Circularly polarized light appears as $\mathbf{S}^T = (S_0, 0, 0, S_3)$ with $S_3 < 0$ for left-circularly polarized light and $S_3 > 0$ for right-circularly polarized light.

Equation (1.60) holds only for the ideal case of monochromatic, perfectly polarized light. In practice, however, there will exist some degree of dispersion in wavelength and the light will be a superposition of waves of different wavelength. The frequency, ω, of the light will be distributed over a finite range of values leading to a complex time dependence of the total electric field. The electric field will generally be quasi-chromatic, and formally represented with the following time dependence [5]:

$$\mathbf{A} = \begin{bmatrix} \tilde{A}_x(t)\, e^{i\delta_x(t)} \\ \tilde{A}_y(t)\, e^{i\delta_y(t)} \end{bmatrix}, \tag{1.61}$$

where the amplitudes, $\tilde{A}_i(t)$, and phases, $\delta_i(t)$, vary slowly in time.

Observable quantities are the result of averaging over a duration, T, long enough to yield time-independent measurements. The average of a quantity $u(t)$ is then:

$$\langle u \rangle = \frac{1}{T}\int_0^T u(t)\, dt\,. \tag{1.62}$$

The intensity, for example, will be

$$I = \langle S_0 \rangle = \langle |A_x(t)|^2 \rangle + \langle |A_y(t)|^2 \rangle\,. \tag{1.63}$$

The consequence of dispersion in wavelength is that the polarization properties of the electric vector will fluctuate randomly in time. The parametric mapping of the electric vector shown in Figure 1.2 will produce blurred contours and the light will be partially polarized. If the light shows no preference towards a particular polarization state, it is referred to as unpolarized, or natural light. The Stokes vector for this case is

$$\mathbf{S}^T = (S_0, 0, 0, 0) \tag{1.64}$$

and would represent "pure sunlight." Because this light is a superposition of electric fields of random amplitude and phase, $|A_x^2| = |A_y^2|$ and $\langle \cos\delta \rangle = \langle \sin\delta \rangle = 0$.

The Stokes vector describing quasi-chromatic light has elements that are time averaged quantities and the equality (1.60) must be replaced by

$$S_0^2 \geq S_1^2 + S_2^2 + S_3^2\,. \tag{1.65}$$

Quantitative treatments of partially polarized light can be found in the texts by Born and Wolf [2], and Azzam and Bashara [5]. In this monograph, the light will be assumed to be perfectly polarized. It should be noted, however, that in many experimental situations depolarization can readily occur and care must be taken to either account for it, or to minimize this possibility. The most common source of depolarization in optical rheometry is multiple scattering by such systems as dense suspensions and liquid crystals.

1.5 Boundary Conditions

The Maxwell equations are subject to boundary conditions at surfaces defining a system of interest. Figure 1.3 is one such interface, separating two regions, 1 and 2. These boundary conditions are developed by performing volume integrals (the volume V in Fig. 1.3) over the first and the last of the Maxwell equations in (1.3) and surface integrals (the surface S in Fig. 1.3) over the second and third equations in (1.3). Applying the divergence theorem and Stokes theorem to the volume and surface integrals, respectively, the boundary conditions are established [1]. To demonstrate this procedure, consider applying this method to the first equation in (1.3). Taking the volume integral of both sides, we have,

$$\int_V d\mathbf{x} \nabla \cdot \mathbf{D} = \int_V d\mathbf{x} \rho(\mathbf{x}) . \tag{1.66}$$

Applying the divergence theorem yields

$$\int_A dA \mathbf{n} \cdot \mathbf{D} = \int_V d\mathbf{x} \rho(\mathbf{x}) , \tag{1.67}$$

where A is the area of the ends of the cylindrical volume V in Figure 1.3. The vector $\mathbf{n}$ is the surface normal and points towards region 2. In the limit that this volume is very small, this equation becomes

$$A\mathbf{n} \cdot (\mathbf{D}_2 - \mathbf{D}_1) = Al\rho(\mathbf{x}) , \tag{1.68}$$

where $\mathbf{D}_i$ is the displacement vector at the interface on the side i, and the density $\rho(\mathbf{x})$ is evaluated at the surface. In other words, the contributions to the integral at the edges of the volume can be neglected. Defining the surface charge density, $\sigma(\mathbf{x}) = \lim_{l \to 0} l\rho(\mathbf{x})$, we have the first equation in the list of boundary conditions given below:

$$\begin{aligned}
(\mathbf{D}_2 - \mathbf{D}_1) \cdot \mathbf{n} &= \sigma(\mathbf{x}) , \\
\mathbf{n} \times (\mathbf{E}_2 - \mathbf{E}_1) &= 0, \\
\mathbf{n} \times (\mathbf{H}_2 - \mathbf{H}_1) &= \mathbf{K}(\mathbf{x}) , \\
(\mathbf{B}_2 - \mathbf{B}_1) \cdot \mathbf{n} &= 0.
\end{aligned} \tag{1.69}$$

The remaining three conditions are found using a similar procedure to the other Maxwell equations. $\mathbf{K}(\mathbf{x})$ is the surface current density.

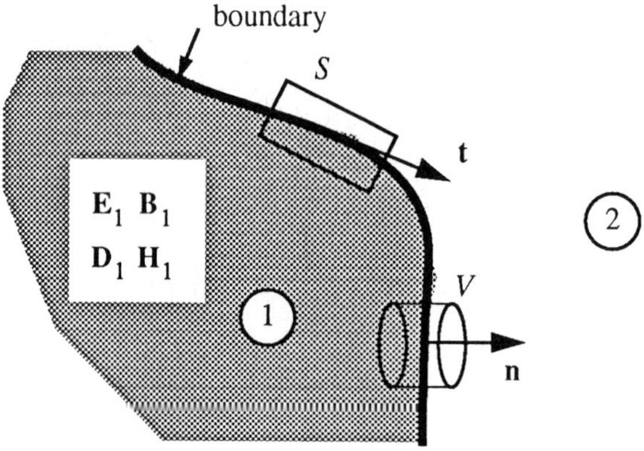

Figure 1.3 Diagram of the boundary separating regions 1 and 2. **n** is the field of surface normal vectors, and **t** is the field of tangent vectors.

1.6 Reflection and Refraction of Plane Waves

A basic interaction of electromagnetic waves with material concerns the reflection and refraction from a single, plane interface separating two dielectric media. This event is pictured in Figure 1.4.

In Figure 1.4, an incident electric field, $\mathbf{E}_1^f$, entering from medium 1, is reflected as $\mathbf{E}_1^r$ into medium 1, and refracted as $\mathbf{E}_2^f$ into medium 2. The superscripts f and r refer to light propagating either forward or backward along the z direction, respectively. The polarization of the light is either parallel to the plane of incidence (denoted as p polarization) or perpendicular to this plane (denoted as s polarization). Angles of refraction and reflection are measured from the unit normal, **n**, defining the orientation of the interface. The two media are isotropic and characterized by dielectric constants and permeabilities ε_i and μ_i, and refractive indices n_i, where $i = 1, 2$.

From equations (1.20) and (1.21), the electric and magnetic fields present in the system are

$$
\begin{aligned}
\mathbf{E}_1^f &= \mathbf{A}_1^f e^{i(\mathbf{k}_1^f \cdot \mathbf{x} - \omega t)}, \quad \mathbf{B}_1^f = \frac{1}{\omega}\mathbf{k}_1^f \times \mathbf{E}_1^f; \\
\mathbf{E}_1^r &= \mathbf{A}_1^r e^{i(\mathbf{k}_1^r \cdot \mathbf{x} - \omega t)}, \quad \mathbf{B}_1^r = \frac{1}{\omega}\mathbf{k}_1^r \times \mathbf{E}_1^r; \\
\mathbf{E}_2^f &= \mathbf{A}_2^f e^{i(\mathbf{k}_2^f \cdot \mathbf{x} - \omega t)}, \quad \mathbf{B}_2^f = \frac{1}{\omega}\mathbf{k}_2^f \times \mathbf{E}_2^f;
\end{aligned}
\tag{1.70}
$$

where $k_1^f = k_1^r = 2\pi n_1/\lambda$ and $k_2^f = 2\pi n_2/\lambda$.

The boundary conditions are applied at the interface defined by the condition z=0. On this plane, the boundary conditions must apply for all times and for all values of the independent variables x and y. Since the boundary conditions are linear and since all the fields have the same frequency, ω, the temporal form of each wave must be $e^{-i\omega t}$. To satisfy the spatial constraints at all positions on the surface, the phases of all the fields must be equal when projected onto the (x,y) plane. This requirement leads to [1]:

$$(\mathbf{k}_1^f \cdot \mathbf{x})_{z=0} = (\mathbf{k}_1^r \cdot \mathbf{x})_{z=0} = (\mathbf{k}_2^f \cdot \mathbf{x})_{z=0}. \tag{1.71}$$

Since $\mathbf{k}_1^f$ resides in the (x,z) plane, so must $\mathbf{k}_1^r$ and $\mathbf{k}_2^f$. Equating the x components, we have Snell's law:

$$n_1 \sin\phi_1^f = n_1 \sin\phi_1^r = n_2 \sin\phi_2^f, \tag{1.72}$$

so that $\phi_1^f = \phi_1^r = \phi_1$. The boundary conditions (1.69) for a system free of charges and currents become

$$\begin{aligned}
&[\varepsilon_2 \mathbf{E}_2^f - \varepsilon_1 (\mathbf{E}_1^f + \mathbf{E}_1^r)] \cdot \mathbf{n} = 0, \\
&[\mathbf{k}_2^f \times \mathbf{E}_2^f - (\mathbf{k}_1^f \times \mathbf{E}_1^f + \mathbf{k}_1^r \times \mathbf{E}_1^r)] \cdot \mathbf{n} = 0, \\
&[\mathbf{E}_2^f - (\mathbf{E}_1^f + \mathbf{E}_1^r)] \times \mathbf{n} = 0, \\
&\left[\frac{1}{\mu_2}\mathbf{k}_2^f \times \mathbf{E}_2^f - \frac{1}{\mu_1}(\mathbf{k}_1^f \times \mathbf{E}_1^f + \mathbf{k}_1^r \times \mathbf{E}_1^r)\right] \times \mathbf{n} = 0.
\end{aligned} \tag{1.73}$$

where the constitutive relations $\mathbf{D}_i = \varepsilon \mathbf{E}_i$ and $\mathbf{B}_i = \mu \mathbf{H}_i$ have been used.

The equations in (1.73) can be used to solve for the ratios E_{1i}^r / E_{1i}^f and E_{2i}^f / E_{1i}^f, where the index $i = (p, s)$ identifies the polarization relative to the plane of incidence. The results are [1]:

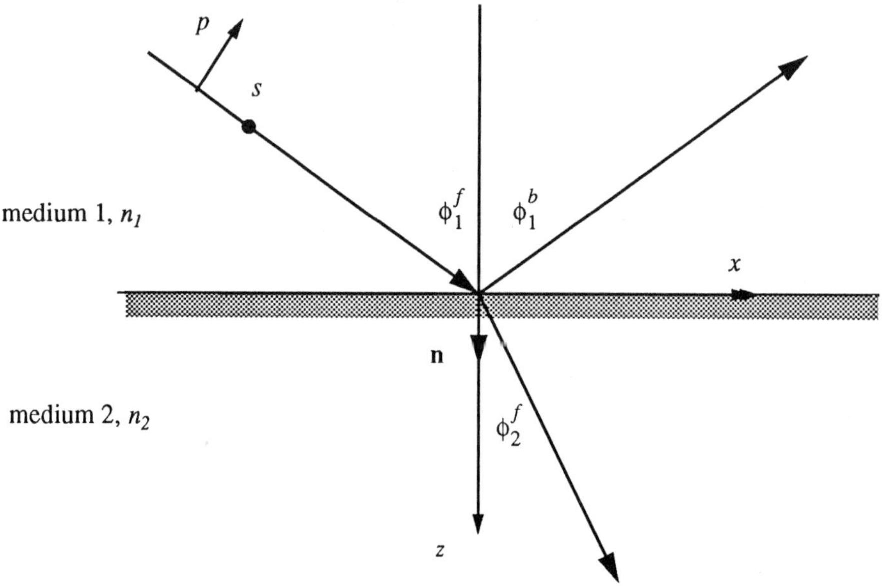

Figure 1.4 Reflection and refraction by a plane interface.

<u>s polarization</u>

$$T_{ss} = \frac{E_{2s}^f}{E_{1s}^f} = \frac{2n_1\cos\phi_1}{n_1\cos\phi_1 + \frac{\mu_1}{\mu_2}\sqrt{n_2^2 - n_1^2\sin^2\phi_1}}, \tag{1.74}$$

$$R_{ss} = \frac{E_{1s}^r}{E_{1s}^f} = \frac{n_1\cos\phi_1 - \frac{\mu_1}{\mu_2}\sqrt{n_2^2 - n_1^2\sin^2\phi_1}}{n_1\cos\phi_1 + \frac{\mu_1}{\mu_2}\sqrt{n_2^2 - n_1^2\sin^2\phi_1}}; \tag{1.75}$$

<u>p polarization</u>

$$T_{pp} = \frac{E_{2p}^f}{E_{1p}^f} = \frac{2n_1 n_2\cos\phi_1}{\frac{\mu_1}{\mu_2}n_2^2\cos\phi_1 + n_1\sqrt{n_2^2 - n_1^2\sin^2\phi_1}}, \tag{1.76}$$

$$R_{pp} = \frac{E_{1p}^r}{E_{1p}^f} = \frac{\frac{\mu_1}{\mu_2}n_2^2\cos\phi_1 - n_1\sqrt{n_2^2 - n_1^2\sin^2\phi_1}}{\frac{\mu_1}{\mu_2}n_2^2\cos\phi_1 + n_1\sqrt{n_2^2 - n_1^2\sin^2\phi_1}}. \tag{1.77}$$

Equations (1.74) to (1.77) are the Fresnel formulae. Examination of equation (1.77) reveals that light with p polarization will not be reflected from a plane interface ($R_{pp} = 0$) at a specific angle of incidence, ϕ_B. This is the *Brewster angle* and, for the case when $\mu_1 = \mu_2$, is

$$\phi_1 = \phi_B = \operatorname{atan}\left(\frac{n_2}{n_1}\right). \tag{1.78}$$

For typical organic materials and most glasses in contact with air, $n_2/n_1 \approx 1.5$ and $\phi_B \approx 56°$. At this angle, $E^r_{1p} = 0$, so that light of arbitrary polarization incident on a surface will be reflected as linearly polarized along the s direction. This phenomena is often used to produced linearly polarized light and Brewster windows are commonly found in laser cavities. This effect can also be used effectively to define the orientation of polarization relative to a reference plane in the laboratory. This latter application will be discussed in Chapter 9.

A second important condition concerns the total internal reflection of light. This occurs when the incident and reflected light reside within a medium of higher refractive index relative to the medium supporting the refracted light ($n_1 > n_2$) and

$$\phi_1 \geq \phi_{TIR} = \operatorname{asin}\left(\frac{n_2}{n_1}\right). \tag{1.79}$$

The plane waves propagating inside medium 2 will have the following spatial dependence:

$$e^{i(\mathbf{k}_2 \cdot \mathbf{x})} = e^{ik_2(x \sin\phi_2 + z\cos\phi_2)}. \tag{1.80}$$

Using Snell's law, the cosine and sin of ϕ_2 are

$$\sin\phi_2 = \frac{\sin\phi_1}{\sin\phi_{TIR}};\quad \cos\phi_2 = i\sqrt{\left(\frac{\sin\phi_1}{\sin\phi_{TIR}}\right)^2 - 1}. \tag{1.81}$$

If equation (1.79) is satisfied, $\sin\phi_2 > 1$ and $\cos\phi_2$ is purely imaginary. When these conditions apply, the plane waves inside medium 2 decay exponentially in the z direction according to:

$$e^{-k_2\sqrt{\left[\left(\frac{\sin\phi_1}{\sin\phi_{TIR}}\right)^2 - 1\right]}z}\, e^{ik_2\left(\frac{\sin\phi_1}{\sin\phi_{TIR}}\right)x}. \tag{1.82}$$

The attenuated, refracted wave, is referred to as an *evanescent* wave and decays with the characteristic length scale:

$$\Lambda = \frac{\lambda}{2\pi n_2\sqrt{\left(\frac{\sin\phi_1}{\sin\phi_{TIR}}\right)^2 - 1}}, \tag{1.83}$$

that can be adjusted by varying the incident angle.

The use of evanescent waves is very valuable to the study of interfacial properties. Techniques such as total internal reflection fluorescence (TIRF) and attenuated transmitted reflectance (ATR) use the energy of evanescent waves to probe thin regions in the vicinity of an interface to determine surface concentrations of interfacial species.

2 Transmission by Anisotropic Media: The Jones and Mueller Calculus

The majority of polarizing elements used to design optical rheometers, and most sample geometries in optical rheometry, involve transmission of light through anisotropic materials. This interaction will generally induce a transformation in the polarization. Throughout this monograph, only linear transformations will be considered, and, in this case, matrix operations conveniently describe the behavior of optical elements. Two calculus procedures have been developed for this purpose for both the Jones and Stokes vectors and are described in this chapter. Using these techniques, the analysis of optical trains can proceed in a systematic, straightforward manner.

2.1 The Jones and Mueller Matrices

The polarization properties of light can be represented by the Jones or Stokes vectors, **A** or **S**, respectively. The latter prescription has the advantages of describing partial polarization and contains directly observable quantities. When light is transmitted through a polarizing element with an incident electric vector $\mathbf{A}_0$ or $\mathbf{S}_0$, the light will exit with polarization $\mathbf{A}_1$ or $\mathbf{S}_1$. This interaction is shown in Figure 2.1. This linear interaction is described by

$$\mathbf{A}_1 = \mathbf{J} \cdot \mathbf{A}_0, \tag{2.1}$$

and

$$\mathbf{S}_1 = \mathbf{M} \cdot \mathbf{S}_0, \tag{2.2}$$

where **J** and **M** are the Jones and Mueller matrices, respectively. It is important to recognize that these are simple matrices and not tensors.

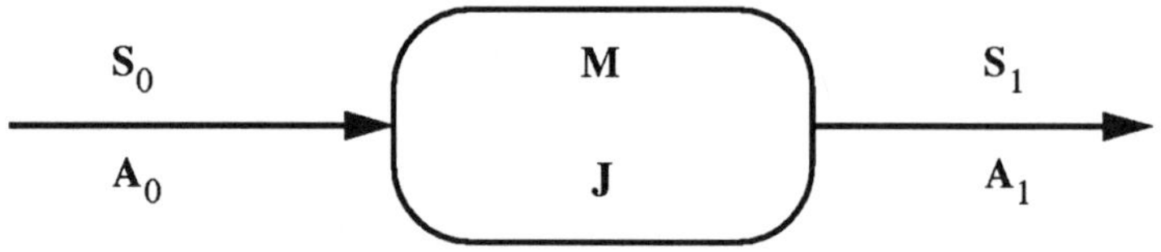

Figure 2.1 Linear transformation of the electric vector.

The connection between the Stokes and Jones vectors, given by equation (1.59) can be used to relate the sixteen-component Mueller matrix to the four-component Jones matrix. Combining equations (2.1), (2.2), and (1.59), we have, using a notation similar to that developed in Azzam and Bashara [5],

$$\mathbf{M} = \begin{bmatrix} \frac{(m_1+m_2+m_3+m_4)}{2} & \frac{(m_1-m_2-m_3+m_4)}{2} & (s_{13}+s_{42}) & -(d_{13}+d_{42}) \\ \frac{(m_1-m_2+m_3-m_4)}{2} & \frac{(m_1+m_2-m_3-m_4)}{2} & (s_{13}-s_{42}) & (d_{42}-d_{13}) \\ (s_{14}+s_{32}) & (s_{14}-s_{32}) & (s_{12}+s_{34}) & (d_{34}-d_{12}) \\ (d_{14}+d_{32}) & (d_{14}-d_{32}) & (d_{12}-d_{34}) & (s_{12}-s_{34}) \end{bmatrix}, \tag{2.3}$$

where

$$\begin{aligned} m_i &= j_i j_i^*, \\ s_{ik} &= Re\,(j_i j_k^*), \\ d_{ik} &= Im\,(j_i^* j_k). \end{aligned} \tag{2.4}$$

Here $(i, k) = 1, 2, 3, 4$ and $(j_1, j_2, j_3, j_4) = (J_{11}, J_{22}, J_{12}, J_{21})$.

For perfectly polarized light, the elements are constrained by the equality of equation (1.60). For that reason, only seven of the elements of the Mueller matrix will be linearly independent. In these circumstances, a sample can be properly characterized by determining a limited set of the components of **M**. For more complex, depolarizing systems, it may be necessary to determine all sixteen Mueller matrix components.

2.2 Analysis of a Series of Polarizing Elements

Most optical experiments consist of a cascade of optical elements, and each will be represented by a Jones or Mueller matrix. Such a series is shown in Figure 2.2. The polarization vector generated by a train of n elements is

$$\mathbf{A}_n = \mathbf{J}_n \cdot \mathbf{J}_{n-1} \cdot \ldots \cdot \mathbf{J}_2 \cdot \mathbf{J}_1 \cdot \mathbf{A}_0, \tag{2.5}$$

or

$$\mathbf{S}_n = \mathbf{M}_n \cdot \mathbf{M}_{n-1} \cdot \ldots \cdot \mathbf{M}_2 \cdot \mathbf{M}_1 \cdot \mathbf{S}_0 . \tag{2.6}$$

Since the Maxwell equations involve the components of the Jones vector, it is normally easier to derive the Jones matrix, **J**, for complex, anisotropic media. Once **J** is obtained, it is generally convenient to transform it to the Mueller matrix representation for the purpose of analyzing the quantities measured in specific optical trains. This is because the components of the Stokes vector are observable, whereas the Jones vector components are not. Since it is the intensity of light that is normally required, only the first element of $\mathbf{S}_n$, S_{n0}, needs to be evaluated.

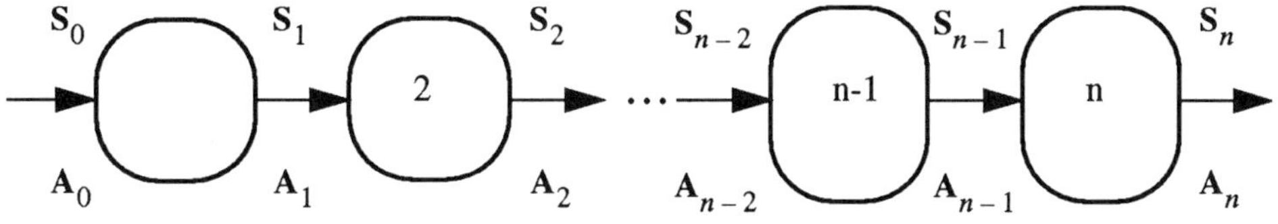

Figure 2.2 Propagation of polarized light through a cascade of optical elements.

2.3 Rotation of Optical Elements

Once a Jones or Mueller matrix of an optical element is obtained for one orthonormal basis set ($\mathbf{e}_1, \mathbf{e}_2$, for example), the corresponding matrices for the element relative to other basis sets can be obtained using standard rotation transformation rules. The action of rotating an optical element through an angle θ and onto a new basis set $\mathbf{e}_1', \mathbf{e}_2'$ is pictured in Figure 2.3. In the nonrotated frame, the exiting polarization vector is:

$$\mathbf{A}_1 = \mathbf{J} \cdot \mathbf{A}_0 ; \mathbf{S}_1 = \mathbf{M} \cdot \mathbf{S}_0 . \tag{2.7}$$

In the rotated frame, the polarization vectors are

$$\mathbf{A'}_i = \mathbf{R}(\theta) \cdot \mathbf{A}_i ; \mathbf{S'}_i = \mathbf{R}_M(\theta) \cdot \mathbf{S}_i ; i = 0, 1 , \tag{2.8}$$

where $\mathbf{R}(\theta)$ is the two-dimensional rotation matrix defined in equation (1.30) and

$$\mathbf{R}_M(\theta) = \begin{bmatrix} 1 & 0 & 0 & 0 \\ 0 & \cos 2\theta & \sin 2\theta & 0 \\ 0 & -\sin 2\theta & \cos 2\theta & 0 \\ 0 & 0 & 0 & 1 \end{bmatrix} . \tag{2.9}$$

The rotation matrices, **R** and $\mathbf{R}_M$, are unitary and have the properties that $\mathbf{R}^{-1}(\theta) = \mathbf{R}(-\theta)$ and $\mathbf{R}_M^{-1}(\theta) = \mathbf{R}_M(-\theta)$. Using equations (2.7) and (2.8), the Jones and Stokes vectors in the rotated frame are:

$$\mathbf{A'}_1 = \mathbf{R}(-\theta) \cdot \mathbf{J} \cdot \mathbf{R}(\theta) \cdot \mathbf{A}_0; \mathbf{S'}_1 = \mathbf{R}_M(-\theta) \cdot \mathbf{M} \cdot \mathbf{R}_M(\theta) \cdot \mathbf{S}_0. \tag{2.10}$$

Therefore, the relationship between the Jones and Mueller matrices in the rotated frame, $\mathbf{J}'$ and $\mathbf{M}'$, are related to their unrotated forms by

$$\mathbf{J}' = \mathbf{R}(-\theta) \cdot \mathbf{J} \cdot \mathbf{R}(\theta); \mathbf{M}' = \mathbf{R}_M(-\theta) \cdot \mathbf{M} \cdot \mathbf{R}_M(\theta). \tag{2.11}$$

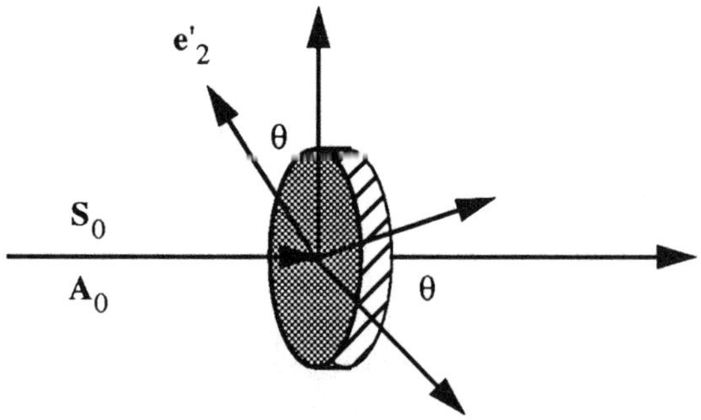

Figure 2.3 Rotation of an optical element by an angle θ.

2.4 Jones Matrices for Simple Polarizing Elements

2.4.1 Isotropic Retarders and Attenuators

Isotropic materials are characterized by a scalar refractive index, $n = n' + in''$, as defined in equation (1.17). The real and imaginary parts of the refractive index induce phase shifts and attenuation of the electric vector, respectively. This is seen by examining the Jones vector, $\mathbf{A}_1$, of light exiting an element of isotropic material with thickness d. If the element is surrounded by a medium of refractive index, n_0, then

$$\begin{aligned}\mathbf{A}_1 &= \tilde{\mathbf{A}}_0 e^{i\frac{2\pi n d}{\lambda} + i\frac{2\pi n_0 z}{\lambda}} \\ &= e^{i\frac{2\pi n' d}{\lambda}} e^{-\frac{2\pi n'' d}{\lambda}} \tilde{\mathbf{A}}_0 e^{i\frac{2\pi n_0 z}{\lambda}} \\ &= e^{i\frac{2\pi n' d}{\lambda}} e^{-\frac{2\pi n'' d}{\lambda}} \mathbf{A}_0,\end{aligned} \tag{2.12}$$

where $\mathbf{A}_0$ is the Jones vector of the incident light and $\tilde{\mathbf{A}}_0$ is its amplitude. It is evident that the Jones matrix of an isotropic retarder/attenuator is

$$\mathbf{J} = e^{i\frac{2\pi n' d}{\lambda}} e^{-\frac{2\pi n'' d}{\lambda}} \mathbf{I}. \tag{2.13}$$

The intensity of the exiting light is

$$I = |\mathbf{A}_1|^2 = e^{-\frac{4\pi n''d}{\lambda}} I_0 = e^{-\nu d} I_0, \tag{2.14}$$

where ν is Beer's law coefficient and I_0 is the incident intensity.

2.4.2 Anisotropic Retarders: Birefringence

This type of material was the subject of the example calculation in section 1.2.1. The refractive index tensor is anisotropic with principal values $(n'_{xx}, n'_{yy}, n'_{zz})$, defined in equation (1.29). As discussed in that example, light propagating along the z axis will be retarded in phase differently, depending on the relative orientation of the polarization vector with the principal axes of the refractive index tensor, $\mathbf{n}'$. The Jones matrix for an element of thickness d is

$$\mathbf{J} = \begin{bmatrix} e^{i\frac{2\pi n'_{xx}d}{\lambda}} & 0 \\ 0 & e^{i\frac{2\pi n'_{yy}d}{\lambda}} \end{bmatrix}$$

$$= e^{i\frac{2\pi n'_{xx}d}{\lambda}} \begin{bmatrix} 1 & 0 \\ 0 & e^{-i\frac{2\pi \Delta n' d}{\lambda}} \end{bmatrix} \tag{2.15}$$

$$= e^{i\frac{2\pi n'_{xx}d}{\lambda}} \begin{bmatrix} 1 & 0 \\ 0 & e^{-i\delta'} \end{bmatrix},$$

where

$$\Delta n' = n'_{xx} - n'_{yy} \tag{2.16}$$

is the linear birefringence in the (x, y) plane and

$$\delta' = \frac{2\pi \Delta n' d}{\lambda} \tag{2.17}$$

is the retardance. The prefactor $\exp[i(2\pi n'_{xx}d/\lambda)]$ is normally omitted in the expression for the Jones matrix since it does not contribute in the calculation of the intensity of light.

Two special retardation devices find wide application in the design of optical polarimeters. These are the quarter-wave plate with $\delta' = \pi/2$ and the half wave plate with $\delta' = \pi$. The utility of a quarter-wave plate can be demonstrated by observing its effect on linearly polarized light oriented at 45° with respect to the principal axes of this device. The incident light is then $\mathbf{A}_0^T = (\tilde{A}_0/\sqrt{2})(1, 1)$. This is shown in Figure 2.4.

The electric vector, $\mathbf{A}_1$, of the light generated by the quarter-wave plate is

$$\mathbf{A}_1 = \begin{bmatrix} 1 & 0 \\ 0 & e^{i\pi/2} \end{bmatrix} \cdot \frac{\tilde{A}_0}{\sqrt{2}} \begin{bmatrix} 1 \\ 1 \end{bmatrix} = \frac{\tilde{A}_0}{\sqrt{2}} \begin{bmatrix} 1 \\ i \end{bmatrix}, \tag{2.18}$$

which is right-circularly polarized light. Left-circularly polarized light is produced when the incident electric vector and the axis defining n'_{xx} have a relative orientation of -45°.

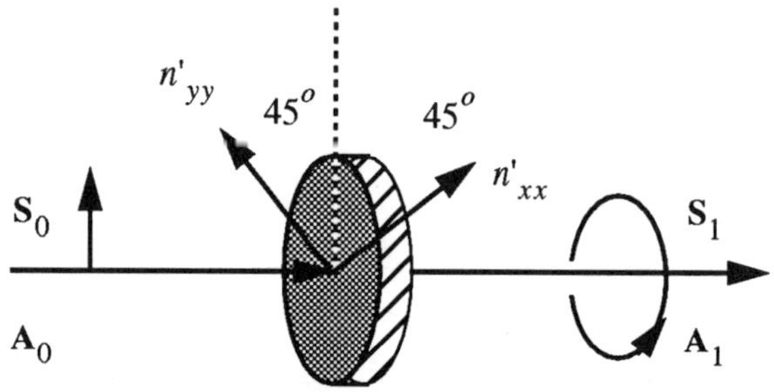

Figure 2.4 The production of circularly polarized light using a quarter-wave plate.

Half-wave plates are used to rotate the electric vector of light. Incident light with its polarization vector oriented at an angle θ relative to the principal axes of $\mathbf{n}'$ has an electric vector $\mathbf{A}_0^T = \tilde{A}_0(\cos\theta, \sin\theta)$. Using equation (2.15) with $\delta' = \pi$, the incident electric vector will be transformed to

$$\mathbf{A}_1 = \tilde{A}_0 \begin{bmatrix} 1 & 0 \\ 0 & -1 \end{bmatrix} \cdot \begin{bmatrix} \cos\theta \\ \sin\theta \end{bmatrix} = \tilde{A}_0 \begin{bmatrix} \cos\theta \\ -\sin\theta \end{bmatrix} = \tilde{A}_0 \begin{bmatrix} \cos(-\theta) \\ \sin(-\theta) \end{bmatrix}. \tag{2.19}$$

Therefore, the exiting light is still linearly polarized but is rotated by an amount 2θ from its original orientation. This operation is shown in Figure 2.5.

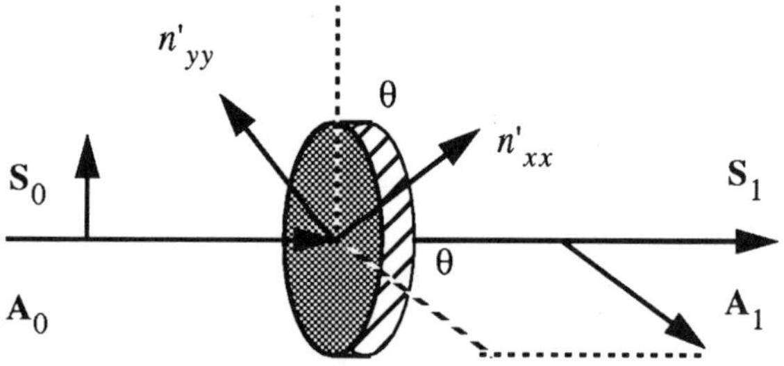

Figure 2.5 The rotation of linearly polarized light using a half-wave plate.

2.4.3 Anisotropic Attenuators: Dichroism

A material with an anisotropic, imaginary refractive index:

$$\mathbf{n}'' = \begin{bmatrix} n''_{xx} & 0 & 0 \\ 0 & n''_{yy} & 0 \\ 0 & 0 & n''_{zz} \end{bmatrix}, \tag{2.20}$$

has a Jones matrix of the form:

$$\begin{aligned} \mathbf{J} &= \begin{bmatrix} e^{-(2\pi n''_{xx}d)/\lambda} & 0 \\ 0 & e^{-(2\pi n''_{yy}d)/\lambda} \end{bmatrix} \\ &= e^{-(2\pi n''_{xx}d)/\lambda} \begin{bmatrix} 1 & 0 \\ 0 & e^{(2\pi\Delta n''d)/\lambda} \end{bmatrix} \\ &= e^{-(2\pi n''_{xx}d)/\lambda} \begin{bmatrix} 1 & 0 \\ 0 & e^{\delta''} \end{bmatrix} \end{aligned} \tag{2.21}$$

for light propagating along the z axis. Here

$$\Delta n'' = n''_{xx} - n''_{yy} \tag{2.22}$$

is the linear dichroism and

$$\delta'' = \frac{2\pi\Delta n''d}{\lambda} \tag{2.23}$$

is the extinction. An ideal polarizer oriented along the x direction is defined as having an infinite, negative dichroism and a Jones matrix:

$$\mathbf{J} = \begin{bmatrix} 1 & 0 \\ 0 & 0 \end{bmatrix}. \tag{2.24}$$

2.4.4 Coaxial Birefringent/Dichroic Materials

Many materials are simultaneously birefringent and dichroic. If the tensors **n**' and **n**'' are coaxial, the Jones matrix will be a simple superposition of these two properties:

$$\mathbf{J} = \begin{bmatrix} 1 & 0 \\ 0 & e^{\delta'' - i\delta'} \end{bmatrix}. \tag{2.25}$$

This as been written in a reference frame coincident with the principal axes of **n**. The case of noncoaxial real and imaginary parts of the refractive index tensor requires a much more complex calculation and is treated in section 2.4.6.

2.4.5 Optically Active Materials: Anisotropy and Circularly Polarized Light

The example presented in section 1.2.2 concerned a material with the following optical characteristics: $\varepsilon_{ij} = \varepsilon\delta_{ij}$, $\mu_{ij} = \mu\delta_{ij}$, $\zeta_{ij} = i\sqrt{\varepsilon\mu}\gamma\delta_{ij}$, and $\zeta'_{ij} = 0$. The presence of a finite optical rotation tensor, ζ_{ij}, couples the electric and magnetic fields and forces a rotation of linearly polarized light as it propagates through the sample. From equation (1.29), the Jones matrix is

$$\mathbf{J} = \mathbf{R}\left(\frac{k\gamma}{2}z\right), \tag{2.26}$$

where the exponential prefactor, $e^{i\sqrt{1+\frac{\gamma^2}{4}}kz}$, has been suppressed. Materials exhibiting this behavior are called optically active and include chiral molecules and molecules with helical structures.

It is of interest to consider the response of circularly polarized light to optically active materials. The action of an optical element represented by a Jones matrix formulated using the basis set for *linearly* polarized light is

$$\mathbf{A}_{linear,\,1} = \mathbf{J}_{linear} \cdot \mathbf{A}_{linear,\,0}, \tag{2.27}$$

where the subscripts $(0, 1)$ refer to the incident and exiting light, respectively. Using equation (1.56), this can be transformed to the *circular* basis set:

$$\mathbf{A}_{circ,\,1} = \mathbf{F}^{-1} \cdot \mathbf{J}_{linear} \cdot \mathbf{F} \cdot \mathbf{A}_{circ,\,0}. \tag{2.28}$$

The Jones matrix in the circular basis set then

$$\mathbf{J}_{circ} = \mathbf{F}^{-1} \cdot \mathbf{J}_{linear} \cdot \mathbf{F}. \tag{2.29}$$

Using this expression, $\mathbf{J}_{circ}$ for an optically active material of thickness d is:

$$\mathbf{J}_{circ} = \begin{bmatrix} e^{-i\frac{\pi\gamma d}{\lambda}} & 0 \\ 0 & e^{i\frac{\pi\gamma d}{\lambda}} \end{bmatrix} = \begin{bmatrix} e^{-i\frac{\pi n_l d}{\lambda}} & 0 \\ 0 & e^{-i\frac{\pi n_r d}{\lambda}} \end{bmatrix} \tag{2.30}$$

In writing equation (2.30), two refractive indices, n_l and n_r, have been defined. For a material with a complex rotary power, $\gamma = \gamma' + i\gamma''$, these are:

$$n_l = \frac{1}{2}(\gamma' + i\gamma'')\,;\; n_r = -\frac{1}{2}(\gamma' + i\gamma'')\,. \tag{2.31}$$

Analogous to the definitions of linear birefringence and linear dichroism following equations (2.15) and (2.21), the form of equation (2.30) suggests the following optical anisotropies for circularly polarized light:

$$\text{circular birefringence: } \Delta n'_{circ} = (n'_l - n'_r) = \gamma', \tag{2.32}$$

$$\text{circular dichroism: } \Delta n''_{circ} = (n''_l - n''_r) = \gamma'' . \tag{2.33}$$

Circular birefringence will induce a differential retardation in the phase of the orthogonal states of circularly polarized light. Circular dichroism, on the other hand, results in anisotropic attenuation of left- and right-circularly polarized light. The Jones matrix of circularly dichroic materials is normally written as:

$$\mathbf{J}_{circ} = \begin{bmatrix} 1 & 0 \\ 0 & e^{(2\pi\Delta n''_{circ} d)/\lambda} \end{bmatrix}, \tag{2.34}$$

in the circular basis set and

$$\mathbf{J}_{linear} = \begin{bmatrix} \cosh(\pi\Delta n''_{circ} d/\lambda) & -i\sinh(\pi\Delta n''_{circ} d/\lambda) \\ i\sinh(\pi\Delta n''_{circ} d/\lambda) & \cosh(\pi\Delta n''_{circ} d/\lambda) \end{bmatrix}, \tag{2.35}$$

in the Cartesian frame. A circular polarizer will perfectly extinguish light of either right- or left-circular polarization and allow the orthogonal polarization to pass. This effect is equivalent to setting $n''_r \to \infty$ while keeping n''_l finite, or vice-versa. Such a device, transmitting only left-circular polarization, has the following Jones matrix:

$$\mathbf{J}_{linear} = \frac{1}{2}\begin{bmatrix} 1 & -i \\ i & 1 \end{bmatrix}, \tag{2.36}$$

whereas a right-circular polarizer has a Jones matrix that is the transpose of (2.36).

2.4.6 Composite Materials and Axially Varying Materials

In an important series of papers [6,7], Jones established an approach for the treatment of materials where the refractive index tensor, $\mathbf{n}(z)$, varies along the propagation direction of the transmitted light. This procedure also lays the foundation for the analysis of complex systems possessing any combination of optical anisotropies.

A material with an axial variation in its refractive index tensor is shown in Figure 2.6. Light propagating through a differential element of thickness Δz is transformed according to

$$\mathbf{A}(z+\Delta z) = \mathbf{P}(z, \Delta z) \cdot \mathbf{A}(z) . \tag{2.37}$$

The Jones matrix, $\mathbf{P}(z, \Delta z)$, describes the linear response associated with the differential element. Subtracting the electric field that is incident on this element, $\mathbf{A}(z)$, from both sides of (2.37), one obtains

$$\mathbf{A}(z+\Delta z) - \mathbf{A}(z) = [\mathbf{P}(z, \Delta z) - \mathbf{I}] \cdot \mathbf{A}(z) . \tag{2.38}$$

Dividing by the differential thickness, Δz, and taking the limit as $\Delta z \to 0$, the following differential equation results:

$$\frac{d\mathbf{A}}{dz} = \mathbf{N}(z) \cdot \mathbf{A}(z) , \tag{2.39}$$

where

$$\mathbf{N}(z) = \lim_{\Delta z \to 0}\left(\frac{\mathbf{P}(z, \Delta z) - \mathbf{I}}{\Delta z}\right), \tag{2.40}$$

is the *differential propagation* Jones matrix.

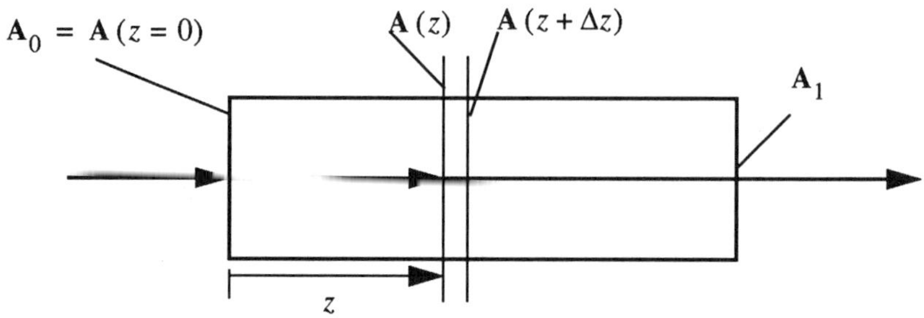

Figure 2.6 Propagation of polarized light through a material with an axial variation in its refractive index tensor.

Alternatively, the Jones vector, $\mathbf{A}(z)$, is related to the Jones vector incident on the front surface of the *total* optical element at $z = 0$, $\mathbf{A}_0 = \mathbf{A}(z = 0)$, by

$$\mathbf{A}(z) = \mathbf{J}(z) \cdot \mathbf{A}_0. \tag{2.41}$$

The matrix $\mathbf{J}(z)$ is the Jones matrix of the material between the plane $z = 0$ and the leading edge of the differential element. Substituting equation (2.41) into (2.37) leads to

$$\mathbf{A}(z + \Delta z) = \mathbf{P}(z, \Delta z) \cdot \mathbf{J}(z) \cdot \mathbf{A}_0 = \mathbf{J}(z + \Delta z) \cdot \mathbf{A}_0, \tag{2.42}$$

from which it is evident that

$$\mathbf{P}(z, \Delta z) = \mathbf{J}(z + \Delta z) \cdot \mathbf{J}^{-1}(z). \tag{2.43}$$

This result can be substituted into equation (2.40) to obtain the following:

$$\begin{aligned} \mathbf{N}(z) &= \lim_{\Delta z \to 0}\left(\frac{\mathbf{J}(z + \Delta z) \cdot \mathbf{J}^{-1}(z) - \mathbf{J}(z) \cdot \mathbf{J}^{-1}(z)}{\Delta z}\right) \\ &= \lim_{\Delta z \to 0}\left(\frac{\mathbf{J}(z + \Delta z) - \mathbf{J}(z)}{\Delta z}\right) \cdot \mathbf{J}^{-1}(z) \\ &= \left(\frac{d}{dz}\mathbf{J}(z)\right) \cdot \mathbf{J}^{-1}(z). \end{aligned} \tag{2.44}$$

The differential equation (2.44) and the boundary condition, $\mathbf{A}(z = 0) = \mathbf{A}_0$, can be used to analyze the propagation of the Jones vector through a material with an axially varying refractive index tensor.

Once obtained, the differential propagation Jones matrix, **N**, can be used to construct the Jones matrix, **J**, for complex materials characterized by a superposition of optical anisotropies. Consider, for example, a *homogeneous* material exhibiting both linear and circular birefringence and dichroism. In such a case, **N** is a constant and independent of z but its form is not immediately evident for a complex material. When **N** is independent of z, equation (2.44) is a set of linear differential equations for the components of $\mathbf{J}(z)$ that can be solved directly, subject to the boundary condition:

$$\mathbf{J}(z \to 0) = \mathbf{I}. \tag{2.45}$$

This boundary condition reflects the fact that the Jones vector will by unchanged by passage through an element of zero thickness. In reference 7, Jones was able to show that the solution to (2.44) is:

$$\mathbf{J}(z) = e^{T_N z} \begin{bmatrix} \cosh Q_N z + \frac{1}{2}(N_{11} - N_{22}) \dfrac{\sinh Q_N z}{Q_N} & N_{12} \dfrac{\sinh Q_N z}{Q_N} \\ N_{21} \dfrac{\sinh Q_N z}{Q_N} & \cosh Q_N z - \frac{1}{2}(N_{11} - N_{22}) \dfrac{\sinh Q_N z}{Q_N} \end{bmatrix}, \tag{2.46}$$

where N_{ij} are the elements of **N** and

$$T_N = \frac{1}{2}(N_{11} + N_{22}),$$
$$Q_N = \sqrt{T_N^2 - D_N},$$
$$D_N = N_{11}N_{22} - N_{12}N_{21}. \tag{2.47}$$

To represent a material containing a combination of optical anisotropies, the construction shown in Figure 2.7 is used. In this figure, the matrix **N** associated with the optical element is composed of a series of lamellae, each one of which introduces a specific optical effect. These include:

1. Isotropic retardation of the phase of light, $\mathbf{N}_1$.
2. Isotropic attenuation of the amplitude of light, $\mathbf{N}_2$.
3. Linear birefringence, $\mathbf{N}_3$.
4. Linear dichroism, $\mathbf{N}_4$.
5. Circular birefringence, $\mathbf{N}_5$.
6. Circular dichroism, $\mathbf{N}_6$.

The Jones matrix, **J**, of the entire series of lamellae shown in Figure 2.7 is the product of the matrices $\mathbf{J}_1, \mathbf{J}_2, \ldots, \mathbf{J}_6$ of the individual optical effects. In other words,

$$\mathbf{J} = \mathbf{J}_1 \cdot \mathbf{J}_2 \cdot \ldots \cdot \mathbf{J}_6. \tag{2.48}$$

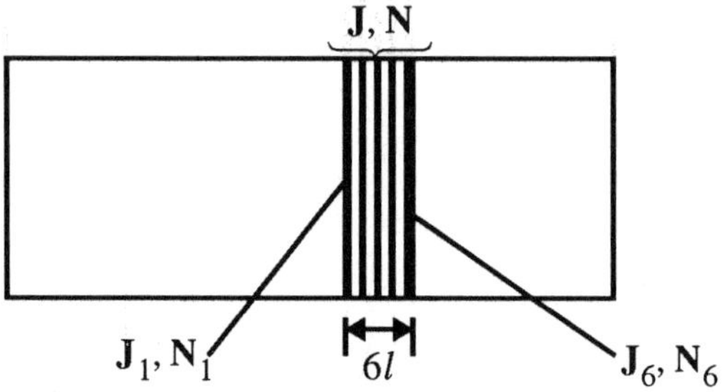

Figure 2.7 Material possessing a superposition of optical anisotropies. Each lamella is characterized by a specific optical effect.

This strategy, however, is only justified if the order of the multiplication in (2.48) is not important. This will be true if the thickness, l, of each lamella is sufficiently small. If the Jones matrix of a lamella is expanded in z about its position, z_0, in the optical element, one obtains

$$\mathbf{J}_i = \mathbf{J}_i(z_0) + \left(\frac{d\mathbf{J}_i}{dz}\right)_{z_0}(z - z_0) + \frac{1}{2}\left(\frac{d^2\mathbf{J}_i}{dz^2}\right)_{z_0}(z - z_0)^2 + \dots \tag{2.49}$$

Equation (2.44) is used to show that the derivatives of $\mathbf{J}_i$ are

$$\frac{d^k\mathbf{J}_i}{dz^k} = \mathbf{N}_i^k \cdot \mathbf{J}_i . \tag{2.50}$$

Taking the thickness of each lamella to be $l = z - z_0$, the Jones matrix, $\mathbf{J}_i$, to order l^2 is

$$\mathbf{J}_i = \mathbf{I} + \mathbf{N}_i l + \mathrm{O}(l^2) . \tag{2.51}$$

Combining this result with equation (2.48) and only retaining terms to $\mathrm{O}(l^2)$, the Jones matrix for the group of six lamellae is

$$\mathbf{J} = \mathbf{I} + \sum_{i=1}^{6} \mathbf{N}_i l + \mathrm{O}(l^2) . \tag{2.52}$$

Equation (2.52) satisfies the requirement that the ordering of the lamellae will not affect the final result. Furthermore, the differential propagation Jones matrix for a composite material is simply the sum of the matrices, $\mathbf{N}_i$, for each of the separate optical effects. The **N** matrices for different optical properties can be derived using equation (2.44). As an example, consider a birefringent material with retardation $\delta' = (2\pi\Delta n'z)/\lambda$. One form of the Jones matrix for this material is

$$\mathbf{J} = \begin{bmatrix} e^{i\frac{\delta'}{2}} & 0 \\ 0 & e^{-i\frac{\delta'}{2}} \end{bmatrix} = \begin{bmatrix} e^{i\frac{\pi\Delta n' z}{\lambda}} & 0 \\ 0 & e^{-i\frac{\pi\Delta n' z}{\lambda}} \end{bmatrix}. \tag{2.53}$$

Using equation (2.44), the differential propagation matrix is

$$\mathbf{N} = \frac{i\pi\Delta n'}{\lambda}\begin{bmatrix} 1 & 0 \\ 0 & -1 \end{bmatrix}. \tag{2.54}$$

If the principal axis of the real part of the refractive index tensor is oriented at an angle θ relative to the laboratory frame, equation (2.11) must be used to rotate the Jones matrices used in the calculation of $\mathbf{N}'$ in the rotated frame to obtain:

$$\begin{aligned} \mathbf{N}' &= \left(\frac{d}{dz}\mathbf{J}'\right) \cdot \mathbf{J}'^{-1} = \frac{d}{dz}[\mathbf{R}^{-1}(\theta) \cdot \mathbf{J} \cdot \mathbf{R}(\theta)] \cdot [\mathbf{R}^{-1}(\theta) \cdot \mathbf{J} \cdot \mathbf{R}(\theta)]^{-1} \\ &= \mathbf{R}^{-1}(\theta) \cdot \left(\frac{d\mathbf{J}}{dz}\right) \cdot \mathbf{J}^{-1} \cdot \mathbf{R}(\theta) \\ &= \mathbf{R}^{-1}(\theta) \cdot \mathbf{N} \cdot \mathbf{R}(\theta). \end{aligned} \tag{2.55}$$

Using the Jones matrices of the optical effects presented earlier in this chapter, along with the rotation property of **N** expressed in (2.55), Table 2.1 listing the different **N** matrices, was constructed.

Table 2.1 **N** matrices for various optical effects.

Optical effect	Definitions	N matrix
isotropic retarder	n' is the scalar, real part of the refractive index	$\frac{i2\pi n'}{\lambda}\begin{bmatrix} 1 & 0 \\ 0 & 1 \end{bmatrix}$
isotropic attenuator	n'' is the scalar, imaginary part of the refractive index	$-\frac{2\pi n''}{\lambda}\begin{bmatrix} 1 & 0 \\ 0 & 1 \end{bmatrix}$
linear birefringent material	$\Delta n'$ is the birefringence and θ is the orientation of the principal axes of the real part of the refractive index tensor	$\frac{i\pi\Delta n'}{\lambda}\begin{bmatrix} \cos 2\theta & \sin 2\theta \\ \sin 2\theta & -\cos 2\theta \end{bmatrix}$
linear dichroic material	$\Delta n''$ is the dichroism and θ is the orientation of the principal axes of the imaginary part of the refractive index tensor	$-\frac{\pi\Delta n''}{\lambda}\begin{bmatrix} \cos 2\theta & \sin 2\theta \\ \sin 2\theta & -\cos 2\theta \end{bmatrix}$

Table 2.1 **N** matrices for various optical effects.

Optical effect	Definitions	N matrix
circular birefringent material	$\Delta n'_{circ}$ is the circular birefringence	$-\frac{\pi \Delta n'_{circ}}{\lambda}\begin{bmatrix} 0 & -1 \\ 1 & 0 \end{bmatrix}$
circular dichroic material	$\Delta n''_{circ}$ is the circular dichroism	$-\frac{\pi \Delta n''_{circ}}{\lambda}\begin{bmatrix} 0 & i \\ -i & 0 \end{bmatrix}$

2.4.7 Combined Birefringent and Dichroic Materials

The results listed in Table 2.1, combined with equations (2.46) and (2.52) can be used to construct the Jones matrices of complex materials displaying any number of anisotropies. One important case is a material containing both birefringence and dichroism. This situation arises frequently in materials deformed by shear flows where constituents comprising the material generate birefringence that may not be oriented coaxially with the constituents giving rise to dichroism. If the angles θ' and θ'' are associated with the orientation of the real and imaginary parts of the refractive index tensor, respectively, the **N** matrix for such a system is

$$\mathbf{N} = \frac{i\pi\Delta n'}{\lambda}\begin{bmatrix} \cos 2\theta' & \sin 2\theta' \\ \sin 2\theta' & -\cos 2\theta' \end{bmatrix} - \frac{\pi\Delta n''}{\lambda}\begin{bmatrix} \cos 2\theta'' & \sin 2\theta'' \\ \sin 2\theta'' & -\cos 2\theta'' \end{bmatrix}. \tag{2.56}$$

Figure 2.8 is a vector diagram describing the principal axes of the refractive index tensor of this material. Using equation (2.47), the Jones matrix is given by (2.46) with parameters:

$$T_N = 0,$$

$$D_N = \frac{\pi^2}{\lambda^2}[\Delta n'^2 - \Delta n''^2 + 2i(\Delta n'\Delta n'')\cos 2(\theta' - \theta'')], \tag{2.57}$$

$$Q_N = i\sqrt{D_N}.$$

If the birefringence and dichroism are coaxial, $(\theta' = \theta'')$, this material is described by equation (2.25) with the Jones matrix rotated by an angle θ'.

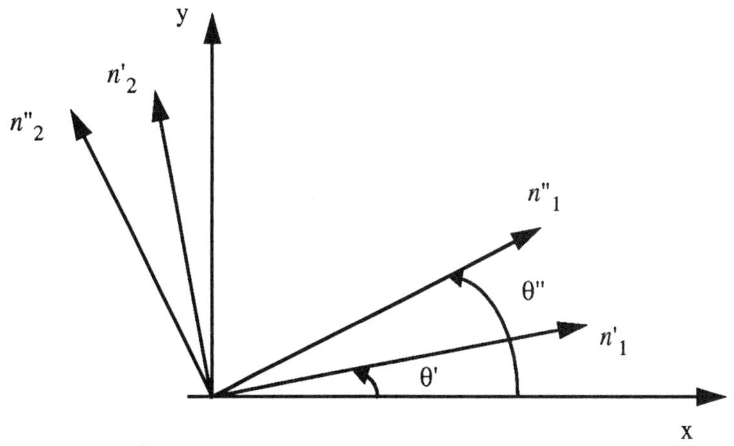

Figure 2.8 Vector diagram of the refractive index tensor for a material with noncoaxial birefringence and dichroism.

2.5 List of Jones and Mueller Matrices

The majority of Jones matrices for transmission polarizing elements have been presented in this chapter. As mentioned earlier, it is generally more convenient to use Mueller matrices and not Jones matrices when analyzing cascades of optical elements making up a particular experiment. For the purpose of such calculations, a list of Jones and Mueller matrices for most of the elements encountered in practice can be found compiled in Appendix I.

2.6 Example Analysis: Crossed Polarizer Experiment

It is illustrative to analyze the simple experimental arrangement shown in Figure 2.9. In this set up, a sample that is both birefringent and dichroic is placed between two crossed polarizers. The real and imaginary parts of the sample's refractive index are coaxial and oriented at an angle θ relative to the laboratory frame (this angle is often referred to as the "extinction angle"). The laboratory frame is specified by the basis set $\mathbf{e}_1, \mathbf{e}_2$. In addition, the pair of crossed polarizers are oriented at an angle α.

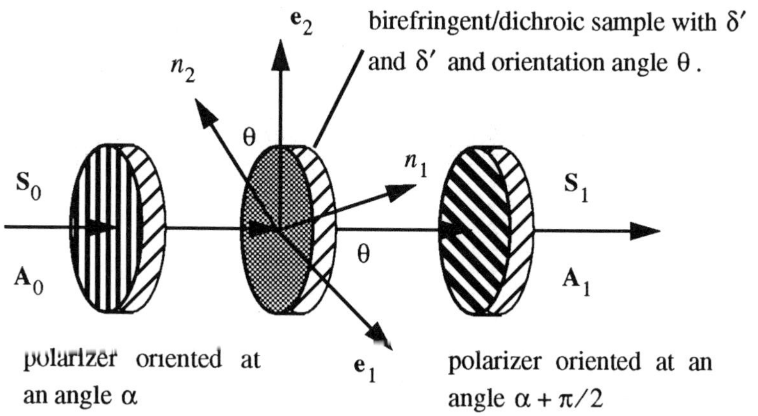

Figure 2.9 Birefringent and dichroic sample sandwiched between crossed polarizers.

The intensity of the light generated in this experiment is easily calculated using equations (2.5) and (2.6) combined with the appropriate Jones and Mueller matrices selected from Appendix I. The Jones and Stokes vectors, $\mathbf{A}_1$ and $\mathbf{S}_1$, exiting this cascade are

$$\mathbf{A}_1 = \mathbf{J}_{11}\left(\alpha + \frac{\pi}{2}\right) \cdot \mathbf{J}_{14}(\delta', \delta'', \theta) \cdot \mathbf{J}_{11}(\alpha) \cdot \mathbf{A}_0, \tag{2.58}$$

and

$$\mathbf{S}_1 = \mathbf{M}_{11}\left(\alpha + \frac{\pi}{2}\right) \cdot \mathbf{M}_{14}(\delta', \delta'', \theta) \cdot \mathbf{M}_{11}(\alpha) \cdot \mathbf{S}_0, \tag{2.59}$$

where the subscripts on the Jones and Mueller matrices refer to the forms found in Appendix I.

Convenient choices for the incident light electric vectors are $\mathbf{A}_0 = (E_0/\sqrt{2})\,(1, i)^T$ and $\mathbf{S}_0 = (I_0, 0, 0, 0)^T$, which represent circularly polarized light and natural light, respectively. Since the Mueller matrix calculation is far easier to accomplish, only that procedure is outlined here in detail. The Stokes vector, $\mathbf{S}_1$, is

$$\mathbf{S}_1 = \frac{1}{2}\begin{bmatrix} 1 & c_{2\alpha+\pi} & s_{2\alpha+\pi} & 0 \\ c_{2\alpha+\pi} & c^2_{2\alpha+\pi} & c_{2\alpha+\pi}s_{2\alpha+\pi} & 0 \\ s_{2\alpha+\pi} & c_{2\alpha+\pi}s_{2\alpha+\pi} & s^2_{2\alpha+\pi} & 0 \\ 0 & 0 & 0 & 0 \end{bmatrix} \cdot$$

$$\begin{bmatrix} C_{\delta''} & -c_{2\theta}S_{\delta''} & -s_{2\theta}S_{\delta''} & 0 \\ -c_{2\theta}S_{\delta''} & [c^2_{2\theta}C_{\delta''} + s^2_{2\theta}c_{\delta'}] & c_{2\theta}s_{2\theta}(C_{\delta''} - c_{\delta'}) & s_{2\theta}s_{\delta'} \\ -s_{2\theta}S_{\delta''} & c_{2\theta}s_{2\theta}(C_{\delta''} - c_{\delta'}) & [s^2_{2\theta}C_{\delta''} + c^2_{2\theta}c_{\delta'}] & -c_{2\theta}s_{\delta'} \\ 0 & s_{2\theta}s_{\delta'} & -c_{2\theta}s_{\delta'} & c_{\delta'} \end{bmatrix} \cdot \tag{2.60}$$

$$\frac{1}{2}\begin{bmatrix} 1 & c_{2\alpha} & s_{2\alpha} & 0 \\ c_{2\alpha} & c^2_{2\alpha} & c_{2\alpha}s_{2\alpha} & 0 \\ s_{2\alpha} & c_{2\alpha}s_{2\alpha} & s^2_{2\alpha} & 0 \\ 0 & 0 & 0 & 0 \end{bmatrix} \cdot \begin{bmatrix} I_0 \\ 0 \\ 0 \\ 0 \end{bmatrix}.$$

Here the notation introduced in Appendix I of $s_\theta = \sin\theta$, $c_\theta = \cos\theta$, $C_\theta = \cosh\theta$, and $S_\theta = \sinh\theta$ is used. Carrying out the multiplications in equation (2.60) one obtains

$$\mathbf{S}_1 = \begin{bmatrix} \frac{I_0}{4}\sin^2[2(\theta-\alpha)]\,(C_{\delta''} - c_{\delta'}) \\ \cdots \\ \cdots \\ \cdots \end{bmatrix}, \tag{2.61}$$

where only the first element has been calculated since the intensity is normally the quantity of interest.

This experiment is among the simplest polarimeters. Upon inspection of equation (2.61), it is evident that the measured intensity will respond to both dichroism and birefringence. It is not possible to separately determine these optical effects. Furthermore, neither the sign of these quantities nor the sign of the orientation angle, θ, can be determined. In Chapter 8, the designs of automated polarimeters capable of independently measuring the dichroism and birefringence are presented. These instruments are also capable of unambiguously determining the signs of these optical anisotropies and their orientation angles.

2.7 Transmission through Homogeneous Materials at Oblique Incidence

The discussion of the transmission of polarized light in previous sections has assumed that the propagation axis is coincident with one of the principal axes of the refractive index tensor. In this section, the analysis of the propagation of light along arbitrary axes is treated, although the sample is still assumed to be homogeneous. The treatment presented here follows the discussion provided in the book by Kong [3] and begins with the development of the "kDB" coordinate system. This is an orthogonal set of axes containing the propagation wavevector, $\mathbf{k}$, the electric displacement vector, $\mathbf{D}$, and the magnetic induction vector, $\mathbf{B}$. The utility of using this system is evident upon examining the set of Maxwell equations listed in equation (1.3) for a media free of charges and currents ($\rho = J = 0$). Taking the time dependence of the fields to be $e^{i\omega t}$, these become

$$\begin{aligned} \nabla\times\mathbf{E} &= i\omega\mathbf{B}, \\ \nabla\times\mathbf{H} &= -i\omega\mathbf{D}, \\ \nabla\cdot\mathbf{B} &= 0, \\ \nabla\cdot\mathbf{D} &= 0. \end{aligned} \tag{2.62}$$

Since the medium is taken to be homogeneous, the solution will be in the form of plane waves, characterized by a phase factor, $e^{i\mathbf{k}\cdot\mathbf{r}}$, which, when attached to each of the field vectors in (2.62), leads to

$$\begin{aligned} \mathbf{k}\times\mathbf{E} &= \omega\mathbf{B}, \\ \mathbf{k}\times\mathbf{H} &= -\omega\mathbf{D}, \\ \mathbf{k}\cdot\mathbf{B} &= 0, \\ \mathbf{k}\cdot\mathbf{D} &= 0. \end{aligned} \tag{2.63}$$

The last two relations in equation (2.63) indicate that the vectors $\mathbf{k}$, $\mathbf{D}$, and $\mathbf{B}$ form a set of orthogonal vectors. For an anisotropic material, characterized by a dielectric tensor ε_{ij}, the electric vector will not necessarily lie in the plane perpendicular to $\mathbf{k}$, and for that reason it is an important advantage to express the polarization properties of the light using the displacement vector. Therefore, the following constitutive relations are used, in place of the ones in equation (1.4),

$$\begin{aligned} E_i &= \kappa_{ij}D_j + \chi_{ij}B_j, \\ H_i &= \nu_{ij}E_j + \gamma_{ij}H_j, \end{aligned} \tag{2.64}$$

where κ_{ij} is the impermittivity tensor and ν_{ij} is the impermeability tensor. In the following discussion, however, only anisotropy in the impermittivity tensor is considered, and $\chi_{ij} = \nu_{ij} = 0$.

The problem at hand is pictured in Figure 2.10. Here light propagates along the wavevector $\mathbf{k}$, which is taken to be parallel to the unit vector $\mathbf{e}_3$. This vector is rotated away from the z axis of the laboratory frame through an angle θ, and its projection onto the (x, y) plane defines the azimuthal angle, ϕ. The vectors $\mathbf{D}$ and $\mathbf{B}$, on the other hand, lie in the plane defined by the unit vectors $\mathbf{e}_1$ and $\mathbf{e}_2$. In terms of the laboratory frame, these unit vectors are

$$\begin{aligned}
\mathbf{e}_1 &= \sin\phi\mathbf{e}_x - \cos\phi\mathbf{e}_y, \\
\mathbf{e}_2 &= \cos\theta\cos\phi\mathbf{e}_x + \cos\theta\sin\phi\mathbf{e}_y - \sin\theta\mathbf{e}_z, \\
\mathbf{e}_3 &= \sin\theta\cos\phi\mathbf{e}_x + \sin\theta\sin\phi\mathbf{e}_y + \cos\theta\mathbf{e}_z,
\end{aligned} \tag{2.65}$$

where the unit vectors $(\mathbf{e}_x, \mathbf{e}_y, \mathbf{e}_z)$ define the laboratory frame. The impermittivity tensor is generally expressed in terms of the laboratory frame, and it is necessary to rotate this tensor onto the kDB frame. It is important to note that it is the impermittivity tensor, and not the dielectric tensor, that is rotated. In the kDB frame, the impermittivity tensor, $\underline{\kappa}_k$, is related to its form in the laboratory frame, $\underline{\kappa}$, according to

$$\underline{\kappa}_k = \mathbf{T} \cdot \underline{\kappa} \cdot \mathbf{T}^{-1}, \tag{2.66}$$

where the rotation tensor is

$$\mathbf{T} = \begin{bmatrix} s_\phi & -c_\theta & 0 \\ (c_\theta c_\phi) & (c_\theta s_\phi) & -s_\theta \\ (s_\theta c_\phi) & (s_\theta s_\phi) & c_\theta \end{bmatrix} \tag{2.67}$$

and the inverse, $\mathbf{T}^{-1}$, is simply the transpose of $\mathbf{T}$.

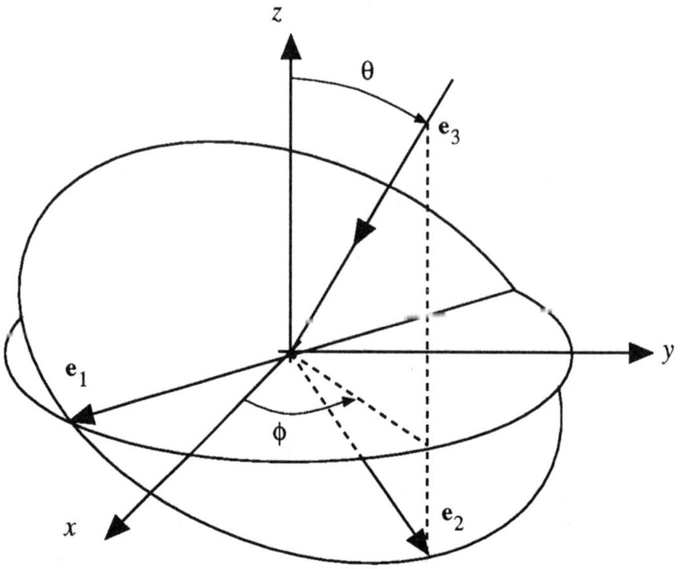

Figure 2.10 Rotation of the impermittivity tensor in the kDB frame.

If the sample possesses a biaxial dielectric tensor of the form,

$$\underline{\varepsilon} = \begin{bmatrix} \varepsilon_1 & 0 & 0 \\ 0 & \varepsilon_2 & 0 \\ 0 & 0 & \varepsilon_3 \end{bmatrix},$$

$$\underline{\kappa} = \begin{bmatrix} \kappa_1 & 0 & 0 \\ 0 & \kappa_2 & 0 \\ 0 & 0 & \kappa_3 \end{bmatrix} = \begin{bmatrix} 1/\varepsilon_1 & 0 & 0 \\ 0 & 1/\varepsilon_2 & 0 \\ 0 & 0 & 1/\varepsilon_3 \end{bmatrix} = \begin{bmatrix} 1/n_1^2 & 0 & 0 \\ 0 & 1/n_2^2 & 0 \\ 0 & 0 & 1/n_3^2 \end{bmatrix}. \tag{2.68}$$

Applying the rotation transformation to equation (2.66) gives

$$\begin{aligned} \kappa_{k11} &= \kappa_1 \sin^2\phi + \kappa_2 \cos^2\phi, \\ \kappa_{k12} = \kappa_{k21} &= (\kappa_1 - \kappa_2)\cos\theta \sin\phi \cos\phi, \\ \kappa_{k22} &= (\kappa_1 \cos^2\phi + \kappa_2 \sin^2\phi)\cos^2\theta + \kappa_3 \sin^2\theta, \\ \kappa_{k13} = \kappa_{k31} &= (\kappa_1 - \kappa_2)\sin\theta \sin\phi \cos\phi, \\ \kappa_{k23} = \kappa_{k32} &= (\kappa_1 \cos^2\phi + \kappa_2 \sin^2\phi - \kappa_3)\sin\theta \cos\theta, \\ \kappa_{k33} &= (\kappa_1 \cos^2\phi + \kappa_2 \sin^2\phi)\sin^2\theta + \kappa_3 \cos^2\theta. \end{aligned} \tag{2.69}$$

Since the light is propagating along the $\mathbf{e}_3$ axis, the displacement vector will sample material properties only in the $(\mathbf{e}_1, \mathbf{e}_2)$ plane. For this reason, the analysis of the polarization of the transmitted light, requires only a consideration of the following submatrix in this plane:

$$\underline{\kappa}_k^* = \begin{bmatrix} \kappa_1 \sin^2\phi + \kappa_2 \cos^2\phi & (\kappa_1 - \kappa_2)\cos\theta\sin\phi\cos\phi \\ (\kappa_1 - \kappa_2)\cos\theta\sin\phi\cos\phi & (\kappa_1\cos^2\phi + \kappa_2\sin^2\phi)\cos^2\theta + \kappa_3\sin^2\theta \end{bmatrix}. \tag{2.70}$$

The procedure to extract the refractive index tensor is to first determine the principal eigenvalues, κ_+^* and κ_-^* of the submatrix $\underline{\kappa}_k^*$, and the orientation angle, α, of its principal axes in the $(\mathbf{e}_1, \mathbf{e}_2)$ plane. Formally, these are

$$\begin{aligned} \kappa_{+,-}^* &= \left[\kappa_{11}^* + \kappa_{22}^* \pm \sqrt{(\kappa_{11}^* - \kappa_{22}^*)^2 + 4\kappa_{12}^{*2}}\right]/2, \\ \tan 2\alpha &= -2\kappa_{12}^{*2}/(\kappa_{11}^* - \kappa_{22}^*). \end{aligned} \tag{2.71}$$

Once these eigenvalues and known, the principal values of the refractive index tensor are calculated as $n_{+,-}^2 = 1/\kappa_{+,-}^*$, and the birefringence measured in the kDB frame, $\Delta n = n_+ - n_-$, can be calculated in terms of the principal values of the refractive index, n_1, n_2 and n_3, defined in equation (2.68).

2.7.1 Example: Oblique Transmission through Parallel Plate Flow

The experimental arrangement analyzed in this section is shown in Figure 2.11. Here a shear flow is applied in the (x, y) plane, with the flow in the x direction. Due to the symmetry of the flow, the principal axes of the refractive index tensor will have n_3 oriented along the z axis and the components n_1 and n_2 will lie in the (x, y) plane, but with the component n_1 oriented at some angle χ relative to the x axis. This situation is represented by simply replacing the angle ϕ in equations (2.69) and (2.70) by $\phi - \chi$. In the limit of small birefringences, $|\Delta n_{12}/n| = |(n_1 - n_2)/n| \ll 1$ and $|\Delta n_{32}/n| = |(n_3 - n_2)/n| \ll 1$, the following approximations can be made:

$$\begin{aligned} \kappa_1 - \kappa_2 &= 1/n_1^2 - 1/n_2^2 \approx -2\Delta n_{12}/n, \\ \kappa_3 - \kappa_2 &= 1/n_3^2 - 1/n_2^2 \approx -2\Delta n_{32}/n, \end{aligned} \tag{2.72}$$

where n is the average refractive index. The refractive index and orientation angle measured in the kDB frame are then

$$\Delta n = \left[\Delta n_{12}^2 (s_{\phi-\chi}^2 + c_{\phi-\chi}^2 c_\theta^2)^2 - 2\Delta n_{12}\Delta n_{32} s_\theta^2 (s_{\phi-\chi}^2 - c_{\phi-\chi}^2 c_\theta^2) + \Delta n_{32}^2 s_\theta^4\right]^{1/2},$$

$$\tan 2\alpha = \frac{2\Delta n_{12} c_\theta s_{\phi-\chi} c_{\phi-\chi}}{\Delta n_{12}(s_{\phi-\chi}^2 - c_{\phi-\chi}^2 c_\theta^2) - \Delta n_{32} s_\theta^2}. \tag{2.73}$$

For a couette flow, the light transmission axis is often directed along the vorticity axis so that $\theta = 0$ and $\phi = \pi/2$. In this case, $\Delta n = \Delta n_{12}$ and $\alpha = \chi$. Another frequently used geometry is the parallel plate cell, where the light propagates along $\theta = \phi = \pi/2$ so that $\Delta n = \Delta n_{12} c_\chi^2 - \Delta n_{32}$ and $\alpha = 0$.

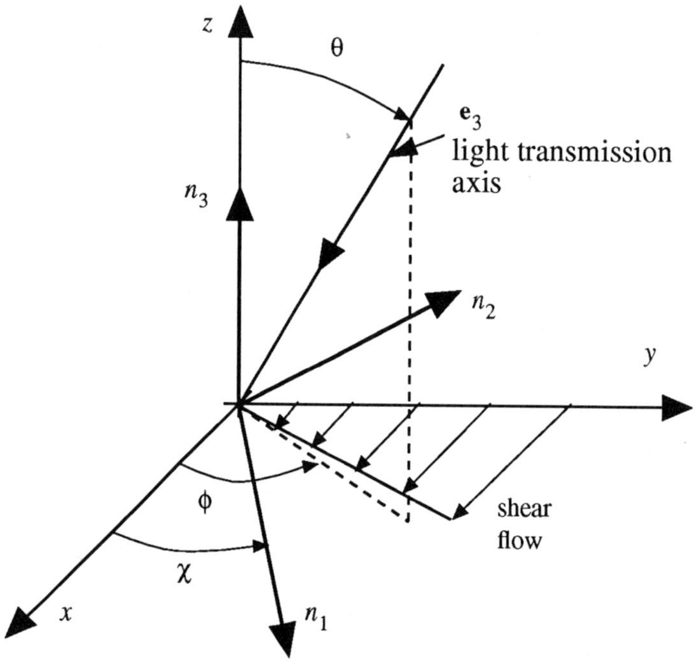

Figure 2.11 Oblique incidence of light through a shear flow. The axes defining the components n_1 and n_2 lie in the (x, y) plane.

3 Reflection and Refraction of Light: Ellipsometry

Light reflected or refracted from a planar interface will be altered in its phase and amplitude. Measurement of the transformation of light following such an interaction can be used to characterize physical properties of the interface. If a thin film is contained at the surface between two media, for example, its refractive index and thickness can be determined. Polarimetry methods used to measure the electric vector emitted in the process of reflection or refraction are commonly referred to as *ellipsometry*, although, in principle, this term can be used interchangeably with polarimetry to describe any technique used to measure the polarization properties of light. In this chapter, the analysis of light interactions with planar interfaces containing any number of stratified thin films is discussed. The treatment of this topic is necessarily brief, and the monograph by Azzam and Bashara [5] is recommended for a more detailed discussion of ellipsometry.

3.1 Reflection and Refraction from a Planar Interface

This problem was treated in section 1.6 of Chapter 1, where the Fresnel coefficients for reflected and refracted light were calculated and presented in equations (1.74) to (1.77). The problem being treated is pictured in Figure 1.4, and it is convenient to represent the electric vector as a Jones vector having orthogonal components that are either parallel (A^i_{jp}) or perpendicular (A^i_{js}) to the plane of incidence (the plane of the figure). Here the superscript $i = f$ and r refers to light propagating in the forward (f) or the reverse (r) direction, relative to the incident light. The subscript $j = 1, 2$ refers to the medium supporting the electric vector. In other words,

$$\mathbf{A}_1^i = \begin{bmatrix} A_{1p}^i \\ A_{1s}^i \end{bmatrix}, \tag{3.1}$$

for the incident, reflected or the refracted light. The Jones matrix describing the transformation of the incident light to reflected light has the general form:

$$\mathbf{J}_R = \begin{bmatrix} R_{pp} & R_{ps} \\ R_{sp} & R_{ss} \end{bmatrix}. \tag{3.2}$$

This matrix contains finite, off diagonal terms, R_{ps} and R_{sp}, when the interface is anisotropic in the plane of the interface; otherwise these terms are zero. For the simple case of isotropic media,

$$\begin{aligned} R_{pp} &= E_{1p}^r / E_{1p}^f, \\ R_{ss} &= E_{1s}^r / E_{1s}^f, \\ R_{ps} &= R_{sp} = 0, \end{aligned} \tag{3.3}$$

where the ratios E_{1i}^r / E_{1i}^f are presented in equations (1.75) and (1.77). In a similar way, the Jones matrix for the refracted light is

$$\mathbf{J}_T = \begin{bmatrix} T_{pp} & T_{ps} \\ T_{sp} & T_{ss} \end{bmatrix}, \tag{3.4}$$

and

$$\begin{aligned} T_{pp} &= E_{2p}^f / E_{1p}^f, \\ T_{ss} &= E_{2s}^f / E_{1s}^f, \\ T_{ps} &= T_{sp} = 0, \end{aligned} \tag{3.5}$$

for isotropic media.

It is useful to define the ellipsometric angles, ψ and Δ, for reflected and refracted light. These are related to the ratios

$$\begin{aligned} \rho_r &= \frac{R_{pp}}{R_{ss}} = \tan\psi_r e^{i\Delta_r}, \\ \rho_t &= \frac{T_{pp}}{T_{ss}} = \tan\psi_t e^{i\Delta_t}, \end{aligned} \tag{3.6}$$

and are often used to represent ellipsometric data since they are directly determined using some ellipsometer designs.

3.2 Stratified, Isotropic Thin Films

The presence of a thin film, or a stack of films, at an interface will affect the polarization properties of reflected and transmitted light. The analysis for isotropic materials is simplified by the fact that the Jones matrix will not contain off-diagonal elements. This means that the reflection and transmission of *p* and *s* polarized light can be treated separately and these subscripts can be dropped in the description of the components of the electric vector.

The physical problem to be solved is described in Figure 3.1. Here a stack of *m* films is contained between two semi-infinite media. The upper medium has refractive index n_0 and supports the incident field, $\mathbf{E}_0$. The lower medium has refractive index n_{m+1} and transmits the refracted light, $\mathbf{E}_{m+1}$. Each film is characterized by a refractive index n_i and thickness d_i, where $i = 1$ to m.

The analysis used here was originally developed by Abeles and closely follows the description offered by Azzam and Bashara [5]. Since the films are isotropic, the orthogonal polarization states, *p* and *s*, can be treated separately, and it is convenient to define a 2×2 vector that recognizes the *direction* of propagation of the electric fields throughout the system. As in section 1.6, electric fields propagating in the forward, or positive z direction, are denoted by a superscript f. A superscript r refers to light propagating in the reverse direction. This vector is

$$\tilde{\mathbf{A}}_i = \begin{bmatrix} E_i^f \\ E_i^r \end{bmatrix}, \tag{3.7}$$

where the index, $i = 1$ to m, refers to the location of the field in the stratified layer.

As light propagates through the stratified layers, it is affected by two basic mechanisms, shown in Figure 3.2. The first mechanism concerns the transmission of the light through a film, and the second arises from reflection and refraction at an interface. An additional subscript, (0,1), has been added to denote whether the electric field exists either at the inside *front* surface of a film, or at the inside *back* surface of a film, respectively. In both cases, these interactions induce the following linear transformations:

$$\tilde{\mathbf{A}}_{i,0} = \mathbf{L}_i \cdot \tilde{\mathbf{A}}_{i,1}; \ \tilde{\mathbf{A}}_{i,1} = \mathbf{K}_{i,i+1} \cdot \tilde{\mathbf{A}}_{i+1,0}, \tag{3.8}$$

where the 2x2 matrices, $\mathbf{L}_i$ and $\mathbf{K}_{i,i+1}$, describe the optical properties of the film i and the interface $i, i+1$, respectively. The total effect of the stratified structure is then

$$\tilde{\mathbf{A}}_0 = \mathbf{K}_{0,1} \cdot \mathbf{L}_1 \cdot \mathbf{K}_{1,2} \cdot \ldots \cdot \mathbf{L}_{m-1} \cdot \mathbf{K}_{m-1,m} \cdot \mathbf{L}_m \cdot \mathbf{K}_{m,m+1} \cdot \tilde{\mathbf{A}}_{m+1}. \tag{3.9}$$

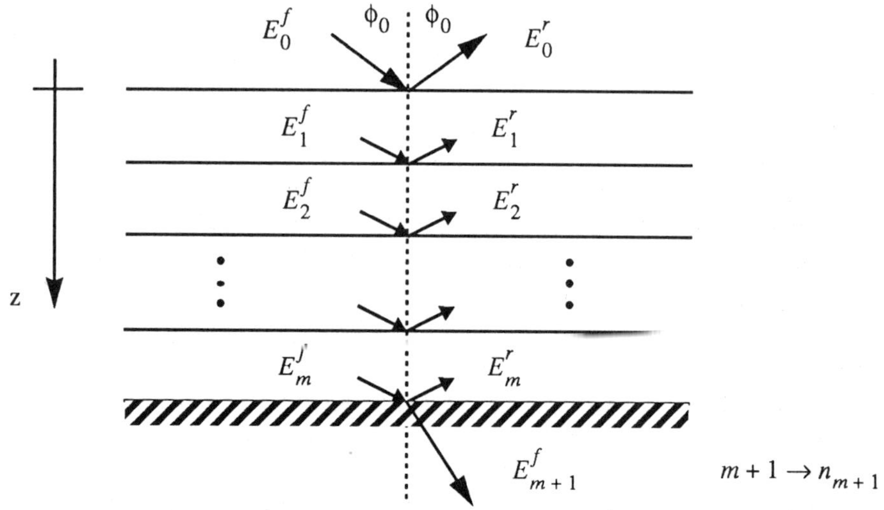

Figure 3.1 Stratified films of isotropic materials. Each film is characterized by a refractive index, n_i, and thickness, d_i.

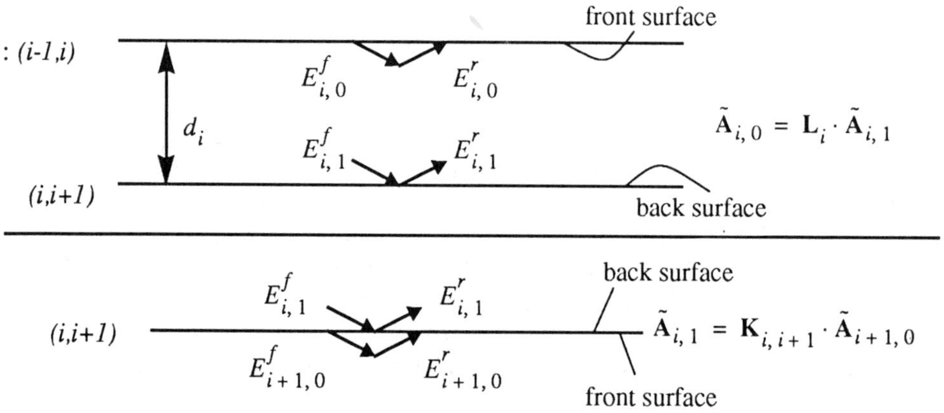

Figure 3.2 Top: Interaction of the electric field through film i. Bottom: Interaction of the electric field across the interface $(i,i+1)$.

The components of the matrices $\mathbf{L}_i$ and $\mathbf{K}_{i,i+1}$ are found by considering the basic processes of transmission and reflection from single slabs and interfaces of material. For example, in Figure 3.2a, light is transmitted through an isotropic slab of material with refractive index n_i and thickness d_i. This situation is described as an isotropic attenuator/retarder and

$$\begin{bmatrix} E^f_{i,0} \\ E^r_{i,0} \end{bmatrix} = \begin{bmatrix} e^{i\delta_i} & 0 \\ 0 & e^{-i\delta_i} \end{bmatrix} \cdot \begin{bmatrix} E^f_{i,1} \\ E^r_{i,1} \end{bmatrix}, \tag{3.10}$$

where

$$\delta_i = \frac{2\pi n_i d_i \cos\phi_i}{\lambda}, \tag{3.11}$$

and the angle ϕ_i is the angle of incidence within the film i. In other words,

$$\mathbf{L}_i = \begin{bmatrix} e^{i\delta_i} & 0 \\ 0 & e^{-i\delta_i} \end{bmatrix}. \tag{3.12}$$

The interface matrix, $\mathbf{K}_{i,\,i+1}$ can be constructed by considering the two elemental process pictured in Figure 3.3. These are simply the reflection/refraction processes described in section 1.6, and it is known that

$$\begin{aligned} E^r_{i,1} &= R_{i,\,i+1} E^f_{i,1}, \\ E^f_{i+1,0} &= T_{i,\,i+1} E^f_{i,1}, \\ E^f_{i+1,0} &= R_{i+1,\,i} E^r_{i+1,0}, \\ E^r_{i,1} &= T_{i+1,\,i} E^r_{i+1,0}. \end{aligned} \tag{3.13}$$

Here $R_{i,\,i+1}$ and $T_{i,\,i+1}$ are the reflection and refraction coefficients defined in equations (1.74) to (1.77) when the incident light travels forward from material i to material $i+1$. The coefficients $R_{i+1,\,i}$ and $T_{i+1,\,i}$ describe the case where the incident light travels in the reverse direction, from material $i+1$ to material i. Examination of equations (1.74) to (1.77) leads to the following relationships:

$$R_{i,\,i+1} = -R_{i+1,\,i}, \tag{3.14}$$

and

$$T_{i+1,\,i} = \frac{1 - R^2_{i,\,i+1}}{T_{i,\,i+1}}. \tag{3.15}$$

Combining the above equations, the matrix $\mathbf{K}_{i,\,i+1}$ is

$$\mathbf{K}_{i,i+1} = \frac{1}{T_{i,i+1}} \begin{bmatrix} 1 & R_{i,i+1} \\ R_{i,i+1} & 1 \end{bmatrix}. \tag{3.16}$$

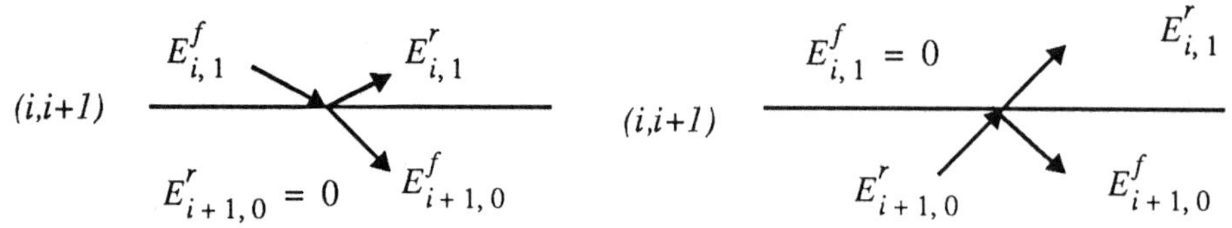

Figure 3.3 Basic reflection and refraction processes making up the matrix $\mathbf{K}_{i,i+1}$.

Equations (3.9), (3.12), and (3.16) are sufficient to calculate the reflection and refraction coefficients for a stratified structure of thin films. In writing the phase factors, δ_i, and the coefficients $R_{i,i+1}$ and $T_{i,i+1}$, however, care must be made to use the angles ϕ_i associated with each film. These are obtained from Snell's law:

$$n_0 \sin\phi_0 = n_1 \sin\phi_1 = \ldots = n_i \sin\phi_i = \ldots = n_{m+1} \sin\phi_{m+1}. \tag{3.17}$$

3.2.1 Example Calculation: Single Isotropic Thin Film

With a single film of refractive index n_1 and thickness d_1 present, the $\mathbf{L}$ and $\mathbf{K}$ matrices are

$$\mathbf{L} = \begin{bmatrix} e^{i\delta} & 0 \\ 0 & e^{-i\delta} \end{bmatrix}; \mathbf{K}_{01} = \frac{1}{T_{01}} \begin{bmatrix} 1 & R_{01} \\ R_{01} & 1 \end{bmatrix}; \mathbf{K}_{12} = \frac{1}{T_{12}} \begin{bmatrix} 1 & R_{12} \\ R_{12} & 1 \end{bmatrix}, \tag{3.18}$$

with $\delta = (2\pi n_1 d_1 \cos\phi_1)/\lambda$. Applying equation (3.9), one obtains

$$\tilde{\mathbf{A}}_0 = \frac{1}{T_{01}T_{12}} \begin{bmatrix} 1 & R_{01} \\ R_{01} & 1 \end{bmatrix} \cdot \begin{bmatrix} e^{i\delta} & 0 \\ 0 & e^{-i\delta} \end{bmatrix} \cdot \begin{bmatrix} 1 & R_{12} \\ R_{12} & 1 \end{bmatrix} \cdot \tilde{\mathbf{A}}_2, \tag{3.19}$$

or

$$\begin{bmatrix} E^f_0 \\ E^r_0 \end{bmatrix} = \frac{e^{i\delta}}{T_{01}T_{12}} \begin{bmatrix} 1 + R_{01}R_{12}e^{-i2\delta} & R_{12} + R_{01}e^{-i2\delta} \\ R_{01} + R_{12}e^{-i2\delta} & R_{01}R_{12} + e^{-i2\delta} \end{bmatrix} \cdot \begin{bmatrix} E^f_2 \\ 0 \end{bmatrix}. \tag{3.20}$$

From this result, the reflection and transmission coefficients for an interface containing a single film are

$$T = \frac{E_0^f}{E_0^f} = \frac{T_{01}T_{12}e^{-i\delta}}{1 + R_{01}R_{12}e^{-i2\delta}}, \tag{3.21}$$

and

$$R = \frac{E_2^r}{E_0^f} = \frac{R_{01} + R_{12}e^{-i2\delta}}{1 + R_{01}R_{12}e^{-i2\delta}}. \tag{3.22}$$

Equations (3.21) and (3.22) can be used for either s or p polarized light. To use them for a specific polarization, the appropriate forms of the reflection and transmission coefficients, $R_{i,i+1,pp}$, $R_{i,i+1,ss}$, $T_{i,i+1,pp}$, and $T_{i,i+1,ss}$, must be used.

The results of this model calculation can be used to extract unknown film properties, such as the film refractive index and thickness. Normally, however, an ellipsometer will yield only two independent observables (the two ellipsometric angles defined in equation (3.6), for example). For that reason, only two film properties can be extracted from a single measurement. If the film is nonabsorbing with a real refractive index, an ellipsometric measurement can provide the real part of the refractive index, n', and the thickness. For an absorbing film, however, there can be up to three unknowns: n', n'', and d. In this case, multiple measurements made at different incident angles, ϕ_0, may be required to specify all three unknowns.

Another important point is that reflection ellipsometers normally yield ratios of the reflection coefficients, R_{pp} and R_{ss}. The equations for these coefficients are nonlinear, transcendental, algebraic equations that must be solved simultaneously for the desired unknowns in an experiment. Techniques to solve these equations are presented in the monograph by Azzam and Bashara [5].

4 Total Intensity Light Scattering

When light intercepts an obstacle it will be scattered, and measurement of the amplitude and polarization properties of the scattered light at various angles relative to the incident beam can provide important structural and dynamic information about a material. Figure 4.1 describes the basic light scattering experiment. Incident light with an electric field $\mathbf{E}_i$ is incident on a scattering object and scatters light $\mathbf{E}_s$ at a scattering angle θ relative to the incident light direction. The wave vectors $\mathbf{k}_i$ and $\mathbf{k}_s$ define the propagation direction of the incident and scattering fields. The vector $\mathbf{x}$ specifies the point of measurement. The vector $\mathbf{x}'$ scales with the size of the scattering object.

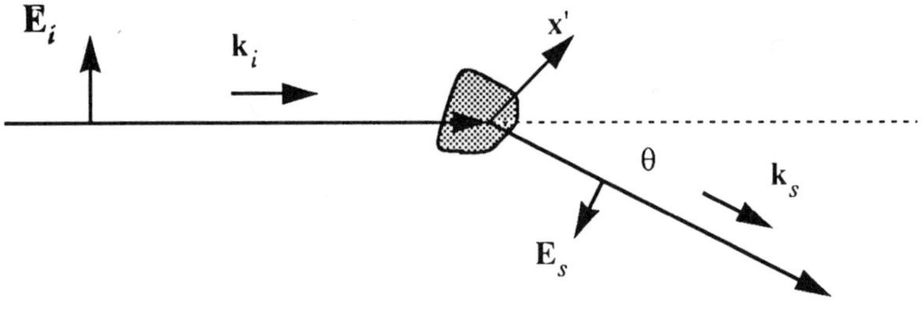

Figure 4.1 The basic light scattering experiment.

The treatment of light scattering in this monograph is found in three chapters. The present chapter discusses total intensity light scattering where the time-averaged intensity of scattered light is measured and $|\mathbf{k}_i| = |\mathbf{k}_s|$. Dynamic light scattering, where time correlations of the light scattered from fluctuations in the sample are measured, is covered in Chapter 6. Raman scattering, which results from vibrational modes in the sample is distinguished from the latter two processes in that $|\mathbf{k}_i| \neq |\mathbf{k}_s|$ and can be used to determine spectroscopic measurements of orientation dynamics. This phenomena is discussed in section 5.4.

4.1 Light Scattering in the Far Field: The Born Approximation

Equation (1.35) provides a general solution to the Maxwell equations and is particularly useful for problems in scattering. Rewriting the results for the scattered field, we have

$$\mathbf{E}_s(\mathbf{x}) = k_0^2 \int_V d\mathbf{x}' \mathbf{G}(\mathbf{x}, \mathbf{x}') \cdot (n^2(\mathbf{x}') - \mathbf{I}) \cdot \mathbf{E}(\mathbf{x}') , \tag{4.1}$$

where the Green's function is

$$\mathbf{G}(\mathbf{x}, \mathbf{x}') = (\mathbf{I} + k_0^{-2} \nabla\nabla) \frac{e^{ik_0|\mathbf{x}-\mathbf{x}'|}}{4\pi|\mathbf{x}-\mathbf{x}'|} . \tag{4.2}$$

In this expression, the vector $\mathbf{x}$ defines the position at which the scattered light is detected, and $\mathbf{x}'$ is an integration variable over a volume, V, containing the scattering element. Since the scattered light is normally measured at a great distance relative to length scales associated with the sample $(|\mathbf{x}| \gg |\mathbf{x}'|)$, the following approximation can be used for $|\mathbf{x} - \mathbf{x}'|$:[1,8]

$$|\mathbf{x} - \mathbf{x}'| = r\left(1 - 2\frac{\mathbf{x} \cdot \mathbf{x}'}{r^2} + \frac{r'^2}{r^2}\right)^{1/2} \approx r - \frac{\mathbf{x} \cdot \mathbf{x}'}{r} , \tag{4.3}$$

where $r = |\mathbf{x}|$ and $r' = |\mathbf{x}'|$. Defining $\mathbf{u} = \mathbf{x}/r$, this results in the following simpler form of the Green's function,

$$\mathbf{G}(\mathbf{x}, \mathbf{x}') = (\mathbf{I} - \mathbf{u}\mathbf{u}) \frac{e^{ik_0 r}}{4\pi r} e^{-ik_0 \mathbf{u} \cdot \mathbf{x}'} . \tag{4.4}$$

The unit vector, $\mathbf{u}$, defines the direction of propagation of the scattered light and $\mathbf{k}_s = k_0 \mathbf{u}$. Inserting this result into equation (4.1) leads to

$$\mathbf{E}_s = k_0^2 \int_V d\mathbf{x}' \frac{e^{ikr}}{4\pi r} e^{-i\mathbf{k}_s \cdot \mathbf{x}'} (\mathbf{I} - \mathbf{u}\mathbf{u}) \cdot (n^2(\mathbf{x}') - \mathbf{I}) \cdot \mathbf{E}(\mathbf{x}') . \tag{4.5}$$

4.1.1 Dipole or Rayleigh Scattering

An electric field will polarize and distort the charge distribution, $\rho(\mathbf{x})$, within a material. The first moment of this distribution is the charge dipole [1],

$$\mathbf{p} = \int d\mathbf{x}\, \mathbf{x} \rho(\mathbf{x}) . \tag{4.6}$$

Scattering from a discrete, small particle with a characteristic dimension much smaller than the wavelength of light, $a \ll \lambda$, can be approximated as scattering from a dipole. In this case, the term $(\boldsymbol{n}^2(\mathbf{x}') - \mathbf{I}) \cdot \mathbf{E}(\mathbf{x}')$ in equation (4.5) is replaced by

$$(\boldsymbol{n}^2(\mathbf{x}') - \mathbf{I}) \cdot \mathbf{E}(\mathbf{x}') \rightarrow \mathbf{p}\delta(\mathbf{x}') , \tag{4.7}$$

for a dipole situated at the origin. The scattered light electric field is

$$\mathbf{E}_s = \mathbf{k}_s \times (\mathbf{k}_s \times \mathbf{p}) \frac{e^{ik_0 r}}{4\pi r} , \tag{4.8}$$

where the identity, $k_0^2(\mathbf{I} - \mathbf{uu}) \cdot \mathbf{p} = \mathbf{k}_s \times (\mathbf{k}_s \times \mathbf{p})$, has been used.

As an example, consider the case pictured in Figure 4.2. Here a dipole, $\mathbf{p}$, is oriented normal to the incident wave vector, $\mathbf{k}_i$, and the scattering angle is θ. The double cross product in (4.8) ensures that the scattered electric vector is orthogonal to the scattered light wave vector, $\mathbf{k}_s$. The magnitude of $\mathbf{E}_s$ is

$$E_s = k_s^2 p \cos\theta \frac{e^{ik_0 r}}{4\pi r} , \tag{4.9}$$

so that the scattered light will produce isointensity contours, $I = |E_s|^2$, as shown in Figure 4.3. The intensity of light observed at $\theta = \pi/2$, therefore, will be zero for a dipole oriented along this direction.

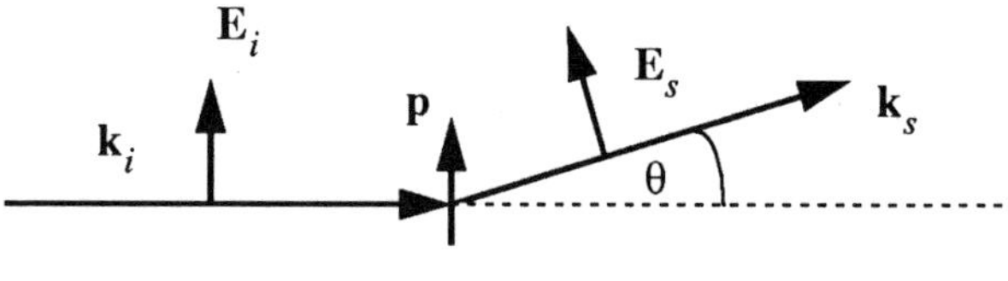

Figure 4.2 Scattering by a dipole.

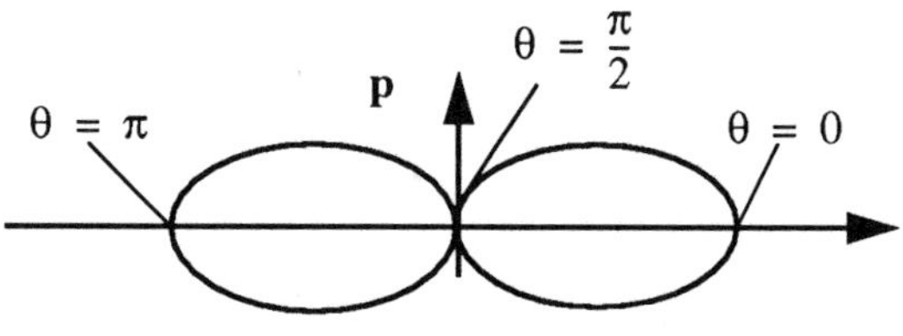

Figure 4.3 Isointensity contour for the dipole pictured in Figure 4.2.

4.1.2 The Polarizability Tensor

At optical frequencies, it is the dipole *induced* by the incident electric field that is normally important. The induced dipole is calculated from the following empirical relationship:[9]

$$\mathbf{p} = \underline{\alpha} \cdot \mathbf{E}, \tag{4.10}$$

where $\underline{\alpha}$ is the polarizability tensor. The induced dipole, therefore, is a measure of how facile the electron distribution is to distortion by an electric field. If the material is isotropic, the polarizability is a simple scalar and the induced dipole is oriented parallel to the incident electric field as described in Figure 4.2.

For an anisotropic material, the polarizability will be a tensor that will depend on the relative orientation of the principal axes of the material and the incident electric field. If the polarizability is biaxial and of the form,

$$\underline{\alpha} = \begin{bmatrix} \alpha_3 & 0 & 0 \\ 0 & \alpha_2 & 0 \\ 0 & 0 & \alpha_1 \end{bmatrix}, \tag{4.11}$$

in a frame coaxial with its principal directions, it must be rotated to represent a system oriented at the polar and azimuthal angles, θ' and φ', shown in Figure 4.4.

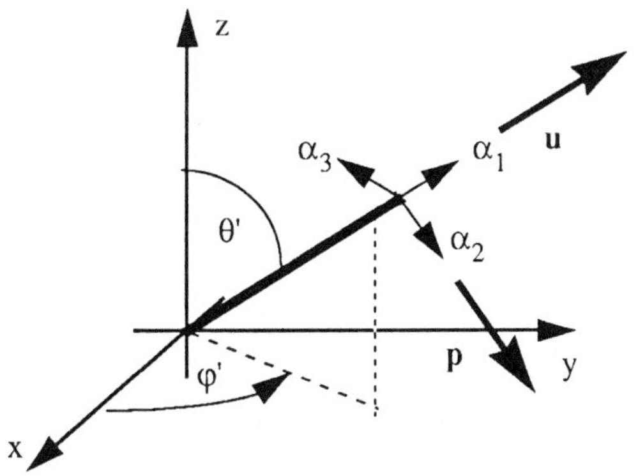

Figure 4.4 Biaxial polarizability tensor rotated by (θ', φ').

Applying rotation matrices defined in (2.67) to this tensor and assuming a *uniaxial* form ($\alpha_2 = \alpha_3$) gives

$$\underline{\alpha}(\theta, \phi) = \begin{bmatrix} s_{\varphi'} & (c_{\theta'}c_{\varphi'}) & (s_{\theta'}c_{\varphi'}) \\ -c_{\varphi'} & (c_{\theta'}s_{\varphi'}) & (s_{\theta'}s_{\varphi'}) \\ 0 & -s_{\theta'} & c_{\theta'} \end{bmatrix} \cdot \begin{bmatrix} \alpha_2 & 0 & 0 \\ 0 & \alpha_2 & 0 \\ 0 & 0 & \alpha_1 \end{bmatrix} \cdot \begin{bmatrix} s_{\varphi'} & -c_{\theta'} & 0 \\ (c_{\theta'}c_{\varphi'}) & (c_{\theta'}s_{\varphi'}) & -s_{\theta'} \\ (s_{\theta'}c_{\varphi'}) & (s_{\theta'}s_{\varphi'}) & c_{\theta'} \end{bmatrix}$$

$$= \alpha_2\mathbf{I} + (\alpha_1 - \alpha_2)\begin{bmatrix} (c_{\varphi'}^2 s_{\theta'}^2) & (c_{\varphi'}s_{\varphi'}s_{\theta'}^2) & (c_{\varphi'}s_{\theta'}c_{\theta'}) \\ (c_{\varphi'}s_{\varphi'}s_{\theta'}^2) & (s_{\varphi'}^2 s_{\theta'}^2) & (s_{\varphi'}s_{\theta'}c_{\theta'}) \\ (c_{\varphi'}s_{\theta'}c_{\theta'}) & (s_{\varphi'}s_{\theta'}c_{\theta'}) & c_{\theta'}^2 \end{bmatrix} \tag{4.12}$$

$$= \alpha_2\mathbf{I} + (\alpha_1 - \alpha_2)(\mathbf{uu}).$$

In writing equation (4.12), the shorthand $s_\theta = \sin\theta$ and $c_\theta = \cos\theta$ has been used, and the unit vector $\mathbf{u}$ defines the orientation of the symmetry axis of the polarizability tensor (see Fig. 4.4).

Use of this form for the polarizability tensor in equation (4.10) leads to the result that the dipole will not be oriented parallel to the incident electric field, as it is when the polarizability is isotropic. Additionally, inserting the resulting expression for the dipole into equation (4.8) will produce a much more complex angular dependence for the scattered light electric field.

An alternative and simpler approach to deriving the result in equation (4.12) is to express the polarizability tensor as a general expansion in the two orthogonal unit vectors, $\mathbf{u}$ and $\mathbf{p}$, embedded on the principal axes shown in Figure 4.4. Evidently, using Einstein notation, the polarizability can be written as

$$\alpha_{ij} = a\delta_{ij} + bu_iu_j + cp_ip_j + d(u_ip_j + p_iu_j), \tag{4.13}$$

where the constants a, b, c, and d are independent of orientation. These constants can be determined by rotating the polarizability tensor to be coaxial with the laboratory frame ($u_i = (0, 0, 1)$ and $p_i = (1, 0, 0)$), so that,

$$\begin{bmatrix} \alpha_3 & 0 & 0 \\ 0 & \alpha_2 & 0 \\ 0 & 0 & \alpha_1 \end{bmatrix} = a\begin{bmatrix} 1 & 0 & 0 \\ 0 & 1 & 0 \\ 0 & 0 & 1 \end{bmatrix} + b\begin{bmatrix} 0 & 0 & 0 \\ 0 & 0 & 0 \\ 0 & 0 & 1 \end{bmatrix} + c\begin{bmatrix} 1 & 0 & 0 \\ 0 & 0 & 0 \\ 0 & 0 & 0 \end{bmatrix} + d\begin{bmatrix} 0 & 0 & 1 \\ 0 & 0 & 0 \\ 1 & 0 & 0 \end{bmatrix}, \tag{4.14}$$

from which

$$\begin{aligned} a &= \alpha_2, \\ a + c &= \alpha_3, \\ a + b &= \alpha_1, \\ d &= 0. \end{aligned} \tag{4.15}$$

The polarizability for a biaxial material is then

$$\alpha_{ij} = \alpha_2\delta_{ij} + \Delta_{12}u_iu_j + \Delta_{32}p_ip_j, \tag{4.16}$$

where $\Delta_{12} = \alpha_1 - \alpha_2$ and $\Delta_{32} = \alpha_3 - \alpha_2$. This result agrees with equation (4.12) when $\alpha_2 = \alpha_3$.

4.1.3 Polarizability of a Dielectric Sphere

It is instructive to develop the solution for scattering by a small sphere of radius, $a \ll \lambda$. In such a limit the sphere is represented as a point dipole, and to determine its polarizability, the interaction of the sphere with the electric field is modeled as shown in Figure 4.5. The restriction that the sphere is much smaller that the wavelength of light suggests that to a first approximation, the electric field, at an instant in time, appears to the sphere as a uniform field. We must solve the following *time-independent* Maxwell's equations [1],

$$\begin{aligned} \nabla\times\mathbf{E} &= 0, \\ \nabla\cdot\mathbf{E} &= 0, \end{aligned} \tag{4.17}$$

subject to the following boundary conditions,

$$\begin{aligned} (\varepsilon_2\mathbf{E}_2 - \varepsilon_1\mathbf{E}_1)\cdot\mathbf{n} &= 0;\ |\mathbf{x}| = a, \\ (\mathbf{E}_2 - \mathbf{E}_1)\times\mathbf{n} &= 0;\ |\mathbf{x}| = a, \\ \mathbf{E} \to \mathbf{E}_0;\ |\mathbf{x}| &\to \infty, \end{aligned} \tag{4.18}$$

where ε_1 and ε_2 are the dielectric constants of the sphere and matrix, $\mathbf{n}$ is the unit normal on the surface of the sphere, and $\mathbf{E}_0$ is the incident electric field.

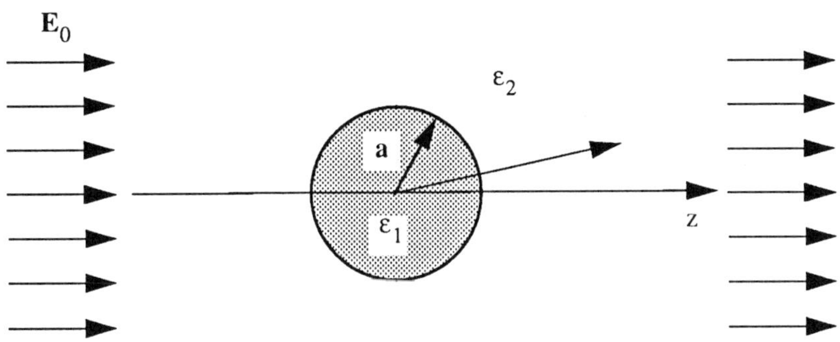

Figure 4.5 Dielectric sphere immersed with a uniform electric field, $\mathbf{E}_0$. The sphere and matrix have dielectric constants, ε_1 and ε_2, respectively.

The requirement that $\nabla\times\mathbf{E} = 0$ is immediately satisfied by seeking a solution in the form of $\mathbf{E} = -\nabla\Phi$, where $\Phi(\mathbf{x})$ is a potential function. The divergence condition, $\nabla\cdot\mathbf{E} = 0$, on the electric field leads to

$$\nabla^2\Phi = 0, \tag{4.19}$$

subject to

$$\begin{aligned}(\varepsilon_2\nabla\Phi_2 - \varepsilon_1\nabla\Phi_1)\cdot\mathbf{n} &= 0;\ |\mathbf{x}| = a,\\ \Phi_2 = \Phi_1;\ |\mathbf{x}| &= a,\\ \Phi_1 \to -E_0 z;\ |\mathbf{x}| &\to \infty.\end{aligned} \tag{4.20}$$

Equation (4.19) is Laplace's equation and has fundamental solutions in the form of harmonic functions. These functions are of two types: growing harmonics, which are appropriate for bounded, interior regions, and decaying harmonics, which apply to unbounded space. These functions are expansions of the Green's function solution to Laplace's equation, $G(\mathbf{x}) = 1/(4\pi|\mathbf{x}|)$, and are

$$\frac{4\pi(-1)^n}{1\cdot 3\cdot\ldots\cdot(2n-1)}\frac{\partial^n}{\partial x_i\partial x_j\ldots\partial x_k}G(\mathbf{x}) = \frac{1}{|\mathbf{x}|}, \frac{\mathbf{x}}{|\mathbf{x}|^3}, \left(\frac{\mathbf{xx}}{|\mathbf{x}|^5} - \frac{\mathbf{I}}{3|x|^3}\right), \ldots \tag{4.21}$$

where $n = 0, 1, 2, \ldots$, for the decaying harmonics. The growing harmonics are

$$|\mathbf{x}|^{2n+1}\frac{4\pi(-1)^n}{1\cdot 3\cdot\ldots\cdot(2n-1)}\frac{\partial^n}{\partial x_i\partial x_j\ldots\partial x_k}G(\mathbf{x}) = 1, \mathbf{x}, \left(\mathbf{xx} - \frac{|x|^2\mathbf{I}}{3}\right), \ldots. \tag{4.22}$$

In general, the solution to Laplace's equation will be a superposition of harmonic functions. For the problem of the sphere, the fields Φ_1 and Φ_2 are

$$\begin{aligned}\Phi_1 &= c_1\mathbf{x}\cdot\mathbf{E}_0,\\ \Phi_2 &= c_2\frac{\mathbf{x}}{|\mathbf{x}|^3}\cdot\mathbf{E}_0 - E_0 z,\end{aligned} \tag{4.23}$$

and where c_1 and c_2 are constants. This form of the solution is motivated by recognizing that the potential is a scalar function that must be linear in the external field, $\mathbf{E}_0$. Applying the boundary conditions gives

$$\begin{aligned}\Phi_1 &= -\frac{3\varepsilon_2}{(\varepsilon_1 + 2\varepsilon_2)}E_0 z,\\ \Phi_2 &= \frac{(\varepsilon_1 - \varepsilon_2)a^3}{(\varepsilon_1 + 2\varepsilon_2)}\frac{E_0 z}{|\mathbf{x}|^3} - E_0 z.\end{aligned} \tag{4.24}$$

To identify the polarizability of the sphere, it is left to compare this solution to the solution of the electric field in the presence of a point dipole. To solve this problem, the simple model of a point dipole displayed in Figure 4.6 is used. Here two opposite point charges are separated by a distance l about the origin. The charge distribution is then

$$\rho(\mathbf{x}) = q\delta\left(\mathbf{x} - \frac{l}{2}\mathbf{u}_z\right) - q\delta\left(\mathbf{x} + \frac{l}{2}\mathbf{u}_z\right), \tag{4.25}$$

where $\mathbf{u}_z$ is the unit vector in the z direction. From the definition in equation (4.6), the point dipole is taken as

$$\mathbf{p} = \lim_{l \to 0} \int d\mathbf{x} \rho(\mathbf{x}) \mathbf{x}. \tag{4.26}$$

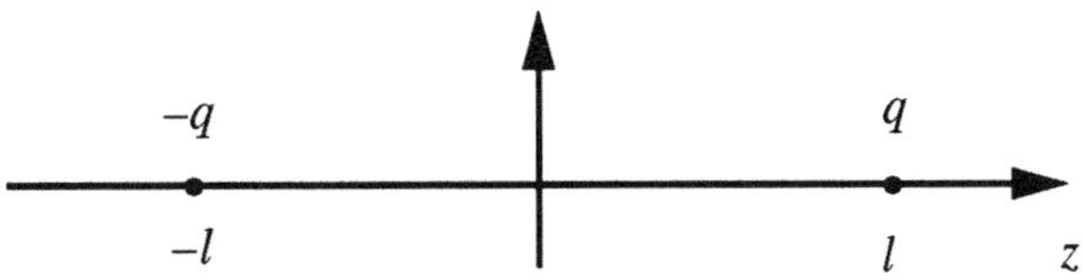

Figure 4.6 Model of a point dipole with charges q and $-q$ located at l and $-l$, respectively.

The time-independent Maxwell's equations to be solved in this case are

$$\nabla \cdot \mathbf{E} = \rho(\mathbf{x}),$$
$$\nabla \times \mathbf{E} = 0. \tag{4.27}$$

Employing the potential function, Φ, we are led to an inhomogeneous form of Laplace's equation with the solution,

$$\Phi(\mathbf{x}) + E_0 z = \lim_{l \to 0} \frac{1}{4\pi} \int d\mathbf{x}' \frac{\rho(\mathbf{x}')}{|\mathbf{x} - \mathbf{x}'|} = \frac{1}{4\pi} \frac{pz}{|\mathbf{x}|^3}. \tag{4.28}$$

Comparing equations (4.24) and (4.28) along with the definition of the polarizability in equation (4.10) gives

$$\underline{\alpha} = \frac{3}{4\pi} \frac{(\varepsilon_1 - \varepsilon_2)}{\varepsilon_1 + 2\varepsilon_2} V\mathbf{I}, \tag{4.29}$$

where V is the volume of the sphere.

In general, the polarizability is related to the volume of an object and the magnitude of the scattered light will be proportional to the contrast in the dielectric constants of the scattering object and its supporting matrix.

4.2 Rayleigh-Debye Scattering

The previous calculations have assumed that the element scattering the light is sufficiently small relative to the wavelength of light so that the element can be approximated as a dipole located at a discrete point in space. The elements are also assumed to be sufficiently well separated so that the light striking them is the incident field and multiple scattering is negligible. Furthermore, the scattered light from separate elements does not interact and simply superimposes to produce the total scattered field.

When a particle becomes comparable in size or larger than the wavelength, several complications may arise. If the index of refraction of the particle is different from its surroundings ($n_p/n_s \neq 1$), refraction can occur at the interface that will contribute an angular dependence to the scattered light. This is referred to as anomalous diffraction. In addition, the total scattered field will be a superposition of light scattered from elements internal to the particle. When these individual contributions are summed, phase differences must be considered and will affect the result in two ways: (a) the incident light striking any element may be changed in phase due to a previous scattering interaction by another element, and (b) there will be phase interference (constructive and destructive) when individual contributions are added to calculate the total field.

Rayleigh-Debye theory operates under two assumptions [9] that allow the problem to be simplified, while still permitting the analysis of large particles. The first assumption is that $|n_p/n_s - 1| \ll 1$, which minimizes the effects of anomalous diffraction. The second restriction is that $2\pi|n_p/n_s - 1|\ (a/\lambda) \ll 1$, where a is a characteristic length scale of the particle. When this is satisfied, phase difference effects *inside* the particle can be neglected. Constructive and destructive interference at the point of measurement, however, will still need to be considered. Indeed, these effects allow light scattering to provide a sensitive measurement of the size and structure of scattering elements.

When these conditions are met, equation (4.5) becomes

$$\mathbf{E}_s = E_0 \mathbf{k}_s \times \mathbf{k}_s \times \mathbf{n}_i \left[\left(\frac{n_p}{n_s} \right)^2 - 1 \right] \frac{e^{ikr}}{4\pi r} \int_V dx' e^{-i(\mathbf{k}_s - \mathbf{k}_i) \cdot \mathbf{x}'} . \tag{4.30}$$

In writing this result, the dielectric properties are taken to be isotropic and the internal field, $\mathbf{E}(\mathbf{x}')$, in equation (4.5) has been replaced by the incident field due to the assumptions mentioned above. In other words,

$$\mathbf{E}(\mathbf{x}') \rightarrow \mathbf{E}_0(\mathbf{x}') = E_0 e^{i\mathbf{k}_i \cdot \mathbf{x}'} \mathbf{n}_i . \tag{4.31}$$

The difference $\mathbf{k}_s - \mathbf{k}_i$ defines the scattering vector,

$$\mathbf{q} = \mathbf{k}_s - \mathbf{k}_i , \tag{4.32}$$

and has a magnitude $2k \sin(\theta/2)$. The orientations of the three wave vectors associated with a scattering experiment are shown in Figure 4.7.

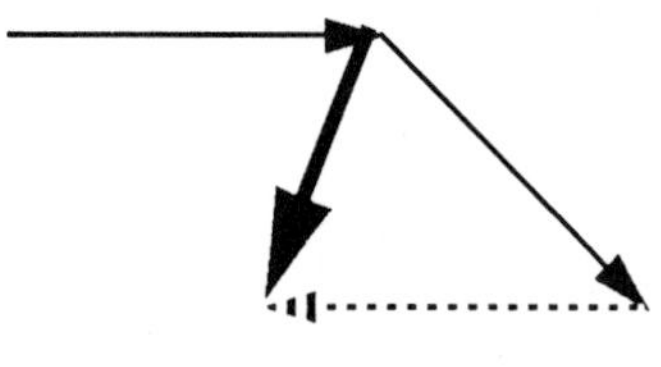

Figure 4.7 The scattering vector, $\mathbf{q}$.

The integral in equation (4.30) describes the interaction of the phase of scattered light emitted over the volume of the particle. Figure 4.8 describes the process where light is scattered by two separate elements within a particle. These elements are separated by a vector, $\mathbf{x}'$, and scatter light along the direction specified by $\mathbf{k}_s$. The incident wave vector is $\mathbf{k}_i$ and the difference in the phase of the scattered light by these two elements is simply [9],

$$\Delta\delta = kl_2 - kl_1 = \mathbf{k}_s \cdot \mathbf{x}' - \mathbf{k}_i \cdot \mathbf{x}' = \mathbf{q} \cdot \mathbf{x}' . \qquad (4.33)$$

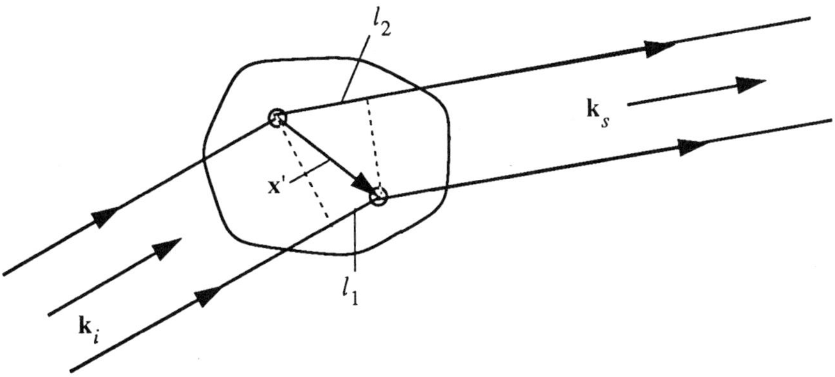

Figure 4.8 The phase difference between the scattered light from two elements within a particle.

The total phase difference of the scattered light from the entire particle is simply the volume integral of the phase factor, $e^{i\mathbf{q} \cdot \mathbf{x}'}$, appearing in equation (4.30) and is referred to as the Rayleigh form factor,

$$R(\theta, \varphi) = \frac{1}{V}\int_V d\mathbf{x}' e^{i\mathbf{q} \cdot \mathbf{x}'} . \qquad (4.34)$$

This factor can be evaluated exactly for many geometries. It is usually an advantage to perform the integration using lamella parallel to the *bisectrix*, $\mathbf{x}''$, defined by

$$\mathbf{x}'' \cdot \mathbf{q} = 0 , \qquad (4.35)$$

since all points within such lamella have a zero phase difference. This construction is shown in Figure 4.9.

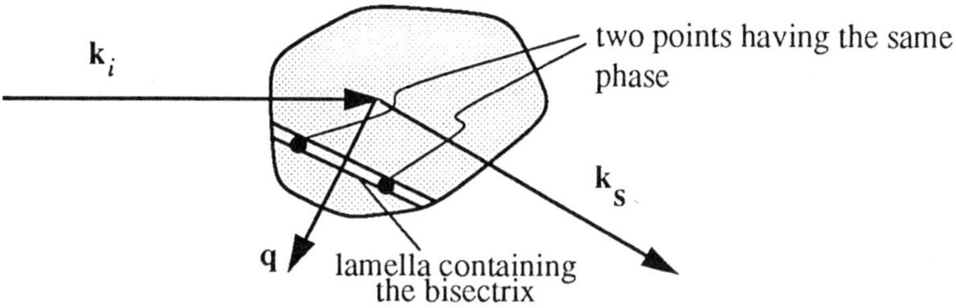

Figure 4.9 Construction of the bisectrix to produce points having the same phase.

4.2.1 Rayleigh Form Factor of a Sphere

A sphere of radius a is pictured in Figure 4.10. The bisectrix disk shown in this figure has a radius $\sqrt{a^2 - b^2}$. All points within the disk have a phase relative to the origin of

$$\delta(b) = qb = 2kb\sin\frac{\theta}{2}. \tag{4.36}$$

The differential volume of the bisectrix disk is $\pi(a^2 - b^2)\,db$ and the form factor is

$$R(\theta, \varphi) = \frac{1}{\frac{4}{3}\pi a^3}\int_{-a}^{a} db\pi(a^2 - b^2)e^{iqb} = \frac{3}{(qa)^3}[\sin(qa) - qa\cos(qa)]. \tag{4.37}$$

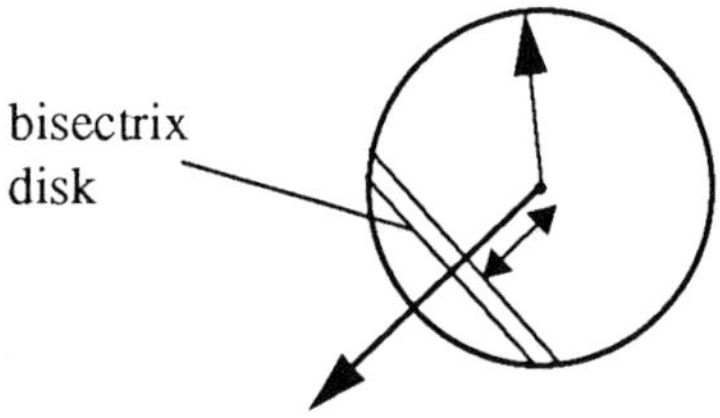

Figure 4.10 Bisectrix construction for a sphere.

The scattered light intensity is

$$I = I_0|\mathbf{k}_s \times (\mathbf{k}_s \times \mathbf{n}_i)|^2\left[\left(\frac{n_p}{n_s}\right)^2 - 1\right]^2\frac{a^6}{9r^2}R^2(\theta), \tag{4.38}$$

where I_0 is the incident light intensity. The factor, $|\mathbf{k}_s \times (\mathbf{k}_s \times \mathbf{n}_i)|^2$ is $k^4\cos^2\theta$ for incident light polarized in the scattering plane (the plane containing the scattering vector, $\mathbf{q}$). When $\mathbf{n}_i$ is orthogonal to the scattering plane, this factor is simply k^4.

Figure 4.11 plots the square of the form factor, $R^2(\theta)$, as a function of the dimensionless scattering vector, qa. Scattering from a sphere will produce a spherically symmetric scattering pattern with a central peak circumscribed by rings with intensity minima located at the roots of $R(\theta) = 0$. Location of the first minimum can therefore be used to determine the size of the scattering sphere. Polydispersity in a sample, however, will smooth this oscillatory pattern and remove the appearance of the minima.

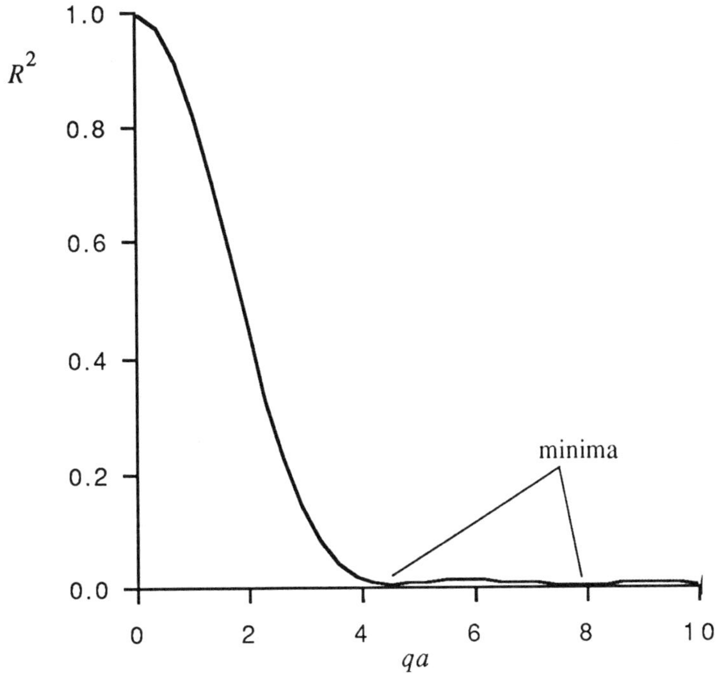

Figure 4.11 Intensity of scattered light as a function of the dimensionless scattering vector.

4.2.2 Rayleigh Form Factor for a Cylinder

The Rayleigh form factor for a cylinder can be easily calculated and the details of the calculation can be found in van de Hulst [9]. The geometry of the scattering problem is shown in Figure 4.12. Here the symmetry axis of the cylinder is oriented according to the spherical angles θ' and ϕ'. The incident light is directed along the positive z direction, and scattered light is observed along the axis defined by the angles θ and ϕ. For a cylinder of length a and diameter b, the Rayleigh form factor is [10]

$$R(\theta, \phi) = 2\frac{J_1(K_b)}{K_b}\frac{\sin K_a}{K_a}, \tag{4.39}$$

where $K_b = kb\sin(\theta/2)\sin\beta$, $K_a = ka\sin(\theta/2)\cos\beta$, and J_1 is a Bessel function of the first kind. The angle β is the angle spanning the particle's symmetry axis and the bisectrix. It is given by

$$\cos\beta = \sin\theta'\cos(\theta/2)\cos(\phi' - \varepsilon) - \cos\theta'\sin(\theta/2) . \tag{4.40}$$

Here the angle ε is the angle between the x axis and the "s" polarization of the scattered light (see Fig. 4.12).

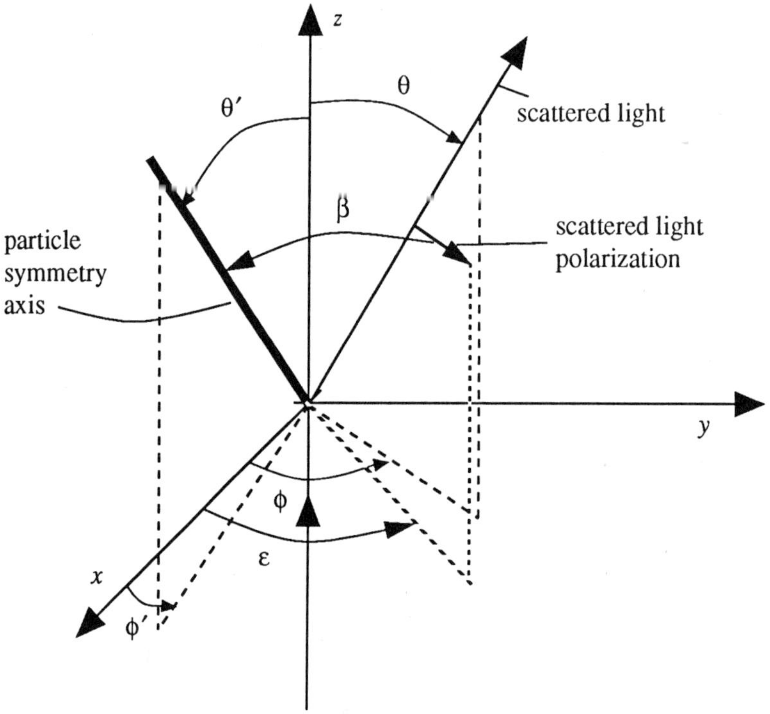

Figure 4.12 Scattering geometry for an axisymmetric particle.

4.2.3 Rayleigh Form Factor for a Spheroid

The Rayleigh form factor for a spheroid of aspect ratio μ and semiminor axis b was calculated by Saito and Ikeda.[11] Using the same geometry as supplied in Figure 4.12, the result is

$$R(\theta, \phi) = \frac{3\sin K}{K^3} - \frac{3\cos K}{K^2} , \tag{4.41}$$

where $K = 2kb\sin(\theta/2)\sqrt{1 + (\mu^2 - 1)\cos^2\beta}$. This result has been used by Meeten [10] and Frattini and Fuller [12] for the calculation of scattering dichroism in the Rayleigh-Debye scattering approximation.

4.3 Light Scattering from Fluctuations and the Structure Factor

Fluctuations in the optical properties of a material can induce spatial and temporal inhomogeneities that scatter light. In general, the dielectric tensor is taken as the following function of space and time,

$$\varepsilon_{ij}(\mathbf{x}, t) = \langle \varepsilon_{ij} \rangle + \delta\varepsilon_{ij}(\mathbf{x}, t) , \tag{4.42}$$

where $\langle \varepsilon_{ij} \rangle$ is the average, homogeneous dielectric tensor and $\delta\varepsilon_{ij}$ is the fluctuation. These fluctuations will, in turn, cause fluctuations in the electric field within the material:

$$\mathbf{E} = \langle \mathbf{E} \rangle + \delta\mathbf{E} , \tag{4.43}$$

with the fluctuating term being the scattered light contribution. The average field, $\langle \mathbf{E} \rangle = \mathbf{E}_0 e^{-\mathbf{k}_i \cdot \mathbf{x}}$, where $\mathbf{k}_i$ is the wave vector of the incident field, is primarily due to transmission of the incident light through the material. For that reason, it is influenced by the mean dielectric tensor of the material.

The light scattered from such a disturbance is found by simply replacing the term $n^2 - \mathbf{I}$ in equation (1.35) by $\delta\underline{\varepsilon}(\mathbf{x})$ to give [8,13],

$$\delta\mathbf{E}(\mathbf{x}, t) = k_0^2 \int d\mathbf{x}' \mathbf{G}(\mathbf{x}, \mathbf{x}') \cdot \delta\underline{\varepsilon}(\mathbf{x}') \cdot \mathbf{E}(\mathbf{x}') . \tag{4.44}$$

Using the result for the Green's function in equations (1.47) and (1.49), the scattered field is

$$\delta\mathbf{E}(\mathbf{x}) = \lim_{\eta \to 0} \frac{1}{(2\pi)^3} \int d\mathbf{x}' \int d\mathbf{k}' e^{-i\mathbf{k}' \cdot (\mathbf{x} - \mathbf{x}')} \frac{k_0^2 \mathbf{I} - \mathbf{k}'\mathbf{k}'}{k'^2 - (k_0 + i\eta)^2} \cdot \delta\underline{\varepsilon}(\mathbf{x}') \cdot \langle \mathbf{E}(\mathbf{x}') \rangle , \tag{4.45}$$

or

$$\delta\mathbf{E}(\mathbf{x}) = \frac{k_0^2}{4\pi} \int d\mathbf{x}' \frac{e^{-ik|\mathbf{x} - \mathbf{x}'|}}{|\mathbf{x} - \mathbf{x}'|} (\mathbf{I} - k_0^{-2} \nabla\nabla) \cdot \delta\underline{\varepsilon}(\mathbf{x}') \cdot \mathbf{E}(\mathbf{x}') . \tag{4.46}$$

In the far field, the approximations leading up to equation (4.30) can be used to yield

$$\delta\mathbf{E}(\mathbf{x}) = \frac{E_0}{4\pi r} \mathbf{k}_s \times \mathbf{k}_s \times \int d\mathbf{x}' e^{-i\mathbf{q} \cdot \mathbf{x}'} \delta\underline{\varepsilon}(\mathbf{x}', t) \cdot \mathbf{n}_i . \tag{4.47}$$

The fluctuations in $\underline{\varepsilon}$ arise from spatial and temporal variations in the temperature, density, and concentration fields throughout the sample. In other words,

$$\delta\underline{\varepsilon}(\mathbf{x}, t) = \left(\frac{d}{dc}\underline{\varepsilon}\right)_{T,\rho} \delta c(\mathbf{x}, t) + \left(\frac{d}{dT}\underline{\varepsilon}\right)_{c,\rho} \delta T(\mathbf{x}, t) + \left(\frac{d}{d\rho}\underline{\varepsilon}\right)_{T,c} \delta\rho(\mathbf{x}, t) . \tag{4.48}$$

Normally, the anisotropic part of the fluctuations will be much smaller than the isotropic contribution. Considering only concentration fluctuations, the isotropic fluctuation in the dielectric tensor is

$$\delta\underline{\varepsilon}(\mathbf{x}, t) = \left(\frac{\partial \varepsilon_0}{\partial c}\right)\delta c(\mathbf{x}, t)\,\mathbf{I}. \tag{4.49}$$

The intensity of the scattered light is

$$I = \langle \mathbf{E}_s \cdot \mathbf{E}_s^* \rangle, \tag{4.50}$$

where the angular brackets refer to an average over the distribution of the fluctuations throughout the sample. Using equations (4.47) and (4.49), the intensity is

$$I = \frac{|E_0|^2}{(4\pi r)^2}|\mathbf{k}_s \times \mathbf{k}_s \times \mathbf{n}_i|^2\left(\frac{\partial \varepsilon_0}{\partial c}\right)^2 \iint d\mathbf{x}' d\mathbf{x}'' e^{-i\mathbf{q}\cdot(\mathbf{x}'-\mathbf{x}'')}\langle \delta c(\mathbf{x}', t)\,\delta c(\mathbf{x}'', t)\rangle. \tag{4.51}$$

For a macroscopically homogeneous system, the above integral can depend only on the difference vector, $\mathbf{x} = \mathbf{x}' - \mathbf{x}''$ and not the specific values of $\mathbf{x}$ and $\mathbf{x}'$ relative to an arbitrary origin. For this reason, equation (4.51) becomes

$$I = \frac{|E_0|^2}{(4\pi r)^2}|\mathbf{k}_s \times \mathbf{k}_s \times \mathbf{n}_i|^2\left(\frac{\partial \varepsilon_0}{\partial c}\right)^2 V\int d\mathbf{x} e^{-i\mathbf{q}\cdot\mathbf{x}}\langle \delta c(\mathbf{x}, t)\,\delta c(0, t)\rangle, \tag{4.52}$$

where V is the system volume.

The spatial autocorrelation function, $C(\mathbf{x}) = \langle \delta c(\mathbf{x}, t)\,\delta c(0, t)\rangle$, appearing in equation (4.52) is a measure of the strength of the correlation of a fluctuation in concentration at an arbitrary origin with a fluctuation some position, $\mathbf{x}$, away. In general, this function will decay from its value at the origin, $C(0) = \langle \delta c^2 \rangle$, to its value, $C(|\mathbf{x}| \to \infty) = \langle \delta c \rangle \langle \delta c \rangle = 0$. A typical decay of the spatial autocorrelation function for an amorphous system is shown in Figure 4.13 and the length scale over which the decay occurs defines the correlation length, ξ.

The intensity is proportional to the Fourier transformation of $C(\mathbf{x})$, and this defines the structure factor, $S(\mathbf{q})$,

$$S(\mathbf{q}) = \frac{1}{c}\int d\mathbf{x} e^{-\mathbf{q}\cdot\mathbf{x}}\langle \delta c(0)\,\delta c(\mathbf{x})\rangle. \tag{4.53}$$

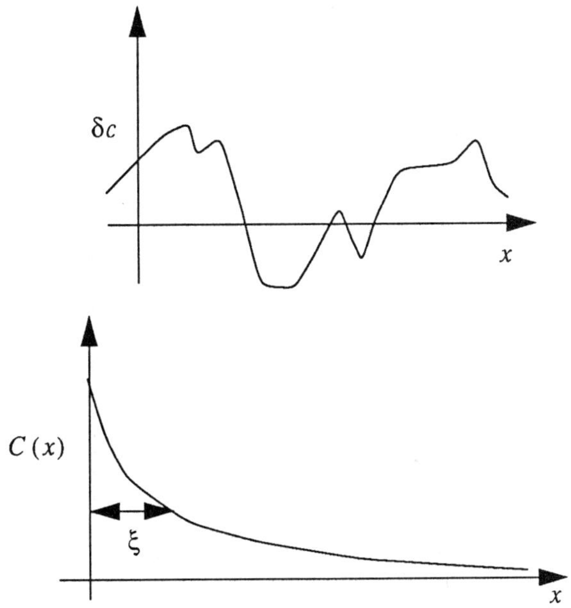

Figure 4.13 Concentration fluctuations and the spatial autocorrelation function.

4.4 Fraunhofer Diffraction from Large Particles

We seek here a description of the scattering from large particles compared with the wavelength of light. In this limit the dominant scattering occurs at small angles. Furthermore, the major contribution to the scattering field arises from interactions with the particle boundary.

The present derivation makes use of two classical results:[9] Babinet's principle and Fresnel's extension of Huygen's principle. The former principal recognizes the equivalence of two scattering experiments. In the first experiment, light impinges on an absorbing disk that has the shape of the area of the scattering particle projected onto the plane that is normal to the incident wave vector. The second experiment involves the interaction of light with an opening in an absorbing screen, and the opening has the shape of the projected particle's area.

The sum of the waves that result in these two experiments must equal the original plane wave incident on the particle, and the two disturbance electric fields due to diffraction are of equal magnitude, but are of opposite sign. Since the intensity is the square modulus of the electric fields, either case will produce the same answer since the sign will not affect the result. It will turn out to be simpler to calculate the scatter light intensity using the second model presented above.

In the limit of small scattering angles, the light diffracted at the particle boundary will not be sensitive to the polarization of the incident light. The difference in phase of the light scattered at various points within the particle's shadow area will, however, need to be taken into account. Consider, for example, light radiating from two points, A and B in plane I of Figure 4.14.

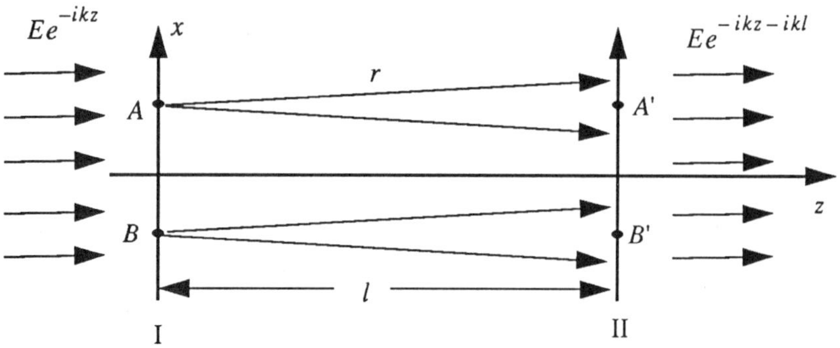

Figure 4.14 Propagation of a plane wave between two planes: Huygen's principle.

Of interest is the light intensity measured on plane II, located a distance l from plane I. Huygen's principle simply states that the field at plane II is the sum of spherical waves emanating from all points in plane I. However, the field at point A' will be primarily influenced by the spherical wave at point A, with a similar relationship between the fields at points B' and B. Fresnel's extension of Huygen's principle quantifies this statement. The spherical wave emanating from a differential area, dS, located at point A is

$$dE_A = C_0 dS \frac{e^{-ikr}}{r} E_I, \tag{4.54}$$

and it is proportional to the incident field, E_I. C_0 is a constant that is determined by first considering the case where the disturbance field emanates uniformly over the entire plane I. In this case, the total field at plane II must be $E_{II} = e^{-ikl}E_I$. Applying Huygen's principle, the field at plane II is also the integration of all the contributions from differential areas located in plane I. In other words,

$$E_{II} = \int_{-\infty}^{\infty}\int_{-\infty}^{\infty} dxdy C_0 \frac{e^{-ikr}}{r} E_I. \tag{4.55}$$

The radial distance, r, locating a single disturbance ray on plane II is approximated as

$$r = \sqrt{l^2 + x^2 + y^2} \approx l + \frac{1}{2l}(x^2 + y^2), \tag{4.56}$$

when the observation distance, l, is large compared with the spatial extent of the incident field. Calculating the field at plane II, we have

$$E_{II} = \frac{E_I C_0 e^{-ikl}}{l} \int_{-\infty}^{\infty}\int_{-\infty}^{\infty} dxdy e^{-\frac{ik}{2l}(x^2 + y^2)} = -i\lambda C_0 e^{-ikl}, \tag{4.57}$$

so that, evidently, $C_0 = i/\lambda$.

Fraunhofer diffraction theory combines the above results to compute the light scattered at small angles from large particles. Such a particle is pictured in Figure 4.15.

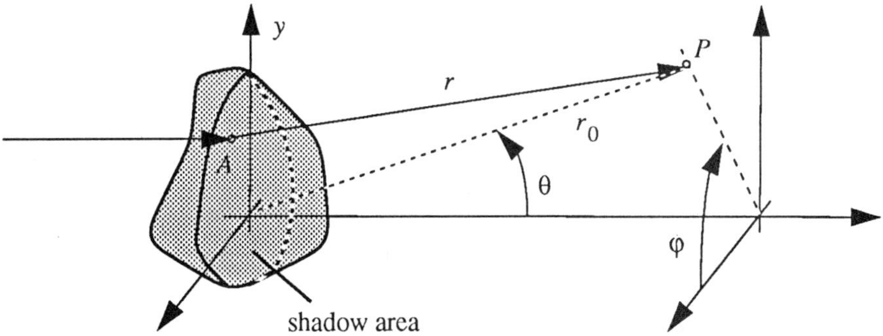

Figure 4.15 Diffraction of light from the shadow area of a particle. Shown is the diffraction of a single ray from position A to a point P.

Light scattered from an element dS located at the position $A = (x, y)$ in the shadow area of the particle is measured at a point P located in a plane some distance away. This point is found at the spherical coordinate (r_0, θ, φ) relative to an arbitrary origin within the shadow area. The spherical disturbance from A reaching P is [9],

$$dE_P = \frac{i}{\lambda} dSE_0 \frac{e^{-ikr}}{r}, \tag{4.58}$$

where E_0 is the incident electric field. The distance r is

$$\begin{aligned} r &= [r_0^2 \cos^2\theta + (x - r_0 \sin\theta\cos\varphi)^2 + (y - r_0 \sin\theta\sin\varphi)^2]^{1/2} \\ &\approx r_0 - (x\cos\varphi + y\sin\varphi)\sin\theta \end{aligned} \tag{4.59}$$

The total field at P is found by integrating this disturbance over the shadow area, A, of the particle so that

$$E_P(\theta, \varphi) = \frac{i}{\lambda} E_0 \iint_A dxdy \frac{e^{-ikr}}{r} = \frac{i}{\lambda} \frac{E_0}{r_0} e^{-ikr_0} \iint_A dxdye^{-ik(x\cos\varphi + y\sin\varphi)\sin\theta}. \tag{4.60}$$

Defining the diffraction function, $D(\theta, \varphi)$, as

$$D(\theta, \varphi) = \frac{1}{A} \iint_A dxdye^{-ik(x\cos\varphi + y\sin\varphi)\sin\theta}, \tag{4.61}$$

where $A = \iint_A dxdy$, the intensity of the diffracted light is

$$I(\theta, \varphi) = \frac{A^2}{\lambda^2 r^2} I_0 |D(\theta, \varphi)|^2, \tag{4.62}$$

and $I_0 = |E_0|^2$ is the incident intensity.

4.4.1 Fraunhofer Diffraction from a Sphere

The function $D(\theta, \varphi)$ is easily calculated for a sphere. For a sphere of radius a, we have

$$D(\theta, \varphi) = \frac{1}{\pi a^2} \int_{-a}^{a} dx \int_{-\sqrt{a^2 - x^2}}^{\sqrt{a^2 - x^2}} dy e^{-ik(x\cos\varphi + y\cos\varphi)\sin\theta} = \frac{2J_1(ka\sin\theta)}{ka\sin\theta}, \tag{4.63}$$

where $J_1(x)$ is a Bessel function of the first kind. As expected, this diffraction pattern produces circular contours of constant intensity. Furthermore, diffraction rings result, located at the zeroes of the Bessel function. As in the case of Rayleigh-Debye scattering, the rings on minimum intensity can be used to determine the size of the scattering particle.

4.4.2 Fraunhofer Diffraction from a Cylinder

The diffraction of light from a cylinder is easily calculated since the projected area of a cylinder will simply be a rectangle. For simplicity, we assume here that the rectangle has a length L and a width d and that the long axis is oriented along the x axis. The diffraction function is then

$$\begin{aligned} D(\theta, \varphi) &= \frac{1}{Ld} \int_{-L/2}^{L/2} dx \int_{-d/2}^{d/2} dy e^{-ik(x\cos\varphi + y\sin\varphi)\sin\theta} \\ &= \frac{\sin[(kL\sin\theta\cos\varphi)/2]}{(kL\sin\theta\cos\varphi)/2} \frac{\sin[(kd\sin\theta\sin\varphi)/2]}{(kd\sin\theta\sin\varphi)/2} \end{aligned} \tag{4.64}$$

Because of the anisotropy in the shape of the particle, this scattering pattern has a dependence on the angle φ.

4.5 The Scattering Jones Matrix

The scattering processes considered in this chapter are linear phenomena and, in general, the scattered electric field can be written as the following linear transformation of the incident field [9],

$$\mathbf{E}_s = \frac{e^{ikr}}{ikr} \mathbf{S}(\theta, \varphi) \cdot \mathbf{E}_0, \tag{4.65}$$

where θ and φ define the angular coordinates of the measurement point of the scattered light. The prefactor, e^{ikr}/r, has been added for convenience and is motivated by its recurrence in all of the previously derived expressions for the scattered light. The factor $1/(ik)$ makes the matrix $\mathbf{S}$ dimensionless.

Depending on the choice made for the components of the electric vector, $\mathbf{S}(\theta, \varphi)$ may not be a true tensor. The convention that is employed here is such a case and for that reason, this function is referred to as the *scattering matrix*.

The convention that is used decomposes the electric fields of both the incident and scattered light into parts that are either parallel (p) or perpendicular (s) to the scattering plane. In other words, the Jones vector for scattering is $\mathbf{A} = (E_p, E_s)^T$. This is analogous to the treatment of reflected and refracted light in Chapter 3. For example, in the case of dipole scattering by an isotropic particle with polarizability, α, the scattering matrix can be derived upon inspection of equation (4.9) and is

$$\mathbf{S}(\theta, \varphi) = ik^3\alpha\begin{bmatrix}\cos\theta & 0\\ 0 & 1\end{bmatrix}. \tag{4.66}$$

Setting the scattering angle $\theta = 90°$ shows that scattering from an isotropic particle at this angle produces the same effect as an ideal polarizer.

4.6 The Optical Theorem: Form Dichroism and Birefringence From Dilute Suspensions

The scattering of light from a suspension of particles will cause the light transmitted through such a material to be changed in phase and amplitude. If particles are nonspherical, and have a net orientation, a birefringence and dichroism will result.

The connection between the scattering properties of the particles and the birefringence and dichroism is easily calculated. We follow here the derivation found in the book by van de Hulst [9]. In Figure 4.16, a layer of particles of thickness l resides in the (x, y) plane. Light propagating along the z axis proceeds through the suspension and is received on the other side, some distance away at point P. Neglecting multiple scattering, to a first approximation, the electric vector of the transmitted light will be the sum of unscattered incident light plus forward scattered light. The latter contribution is calculated using the scattering matrix, $S(\theta, \varphi)$ with the angle $\theta = 0$. In other words,

$$\mathbf{E} = \left[\mathbf{I} + \sum_j \mathbf{S}(0)\frac{e^{-ikr_j}}{ikr_j}\right] \cdot \mathbf{E}_0, \tag{4.67}$$

where the summation is over all the particles in the sample.

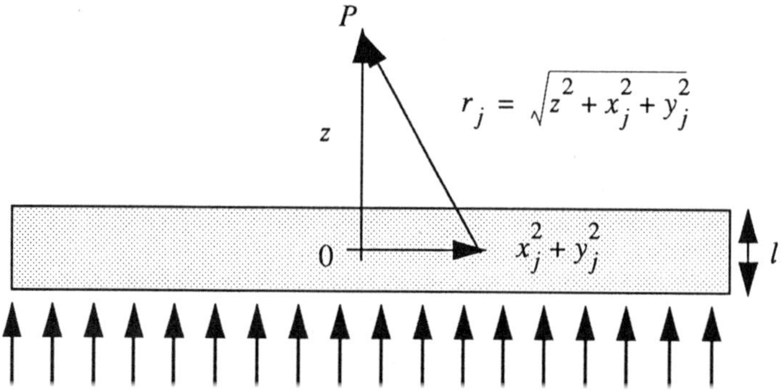

Figure 4.16 Transmission of zero angle scattered light through a slab of particles.

If this field is observed at the point P, it can be expected that the primary contribution will come from particles at distances close to the origin O compared with the distance z. This allows the radial distance r_j to be approximated as

$$r_j = \sqrt{z^2 + x_j^2 + y_j^2} \approx z + \frac{(x_j^2 + y_j^2)}{2z}, \tag{4.68}$$

so that,

$$\mathbf{E} = \left[\mathbf{I} + \frac{lN\mathbf{S}(0)}{ikz} \int_{-\infty}^{\infty} dxdye^{-ik\frac{(x^2+y^2)}{2z}} \right] \cdot \mathbf{E}_0,$$
$$\mathbf{E} = \left[\mathbf{I} - \frac{2\pi lN\mathbf{S}(0)}{k^2} \right] \cdot \mathbf{E}_0. \tag{4.69}$$

where N is the number of particles per unit volume.

For a dilute suspension, this result can be compared with a weakly birefringent and dichroic material. In this case, from equations (2.15) and (2.21),

$$\mathbf{E} = \begin{bmatrix} e^{-i\frac{2\pi n'_{pxx}l}{\lambda} - \frac{2\pi n''_{pxx}l}{\lambda}} & 0 \\ 0 & e^{-i\frac{2\pi n'_{pyy}l}{\lambda} - \frac{2\pi n''_{pyy}l}{\lambda}} \end{bmatrix} \cdot \mathbf{E}_0 \approx \left[\mathbf{I} - \frac{2\pi l}{\lambda}(i\mathbf{n}' + \mathbf{n}'') \right] \cdot \mathbf{E}_0, \tag{4.70}$$

where the subscript p denotes that this is the particle contribution to the refractive index. Equations (4.69) and (4.70) can be compared to produce the result that

$$\mathbf{n}'_p = \frac{2\pi nN}{k^3} Im[\mathbf{S}(0)], \tag{4.71}$$

and

$$\mathbf{n}''_p = \frac{2\pi n N}{k^3} Re\,[\mathbf{S}(0)]\,. \tag{4.72}$$

In the limit of dipole scattering, the zero-angle scattering Jones matrix for a system of anisotropic particles is simply found by extending equation (4.66) to the case were the polarizability is a tensor. In this case,

$$\mathbf{S}(0) = ik^3 \underline{\alpha}\,, \tag{4.73}$$

and

$$\mathbf{n}'_p = (2\pi n N)\,Re\,(\underline{\alpha})\;;\;\; \mathbf{n}''_p = -(2\pi n N)\,Im\,(\underline{\alpha})\,. \tag{4.74}$$

From these expressions, the birefringence from the scattering of oriented, anisotropic particles can be calculated. The dichroism can also be calculated, but in this limit, it is only present if the polarizability has an imaginary component. In other words, only absorbing particles are predicted to be dichroic.

An improved theory for scattering birefringence and dichroism can be obtained by extending the scattering theory to higher order in the ratio of the particle size relative to the wavelength. The details of this calculation will be not repeated here and can be found in van de Hulst. The analysis uses Mie scattering theory that is valid for larger particles and produces the following expansion for the zero-angle scattering matrix,

$$S_{ij}(0) = ik^3 \alpha_{ij} + \frac{2}{3} k^6 \alpha_{ik} \alpha_{kj}\,. \tag{4.75}$$

This result predicts that nonabsorbing particles will produce a real contribution to $S_{ij}(0)$ at higher order in the wavenumber. Using this result, *nonabsorbing* particles produce the following components of the refractive index,

$$\mathbf{n}'_p = 2\pi n N \underline{\alpha}\,, \tag{4.76}$$

and

$$\mathbf{n}''_p = \frac{4\pi n N k^3}{3} \underline{\alpha}^2\,. \tag{4.77}$$

Equation (4.12) connects the polarizability of a particle with its orientation and can be used to derive the following equations for the particle contribution to the refractive index tensor [16]:

$$\mathbf{n}'_p = 2\pi n N\,[\alpha_2 \mathbf{I} + (\alpha_1 - \alpha_2)\,\langle \mathbf{uu} \rangle]\,, \tag{4.78}$$

and

$$\mathbf{n}''_p = \frac{4\pi n N k^3}{3}\,[\alpha_2^2 \mathbf{I} + (\alpha_1^2 - \alpha_2^2)\,\langle \mathbf{uu} \rangle]\,. \tag{4.79}$$

where the angular brackets refer to averages over the orientation distribution of the particles.

Finally, these results can be used to derive expressions for the birefringence, dichroism and orientation angle of a particle suspension. The birefringence and dichroism are simply the differences in the principal eigenvalues of the tensors on the right-hand sides of equations (4.78) and (4.79), respectively. The orientation angle is the angle of rotation that diagonalizes the second-moment tensor, $\langle \mathbf{uu} \rangle$. For light propagating along the z axis, the birefringence and orientation angle are

$$\Delta n' = 2\pi n N (\alpha_1 - \alpha_2) \sqrt{(\langle u_x^2 \rangle - \langle u_y^2 \rangle) + 4 \langle u_x u_y \rangle}\,, \tag{4.80}$$

and

$$\tan 2\theta = \frac{2\langle u_x u_y \rangle}{\langle u_x^2 \rangle - \langle u_y^2 \rangle}\,. \tag{4.81}$$

The ratio of dichroism to birefringence is $(\Delta n'')/(\Delta n') = (4k^3\bar{\alpha})/3$, where $\bar{\alpha}$ is the average polarizability. For this reason, scattering dichroism is generally observable only for particles that are relatively large compared to the wavelength.

4.7 The Onuki-Doi Theory of Form Birefringence and Dichroism

The derivation of form (or scattering) birefringence and dichroism using the optical theorem in section 4.6 is appropriate for systems of discrete scattering centers where dipole, or Rayleigh, scattering is the dominate mechanism. For more complex systems, such as scattering by fluctuations, or in systems where the length scale of the scatterers are comparable to the wavelength, the recent theory of Onuki and Doi [13] should be used.

Starting with equation (1.33), we have

$$\nabla\times(\nabla\times\mathbf{E}) = k_0^2 \varepsilon(\mathbf{X}) \cdot \mathbf{E}\,. \tag{4.82}$$

In a fluctuating system, both the electric field, $\mathbf{E}$, and the dielectric tensor, $\underline{\varepsilon}$, will undergo random oscillations described by equations (4.42) and (4.43). A solution is sought for this equation so that the mean electric field in the sample is of the form $\langle \mathbf{E}(\mathbf{x}) \rangle = \mathbf{E}_0 e^{-i\mathbf{k}_i \cdot \mathbf{x}}$. In other words, the mean field (the total field minus the part due to fluctuations from scattering), propagates in the direction of the incident field, but with an electric vector, $\mathbf{E}_0$, that is altered by interactions with the mean dielectric properties of the material. Substituting this form for the average field into (4.82) leads to the requirement that

$$\mathbf{k}\times(\mathbf{k}\times\mathbf{E}_0) = -k_0^2 \underline{\varepsilon} \cdot \mathbf{E}_0\,, \text{ or } \mathbf{k}\times(\mathbf{k}\times\langle\mathbf{E}\rangle) = -k_0^2 \underline{\varepsilon} \cdot \langle\mathbf{E}\rangle\,. \tag{4.83}$$

Alternatively, equation (4.82) can be averaged and then Fourier transformed to give

$$\mathbf{k}\times(\mathbf{k}\times\langle\mathbf{E}\rangle) = -k_0^2 \langle \underline{\varepsilon} \cdot \mathbf{E} \rangle\,, \tag{4.84}$$

which, when compared with (4.83), gives the result that

$$\underline{\varepsilon} \cdot \langle \mathbf{E} \rangle = \langle \underline{\varepsilon} \cdot \mathbf{E} \rangle . \tag{4.85}$$

Using the definitions of (4.42) and (4.43), evidently,

$$\underline{\varepsilon} \cdot \langle \mathbf{E} \rangle = \langle \underline{\varepsilon} \rangle \cdot \langle \mathbf{E} \rangle + \langle \delta \underline{\varepsilon} \cdot \delta \mathbf{E} \rangle , \tag{4.86}$$

which can be used to relate the total dielectric tensor to its average value and its fluctuation. The electric field can be eliminated from this equation by using the result for $\delta\mathbf{E}$ given in equation (4.45). The dielectric tensor is then

$$\underline{\varepsilon} \cdot \langle \mathbf{E} \rangle = \langle \underline{\varepsilon} \rangle \cdot \langle \mathbf{E} \rangle + \lim_{\eta \to 0} \frac{1}{(2\pi)^3} \int d\mathbf{x}' \int d\mathbf{k}' e^{-i\mathbf{k}' \cdot (\mathbf{x} - \mathbf{x}')} \frac{k_0^2 \mathbf{I} - \mathbf{k}'\mathbf{k}'}{k'^2 - (k_0 + i\eta^2)} \cdot \langle \delta \underline{\varepsilon} (\mathbf{x}) \delta \underline{\varepsilon} (\mathbf{x}') \rangle \cdot \langle \mathbf{E} (\mathbf{x}') \rangle . \tag{4.87}$$

As in section 4.3, the autocorrelation function, $\langle \delta \underline{\varepsilon} (\mathbf{x}) \delta \underline{\varepsilon} (\mathbf{x}') \rangle$, will only depend on the difference vector, $\mathbf{x} - \mathbf{x}'$. Using this fact, and replacing $\langle \mathbf{E} (\mathbf{x}') \rangle$ by $\mathbf{E}_0 e^{-i\mathbf{k}_i \cdot \mathbf{x}'}$, gives the final result for the dielectric tensor,

$$\underline{\varepsilon} = \langle \underline{\varepsilon} \rangle + \lim_{\eta \to 0} \frac{1}{(2\pi)^3} \int d\mathbf{k}' \frac{k_0^2 \mathbf{I} - \mathbf{k}'\mathbf{k}'}{k'^2 - (k_0 + i\eta^2)} \cdot \mathbf{C} (\mathbf{k}_i - \mathbf{k}') , \tag{4.88}$$

where

$$\mathbf{C} (\mathbf{k}) = \int d\mathbf{x}' e^{-i\mathbf{k} \cdot \mathbf{x}} \langle \delta \underline{\varepsilon} (0) \delta \underline{\varepsilon} (\mathbf{x}) \rangle . \tag{4.89}$$

The first term, $\langle \underline{\varepsilon} \rangle$, is the *intrinsic* part of the dielectric tensor, $\underline{\varepsilon}_i$, and will lead to dichroism only if the material is absorbing. The second term in equation (4.88) is the form contribution and is denoted as $\underline{\varepsilon}_f$. It is left to analyze this portion of the dielectric tensor for its real and imaginary parts so that the form birefringence and form dichroism can be calculated. This is accomplished using the relation

$$\frac{1}{k'^2 - (k_0 + i\eta)^2} = P\left[\frac{1}{k'^2 - k_0^2}\right] + i\pi\delta (k'^2 - k_0^2) , \tag{4.90}$$

where $P[\ldots]$ denotes the Cauchy principal value and the real and imaginary parts of the form effect are

$$Re (\underline{\varepsilon}_f) = \frac{1}{(2\pi)^3} \int d\mathbf{k}' P \left[\frac{k_0'^2 \mathbf{I} - \mathbf{k}'\mathbf{k}'}{k'^2 - (k_0 + i\eta)^2}\right] \cdot \mathbf{C} (\mathbf{k}_i - \mathbf{k}') , \tag{4.91}$$

and

$$Im(\varepsilon_f) = \frac{k_0^2}{16\pi^2}\int d\Omega_{\hat{\mathbf{k}}'}(\mathbf{I} - \hat{\mathbf{k}}'\hat{\mathbf{k}}') \cdot \mathbf{C}(\mathbf{k}_i - k_0\hat{\mathbf{k}}') , \tag{4.92}$$

where $\int d\Omega_{\hat{\mathbf{k}}'}$ represents integration over the solid angle of the unit vector, $\hat{\mathbf{k}}' = \mathbf{k}'/k_0$.

If the fluctuations in the dielectric tensor are primarily due to isotropic concentration fluctuations, then using equation (4.49),

$$\varepsilon_f = \lim_{\eta \to 0} \frac{c}{(2\pi)^3}\left(\frac{\partial \varepsilon_0}{\partial c}\right)^2 \int d\mathbf{k}' \frac{k_0^2\mathbf{I} - \mathbf{k}'\mathbf{k}'}{k'^2 - (k_0 + i\eta)^2} S(\mathbf{k}_i - \mathbf{k}') , \tag{4.93}$$

where $S(\mathbf{k})$ is the structure factor defined in equation (4.53).

From the above expressions, the form birefringence and dichroism can be calculated. The form dichroism, in particular, has the simple interpretation of being the anisotropy in the second-moment tensor of the structure factor. It is also evident that the form dichroism appears at a higher order in the wave number than the form birefringence.

5 Spectroscopic Methods

This chapter concerns interactions of light with matter that can be used to identify specific structures at the level of the chemical bonds making up the sample. This is in contrast to the scattering phenomena described in Chapter 4 where information is obtained on length scales comparable to the wavelength of the light itself. The ability to discriminate between different chemical constituents is accomplished by taking advantage of couplings between incident light and various modes of energy contained within the sample. For this reason, the interactions presented here do not conserve energy, as is the case with Rayleigh light scattering.

There are many spectroscopic interactions of light and matter, all of which rely on particular molecular structural changes that affect the electric and magnetic fields of light. In this chapter, dichroism resulting from either electronic or vibrational transitions is discussed, with the most emphasis being given to vibrational absorption spectroscopy in the infrared. That discussion is followed by a second vibrational spectroscopy: Raman scattering. Polarized fluorescence is also presented. In every case, the focus of the discussion is on the spectroscopic measurement of orientation.

5.1 Dichroism in the Ultraviolet, Visible and Infrared

Ultraviolet, visible, and infrared spectroscopies refer to analysis of the absorption characteristics of a sample that are linked to various electronic and vibrational transitions within a molecule [17]. These involve relative displacements of electrons and nuclei that are able to couple to incident light if they induce a dipole in the material. The strength of this coupling is measured by the transition dipole moment [17],

$$\mathbf{M}_{i,j;k,l} = \int d\mathbf{r}_e d\mathbf{r}_n \Psi_{i,j}(\mathbf{r}_e, \mathbf{r}_n)\, \mathbf{P}(\mathbf{r}_e, \mathbf{r}_n)\, \Psi_{k,l}(\mathbf{r}_e, \mathbf{r}_n) \,. \tag{5.1}$$

This integral represents the strength of the dipole that is produced when a molecule undergoes a transition between two states in its configuration. These states are described by the wavefunction, $\Psi_{i,j}(\mathbf{r}_e, \mathbf{r}_n)$, which is an eigenfunction solution of the Schroedinger wave equation. The square of the wavefunction is interpreted as the probability that a molecule will find itself in a state (i, j) with nuclei located at $\mathbf{r}_n$ and electrons located at $\mathbf{r}_e$. The indices i and j refer to the energy levels associated with each state for the nuclei and electrons. These energies are the eigenvalues of the Schroedinger equation. The dipole function, $\mathbf{P}(\mathbf{r}_e, \mathbf{r}_n)$, is given by the classical expression,

$$\mathbf{P}(\mathbf{r}_e, \mathbf{r}_n) = |q| \sum_{m=1}^{N_n} Z_m \mathbf{r}_{nm} - |q| \sum_{n=1}^{N_e} \mathbf{r}_{en}, \tag{5.2}$$

where q is the unit of the charge on an electron, Z_m is the atomic number of nucleus m, and N_n and N_e are the numbers of nuclei and electrons in the molecule.

Absorption will occur only when the light is in resonance with the energy of the transition. This requires that $h\nu = E_{i,j \to k,l}$, where $E_{i,j \to k,l}$ is the discrete energy change involved with the transition from state i, j to k, l, and ν is the frequency of the light. The constant h is Planck's constant. In Figure 5.1, a schematic of an energy level diagram involving different electronic and vibrational states is shown. Two types of energy transitions are shown: vibrational and vibronic. The former transition involves initial and final states that both have the same electronic configuration, but differ in the vibrational modes of the nuclei. The vibronic transition involves changes in both the electronic configuration and the vibrational modes.

Absorption in the ultraviolet ($\lambda \approx 100-400$ nm) and the visible ($\lambda \approx 400-800$ nm) is primarily the result of transitions in the electronic state of the molecule. In such a process, the transition dipole moment would be proportional the overlap in the densities of the charge distributions between the two electron orbitals involved in a transition. The periodic displacement of electrons from one state to another will cause the charge distribution to be anisotropic, with net negative and positive contributions in certain locations within the molecule. The result is the formation of a dipole moment. Very often, dye molecules that absorb in the visible are dispersed within a sample or attached to the molecules of a sample and are used to monitor its degree of alignment. However, since the relative orientation of such a dye molecule to the molecular axes of the constituent sample molecules is often unknown, the interpretation of these measurements can be difficult.

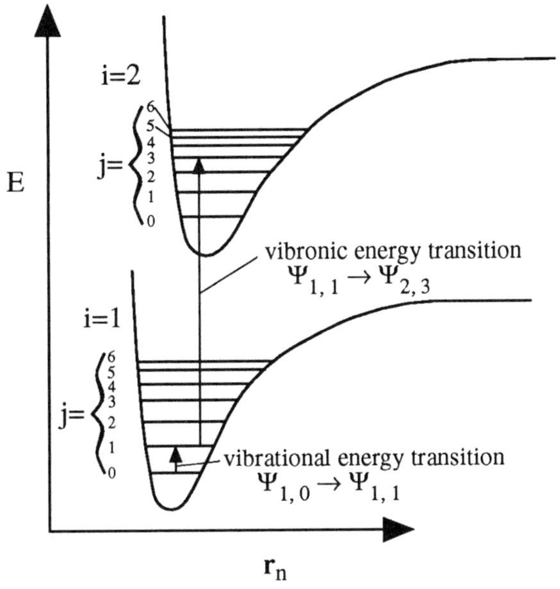

Figure 5.1 Electronic-vibrational energy transitions.

Vibrational absorption spectroscopy occurs in the infrared spectrum. The location of such an absorption will be directly linked to the frequency of vibrational modes. The symmetry of the modes of vibration will also affect the absorption and this is evident upon inspection of equation (5.1). These symmetries are contained in the forms of the wave function themselves, which are symmetric or antisymmetric with respect to planes within a molecule. Upon inspection of the integrand of equation (5.1), it is apparent that certain products of the wave functions with the dipole function can lead to an asymmetric result, causing the integral to vanish. Such transitions are "forbidden" and will not appear in the infrared spectrum.

In many cases the effects of symmetry can be seen from a simple inspection of the particular geometry of a molecule [19]. For diatomic molecules, for example, only heteronuclear systems can be infrared active, since the symmetry of a homonuclear molecule precludes the formation of a dipole moment. In a polymer chain consisting of carbon-carbon bonds along its backbone, this symmetry argument leads to a result that those C-C bonds will generally not contribute to the infrared spectrum. Even in the case of heteronuclear molecules, the particular dynamics of a vibrational mode can negate the formation of a dipole. One such heteronuclear molecule, CO_2, is shown in Figure 5.2. In this figure, two modes of stretching vibrations are shown: the symmetric stretch and the asymmetric stretch. Examination of the symmetric stretch reveals that the dynamics of this type of motion can never produce an asymmetric charge distribution, and therefore cannot lead to the formation of a dipole. The asymmetric stretch, on the other hand, does form a dipole, and is "infrared active". The failure of the symmetric stretch to produce a dipole is a consequence of the linear structure of the CO_2 molecule. In a nonlinear molecule such as CH_2O, the symmetric stretching of the hydrogen atoms relative to the oxygen will lead to a dipole oriented parallel to the CO double bond.

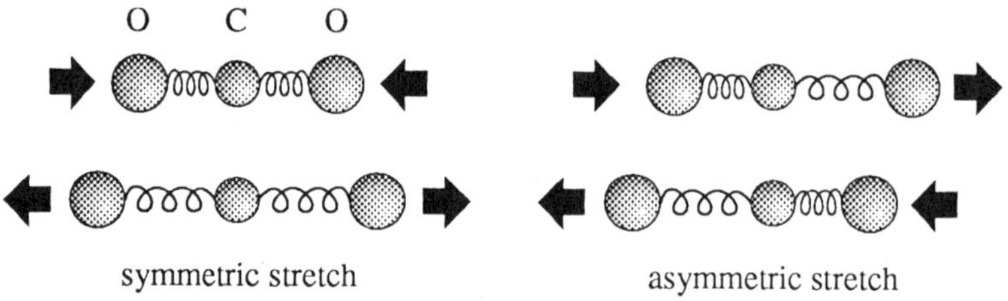

Figure 5.2 Vibrational modes of CO_2.

Polarized absorption spectroscopy is a powerful diagnostic tool for investigating molecular structure and the symmetry considerations discussed above can be extended to the situation where the incident light is polarized along a particular direction. The strength of the interaction of the transition dipole moment, $\mathbf{A}_{i,\,j;k,\,l}$, with the electric vector of the light, $\mathbf{E}$, is proportional to the dot product [17],

$$\mathbf{A}_{i,\,j;k,\,l} = \mathbf{M}_{i,\,j;k,\,l} \cdot \mathbf{E} = |\mathbf{M}_{i,\,j;k,\,l}||\mathbf{E}| \cos\Omega , \tag{5.3}$$

where Ω is the angle between the electric vector and the transition dipole moment. The intensity of an absorption will then be proportional to $|\cos^2\Omega|$ and absorption spectroscopy, performed using various polarizations of light relative to an oriented sample can elucidate the relative orientation of various chemical bonds. In general, the orientation angle, Ω, will be distributed over an orientation distribution function so that the final result will involve the average, $\langle \cos^2\Omega \rangle$.

When an absorption dichroism measurement is analyzed to extract the orientation of a sample, it must be stressed that it is the orientation of a transition dipole moment that is measured, and these will generally have a different orientation distribution than, say, the principal geometric axes of the molecules comprising the sample. If a sample consisting of polymer chains is stretched mechanically, as shown in Figure 5.3, it is the orientation of the backbone of the polymer that is often desired. For the case of a uniaxial extension, this orientation would be related to the average $\langle \cos^2\theta \rangle$, where θ is the polar angle that the polymer segments make with the laboratory axis (see Fig. 5.3). The absorption dichroism measurement, on the other hand, targets the degree of orientation of a transition dipole moment, which is characterized by the ellipse oriented at an angle α with respect to the main chain and an angle Ω with respect to the elongation axis Z. In such an experiment, absorption data would normally be taken using light polarized parallel to the laboratory axes, $X_i = (X, Y, Z)^T$. If the deformation is a uniaxial extension, as will be assumed here, then the orientation distribution describing the polymer chains will be identical along the X and Y axes. In this case absorption data would be collected by transmitting light through the sample along the X axis that is polarized either along the Z or the Y axes. The measured intensities transmitted by the sample will be simple functions of the average $\langle \cos^2 \Omega \rangle$. What is now required is the relationship between this observable, and the desired average, $\langle \cos^2 \theta \rangle$.

The orientation of the segment relative to the laboratory frame containing the transition dipole moment is defined by the angles θ and φ, which are the polar and azimuthal angles associated with the vector (parallel to the z' axis in Fig. 5.3) spanning the two ends of the segment. In addition, a third angle, Φ, is used to measure the rotation of the segment about its own axis. Imbedded within the segment is the reference frame, (x, y, z). In that frame, the orientation of the transition dipole moment is prescribed by the polar and azimuthal angles α and β, respectively. Only two angles are required in this case since the transition dipole moment will be taken to have uniaxial symmetry with principal axes a_1 and a_2 that are parallel and perpendicular to its symmetry axis. These two parameters characterize the strength of light absorption along these respective directions. In the (x, y, z) frame the tensor representing the absorbance by this dipole will have the same form as found for the polarizability tensor in section 4.1.2. In other words,

$$\mathbf{a}' = a_2\mathbf{I} + (a_1 - a_2)\begin{bmatrix} \sin^2\alpha\cos^2\beta & \sin^2\alpha\cos\beta\sin\beta & \sin\alpha\cos\alpha\cos\beta \\ \sin^2\alpha\cos\beta\sin\beta & \sin^2\alpha\sin^2\beta & \sin\alpha\cos\alpha\sin\beta \\ \sin\alpha\cos\alpha\cos\beta & \sin\alpha\cos\alpha\sin\beta & \cos^2\alpha \end{bmatrix}, \qquad (5.4)$$

where the prime on the tensor $\mathbf{a}'$ denotes that it has been written in the (x, y, z) frame.

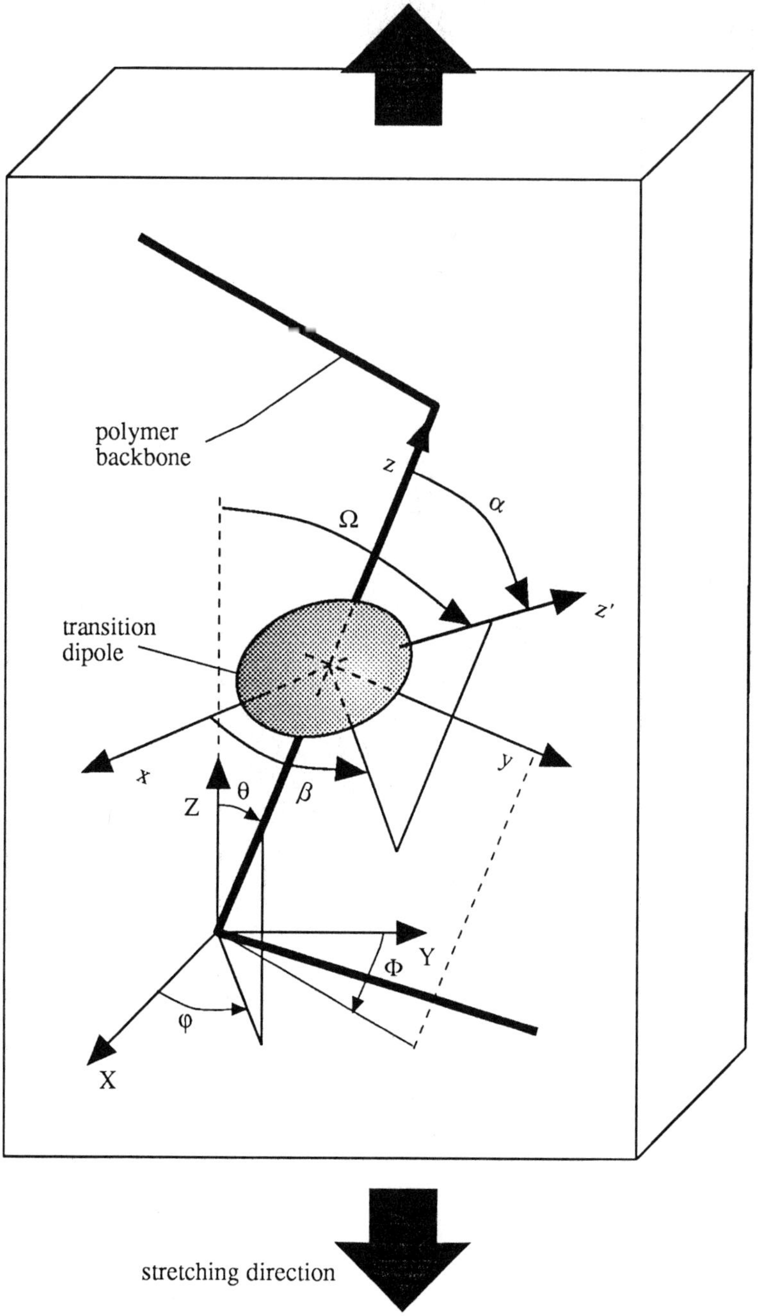

Figure 5.3 Orientation of a transition dipole moment relative to a laboratory frame. The axis z' defines a principal axis of the dipole moment tensor.

The tensor in equation (5.4) must be transformed from the (x, y, z) frame to its form, $\mathbf{a}$, in the laboratory frame, (X, Y, Z). In general, the transformation of a tensor from one frame to another involves the tensor of direction cosines, T_{ij}, describing the relative orientations of each axis of the two frames. The elements of this tensor are the cosines of the angles spanning the axis X_i of the frame (X, Y, Z) and the axis x_j of the frame (x, y, z). This tensor is [17]

$$\mathbf{T} = \begin{bmatrix} (\cos\theta\cos\varphi\cos\Phi - \sin\varphi\sin\Phi) & (-\cos\theta\cos\varphi\sin\Phi - \sin\varphi\cos\Phi) & \sin\theta\cos\varphi \\ (\cos\theta\sin\varphi\cos\Phi + \cos\varphi\sin\Phi) & (-\cos\theta\sin\varphi\sin\Phi + \cos\varphi\cos\Phi) & \sin\theta\sin\varphi \\ -\sin\theta\cos\Phi & \sin\theta\sin\Phi & \cos\theta \end{bmatrix} \tag{5.5}$$

and is a generalization of the rotation tensor given previously in equation (2.67) (the previous result can be recovered by setting $\Phi = -\phi$, $\theta = -\theta$, and $\varphi = \pi/2$).

The linear transformation from $\mathbf{a}'$ to $\mathbf{a}$ is simply

$$\mathbf{a} = \mathbf{T}\cdot\mathbf{a}'\cdot\mathbf{T}^T, \tag{5.6}$$

and for a polymer chain comprised of N identical absorbing units, the total absorption of the macromolecule is,

$$\langle A_{ij}\rangle = N\langle a_{ij}\rangle = N\sum_{m,n}\langle a'_{mn}\rangle\langle T_{im}T_{jn}\rangle. \tag{5.7}$$

The averages in equation (5.7) represent averages over the orientation distribution function characterizing the angles θ, φ, Φ, α, and β. This distribution is assumed to be separable into functions of the sets of angles residing in the two coordinates in Figure 5.3 and is of the form

$$P(\theta, \varphi, \Phi, \alpha, \beta) = p_X(\theta, \varphi, \Phi)\, p_x(\alpha, \beta), \tag{5.8}$$

and this accounts for the factorization of the averaging for a'_{mn} and $T_{im}T_{jn}$. The factorization of the probability distribution function is valid if the deformation of the sample does not affect the local orientation of the transition dipole moments with respect to the polymer backbone.

The distribution function $p_x(\alpha, \beta)$ is usually taken to be uniform in the angle β, and this is justified by imagining that the dipoles are able to sample all azimuthal orientations along the chain axis. The average dipolar absorbance tensor then takes on the simplified form,

$$\mathbf{a}' = a_2\mathbf{I} + (a_1 - a_2)\begin{bmatrix} \frac{1}{2}\sin^2\alpha & 0 & 0 \\ 0 & \frac{1}{2}\sin^2\alpha & 0 \\ 0 & 0 & \cos^2\alpha \end{bmatrix}. \tag{5.9}$$

It remains to carry out the averaging over the direction cosines, T_{ij}. For a uniaxial deformation, it is evident that the bulk absorbance tensor, $\langle A_{ij} \rangle$, must be a diagonal tensor and the distribution function $p_X(\theta, \varphi, \Phi)$ must be uniform in the angle φ. Furthermore, by assuming that the dipoles are uniformily distributed over the angle β, the angle Φ has no meaning, and the final results should be expected to be independent of this angle. Carrying out the averages over φ, the following results are obtained:

$$A_{11}/N = A_{22}/N = a_2 + (a_1 - a_2)\left(\frac{1}{4}\langle \sin^2\alpha \rangle (1 + \langle \cos^2\theta \rangle) + \frac{1}{2}\langle \cos^2\alpha \rangle \langle \sin^2\theta \rangle\right),$$

$$A_{33}/N = a_2 + (a_1 - a_2)\left(\frac{1}{2}\langle \sin^2\alpha \rangle \langle \sin^2\theta \rangle + \langle \cos^2\alpha \rangle \langle \cos^2\theta \rangle\right).$$

(5.10)

The dichroism measured for light propagating along the X axis is calculated as

$$\Delta n'' \approx A_{33} - A_{22} = N(a_1 - a_2)\left(\frac{3\langle \cos^2\alpha \rangle - 1}{2}\right)\left(\frac{3\langle \cos^2\theta \rangle - 1}{2}\right)$$

$$= N(a_1 - a_2)\, S(\langle \cos^2\alpha \rangle)\, S(\langle \cos^2\theta \rangle), \tag{5.11}$$

where the function $S(\langle \cos^2\theta \rangle) = (3\langle \cos^2\theta \rangle - 1)/2$ is the Herman's orientation function [17]. These functions are also equivalent to the Legendre polynomial, $P_2(\langle \cos^2\theta \rangle)$, and are the diagonal terms of the more general Saupe orientation tensor that is widely used in the description of the director field in liquid crystals. The Saupe orientation tensors are defined as

$$S_{ij} = \frac{1}{2}(3\langle \cos\theta_i \cos\theta_j \rangle - \delta_{ij}). \tag{5.12}$$

Polarized absorption spectra are often expressed in terms of the dichroic ratio,

$$R = \frac{A_{33}}{A_{22}} = \frac{a_1 + 2a_2 + 2(a_1 - a_2)\, S(\langle \cos^2\theta \rangle)\, S(\langle \cos^2\alpha \rangle)}{a_1 + 2a_2 - (a_1 - a_2)\, S(\langle \cos^2\theta \rangle)\, S(\langle \cos^2\alpha \rangle)}. \tag{5.13}$$

Assuming that the transition dipole does not absorb light polarized perpendicular to its principal axis ($a_2 = 0$), the dichroic ratio takes on the simple form,

$$R = \frac{1 + 2S(\langle \cos^2\theta \rangle)\, S(\langle \cos^2\alpha \rangle)}{1 - S(\langle \cos^2\theta \rangle)\, S(\langle \cos^2\alpha \rangle)}. \tag{5.14}$$

The dichroic ratio is also the following function of the angle Ω that the dipole makes with the stretching direction,

$$R = \frac{\langle \cos^2 \Omega \rangle}{(1 - \langle \cos^2 \Omega \rangle)/2} . \tag{5.15}$$

Rearranging the dichroic ratio, one obtains

$$\frac{R-1}{R+2} = S(\langle \cos^2 \Omega \rangle) = S(\langle \cos^2 \theta \rangle)\, S(\langle \cos^2 \alpha \rangle) . \tag{5.16}$$

Finally, if the orientation angle α is fixed at a value α_0, the desired quantity $S(\langle \cos^2 \theta \rangle)$, is related to the observable, R, by

$$S(\langle \cos^2 \theta \rangle) = \left(\frac{R_0 + 2}{R_0 - 1}\right)\left(\frac{R-1}{R+2}\right), \tag{5.17}$$

where $R_0 = 2\cot^2 \alpha_0$.

An alternative approach to derive the absorption cross section tensor, a_{ij}, begins by expressing it the following form, in the frame of the tensor itself:

$$a_{ij} = a_2 \delta_{ij} + \Delta'_{12} u_i u_j + \Delta'_{32} p_i p_j , \tag{5.18}$$

where $\Delta'_{12} = a_1 - a_2$ and $\Delta'_{32} = a_3 - a_2$. The orthogonal vectors u_i and p_i define the major axes of the absorpotion cross section tensor and are shown in Figure 5.4. The principal values, a_1 and a_3, are aligned along the directions u_i and p_i, respectively. These vectors can be written in terms of the unit vectors, r_i and n_i, defining the orientation of the molecular segment as

$$\begin{aligned} u_i &= \cos\alpha r_i + \sin\alpha n_i, \\ p_i &= -\sin\alpha r_i + \cos\alpha n_i. \end{aligned} \tag{5.19}$$

so that the tensor a_{ij} becomes

$$\begin{aligned} a_{ij} = a_2 \delta_{ij} &+ \Delta'_{12} [c_\alpha^2 r_i r_j + s_\alpha^2 n_i n_j + c_\alpha s_\alpha (r_i n_j + n_i r_j)] \\ &+ \Delta'_{32} (s_\alpha^2 r_i r_j + c_\alpha^2 n_i n_j - c_\alpha s_\alpha (r_i n_j + n_i r_j)) . \end{aligned} \tag{5.20}$$

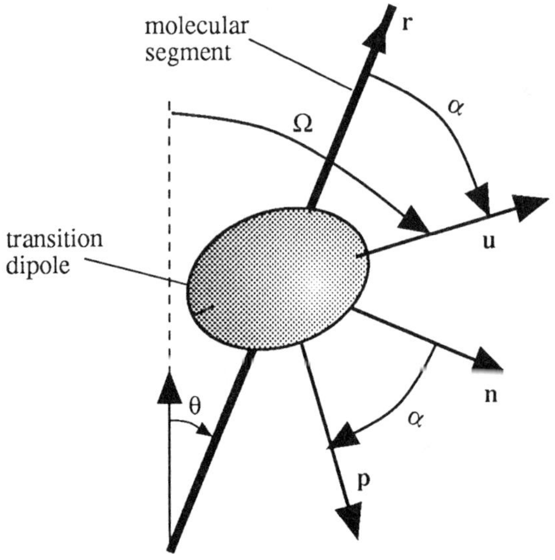

Figure 5.4 Vector diagram describing the orientation of the absorption cross section tensor relative to the molecular axis.

As before, the molecule is allowed to freely spin about the segment axis, r_i, causing the absorption cross section tensor to be tranversely isotropic. This process involves computing the averages $\langle n_i n_j \rangle$ and $\langle n_i \rangle$ on the unit circle normal to the vector r_i. Clearly, the second average will be zero and the first average will have the general form

$$\langle n_i n_j \rangle = A\delta_{ij} + Br_i r_j . \tag{5.21}$$

The trace of (5.21) simply returns the magnitude of the unit vector, $n_i n_i = 1$ so that $3A + B = 1$. Furthermore, if r_i is taken to lie parallel to the "3" axis, the component $\langle n_3 n_3 \rangle$ must be zero since n_i is orthogonal to r_i. This means that $A + B = 0$ and

$$\langle n_i n_j \rangle = \frac{1}{2}(\delta_{ij} - r_i r_j) . \tag{5.22}$$

Using this result, the absorption cross section tensor becomes

$$a_{ij} = \frac{1}{2}(2a_2 + \Delta'_{12}s_\alpha^2 + \Delta_{32}c_\alpha^2)\,\delta_{ij} + \left[\Delta'_{12}P_2(\cos\alpha) + \Delta'_{32}\left(\frac{1}{2} - P_2(\cos\alpha)\right)\right] r_i r_j \tag{5.23}$$

If the absorption dipole is uniaxial ($a_2 = a_3$) then the dichroism for a uniaxial deformation along the "3" axis will be

$$\langle a_{33} - a_{11} \rangle = \Delta'_{12} \langle P_2(\cos\Omega) \rangle$$
$$= \Delta'_{12} P_2(\cos\alpha) \langle r_3^2 - r_1^2 \rangle = \Delta'_{12} P_2(\cos\alpha) \langle P_2(\cos\theta) \rangle, \tag{5.24}$$

which agrees with equation (5.11).

5.2 Raman Scattering

The dominant scattering mechanism is Rayleigh scattering, generating light with the same wavelength as the incident beam. Vibrational and rotational motions within the material, however, will cause a small fraction of the scattered light to be shifted in frequency by discrete amounts. This phenomena is referred to as Raman scattering and provides the capability of spectroscopic identification of orientation dynamics. Furthermore, unlike intrinsic dichroism and birefringence, Raman scattering can provide information concerning *both* the second and fourth moments of the orientation distribution.

Because Raman scattering is a vibrational spectroscopy, its utility is often compared against infrared spectroscopy, which is also a response to vibrational modes within a sample. Certainly, in the years that immediately followed the discovery of the Raman effect in 1923 by C. V. Raman and K. S. Krishnan [20], its application in the laboratory was far easier than for infrared techniques available at that time. Consequently, Raman scattering enjoyed considerable use as a method of structural characterization until the 1950s. That time period marked the beginning of rapid advancement in techniques for infrared spectroscopy, and subsequent developments in Raman scattering methods had to await the invention of the laser. The laser, as a source of powerful, monochromatic radiation, has made Raman scattering a viable complement, and in some cases, an alternative to infrared spectroscopy. In this chapter the physical basis for the effect is explained and the use of polarized Raman light scattering to measure orientation is developed.

5.2.1 Theory of Raman Scattering

Because Raman scattering involves vibrational and rotational modes within a sample, its explanation must necessarily involve a quantum mechanical treatment [21]. This is certainly true when the incident light corresponds to an intrinsic region of absorption in the sample, but it is also required for a quantitative analysis of the simpler Stokes and anti-Stokes Raman scattering, which is the subject of the discussion in this chapter. A detailed quantum mechanical understanding of Raman scattering, however, is not necessary for the applications that are of interest in this book, and for that reason, only a brief account is offered here.

Although both Raman scattering and infrared spectroscopy involve vibrational and rotational motions within a sample, the two processes are quite distinct. Infrared spectroscopy utilizes the absorption of light so that the incident beam must contain frequencies that correspond to the frequencies of those motions. For that reason, infrared spectroscopy requires light sources that generate light in the range of 10,000 to 10 cm^{-1}. Although some lasers can be used for that purpose, black body sources and glow bars are most frequently used. Raman scattering, however, is an inelastic light scattering phenomena that occurs at any region of the spectrum.

When light having a frequency that is removed from any absorption band impinges on a sample, the exiting radiation will be the result of several possible processes. Since, in this discussion, the material is transparent, the majority of the light will be simply transmitted, reflected or refracted. In these cases, the frequency of the light is unaffected, although the polarization vector may have changed. The orientation of wavevector of the light associated with these processes is also well defined. A small amount of light will also be scattered, as a result of interactions with the polarizability tensor of the material. Most of the scattered radiation is the result of an elastic interaction and will also have the same frequency as the incident beam, although it will be directed outward at all angles. This is the process of Rayleigh scattering that was discussed in Chapter 4. Here a scattering element that originally possessed a certain amount of energy will retain that energy at the end of the interaction with the light. This would be true whether the element was in its ground state, or in an excited energy state. For example, if the element was in its first excited state above the ground state with energy $h\nu_1$, following a Rayleigh scattering process, it would be returned to this energy level. Here h is Planck's constant and ν_1 is the frequency associated with that excited state. Pictorially, a quantum mechanical description of this interaction is described in Figure 5.5. In this schematic, light of frequency ν and energy $h\nu$ strikes a sample and elevates the scattering element to a "virtual state" energy level. In Rayleigh scattering, the element quickly returns to its original energy state and radiates light at the frequency of the incident beam.

In Raman scattering, the interaction of light with a scattering element is inelastic and an amount of energy is transferred between the sample and the light. As shown in Figure 5.5, two possibilities exist. If the element is originally in its ground state, upon returning from the virtual state, it can be deposited in the first excited state, thereby producing light with a frequency $\nu - \nu_1$. Light radiated at this lower frequency is referred to as "Stokes" Raman scattered light. Alternatively, elements that originated from the first excited state could be returned to the ground state after radiating light of frequency $\nu + \nu_1$. This process is called "anti-Stokes" Raman scattered light. The relative intensities of Stokes and anti-Stokes Raman scattered light will be proportional to the ratio of the populations of elements in the ground and excited states. Assuming a Boltzmann distribution, this will be proportional to $exp(h\nu_1/kT)$ and for that reason the intensity of the Stokes spectrum will be greater than the anti-Stokes spectrum.

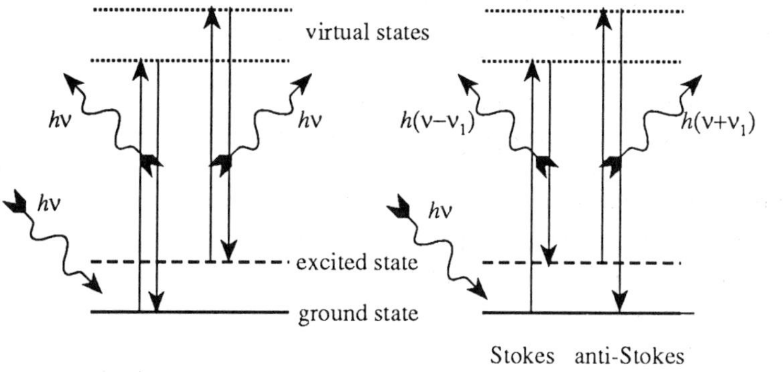

Figure 5.5 Energy level diagram for Rayleigh (left) and Raman (right) scattering.

5.2.2 Classical Theory of Raman Scattering

The following classical description [21], while failing to explain many of the quantitative aspects of the Raman effect, is able to capture its qualitative behavior and provides a useful basis for understanding this phenomena. As in the case of Rayleigh scattering theory, the source of the scattering is the oscillating dipole moment,

$$\mathbf{p} = \underline{\alpha} \cdot \mathbf{E}. \tag{5.25}$$

The polarizability tensor, $\underline{\alpha}$, introduced in section 4.1.2, is a measure of the facility of the electron distribution to distortion by an imposed electric field. The structure of the electron distribution will generally be anisotropic, giving rise to intrinsic birefringence. This optical anisotropy reflects the average electron distribution whereas vibrational and rotational modes of the molecules making up a sample will cause the polarizability to fluctuate in time. These modes are discrete, and considering a particular vibrational frequency, ν_k, the oscillating polarizability can be modeled as

$$\alpha_{ij} = \alpha_{0ij} + \left(\frac{\partial \alpha_{ij}}{\partial Q_k}\right)_0 Q_k \sin 2\pi\nu_k t, \tag{5.26}$$

where Q_k is the amplitude of the mode. The rate of change of the polarizability tensor with the kth vibrational mode defines the derived polarizability tensor,

$$\alpha'_{ij} = \left(\frac{\partial \alpha_{ij}}{\partial Q_k}\right)_0 Q_k. \tag{5.27}$$

Taking the electric field to be $E_{0i} \sin 2\pi\nu t$, the induced dipole moment is then

$$\begin{aligned} p_i &= (\alpha_{0ij} + \alpha'_{ij} \sin 2\pi\nu_k t)\, E_{0j} \sin 2\pi\nu t \\ &= E_{0j}\left[\alpha_{0ij} \sin 2\pi\nu t + \frac{1}{2}\alpha'_{ij}\{\cos 2\pi(\nu - \nu_k)t - \cos 2\pi(\nu + \nu_k)t\}\right], \end{aligned} \tag{5.28}$$

where ν is the frequency of the applied, incident light. This simple analysis reveals that the dipole will oscillate at frequencies $\nu \pm \nu_k$ in addition to the incident frequency. It cannot, however, be used to determine the amplitudes of the Stokes and anti-Stokes Raman bands at the frequencies $\nu - \nu_k$ and $\nu + \nu_k$, respectively.

Using this model, the intensity of light generated by a specific Raman scattering process will simply by the modulus of the electric vector it produces:

$$I = |E_0|^2 \langle (n_i^i \alpha'_{ij} n_j^f)^2 \rangle, \tag{5.29}$$

where n_i^i and n_j^f are unit vectors prescribing the orientations of the polarization vectors of the incident and scattered electric vectors.The angular brackets in equation (5.29) refer to averages over the distribution function specifying configurations of the Raman scatterers in the system. The fourth-rank tensor, $\langle \alpha'_{ij}\alpha'_{kl}\rangle$, is referred to as the *Raman tensor.*

5.2.3 The Depolarization Ratio

The experiment shown in Figure 5.6 shows the basic experiment defining the depolarization ratio. Here incident light, $\mathbf{E}_i$ polarized parallel to the y axis is brought to the sample along the z axis. Raman scattered light is collected either along the x or the y axes and two separate measurements are conducted. With the analyzing polarizer oriented along the y axis, the electric vector of scattered light with polarization parallel to the incident beam, $\mathbf{E}_{r,p}$, is measured. Orientation of the polarizer parallel to the z axis when the measurement is along the x axis (or parallel to the y axis if the measurement is made parallel to the z axis) analyzes the perpendicular polarization state, $\mathbf{E}_{r,s}$. The ratio of the intensities measured in this experiment defines the depolarization ratio,

$$\rho = \frac{|\mathbf{E}_{r,s}|^2}{|\mathbf{E}_{r,p}|^2} = \frac{I_s}{I_p}. \tag{5.30}$$

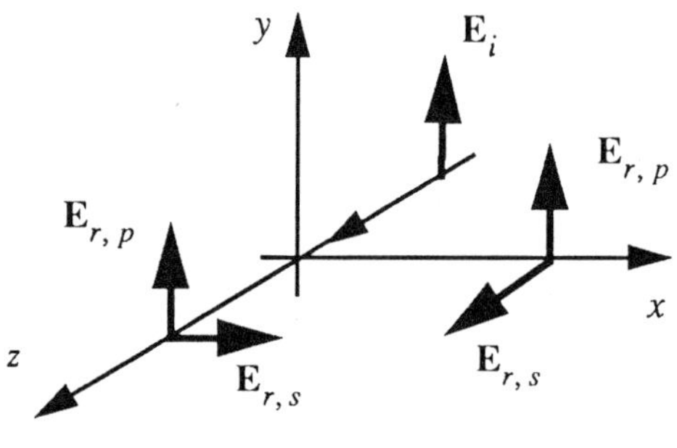

Figure 5.6 Depolarization ratio experiment. $\mathbf{E}_i$ is the incident light and is polarized parallel to the y axis. The Raman scattered light is measured either using a polarizer oriented parallel to the incident polarization ($\mathbf{E}_{r,p}$), or normal to that polarization ($\mathbf{E}_{r,s}$).

For a purely isotropic Raman scattering tensor, the depolarization ratio will be zero since the scattered light from such a system must have a polarization parallel to the incident beam. This is evident when equation equation (5.29) is used in equation (5.30) so that [21],

$$\rho = \frac{\langle (\alpha'_{yz})^2\rangle}{\langle (\alpha'_{yy})^2\rangle}; \text{ or } \rho = \frac{\langle (\alpha'_{xy})^2\rangle}{\langle (\alpha'_{yy})^2\rangle}. \tag{5.31}$$

The off diagonal elements, α'_{yz} and α'_{xy}, will be zero for an isotropic Raman scattering process. For this to occur, not only must the structure of the electron distribution be spherically symmetric, but the vibrational mode under consideration must be dynamically isotropic. These two requirements can be appreciated by inspecting the case of carbon tetrachloride, a molecule that is structurally symmetric, as pictured below in Figure 5.7. Although the polarizability tensor is isotropic for this molecule, gradients in this tensor with respect to vibrational modes can be anisotropic. This will be the case for the asymmetric stretching vibration in Figure 5.7, and this will result in a Raman scattering tensor that is uniaxial, with a principal axis parallel to the axis of the asymmetric vibration. The symmetric stretch, on the other hand, will produce a symmetric form for α'_{ij}.

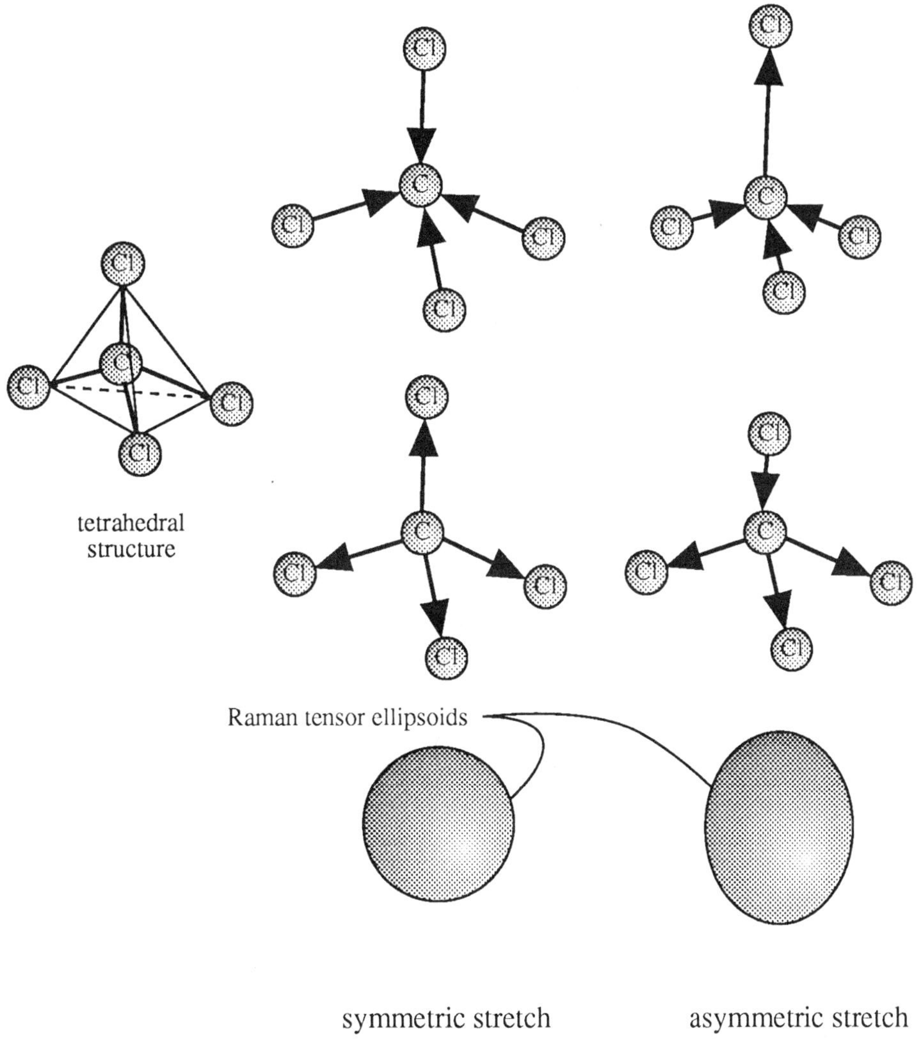

Figure 5.7 The structure of carbon tetrachloride and its symmetric and asymmetric stretching vibrations.

In a system comprised of many molecules, the measured intensity will be an average over all the Raman scatterers in the sample, and this is accounted for by the angular brackets in equation (5.31). Since this average is over the *square* of the derived polarizability tensor, the net result can be nonzero, even for a system of random oriented scatterers, and the depolarization ratio will be finite. As a simple example, consider a material consisting of Raman scattering centers characterized by a uniaxial derived polarizability tensor. Assigning the orientation of the principal axis of a representative scattering center to be parallel to the unit vector, $\mathbf{u}$, to a first approximation, the tensor α'_{ij} can be written as [22]

$$\alpha'_{ij} = \alpha'_0 (\delta_{ij} + \varepsilon u_i u_j) , \tag{5.32}$$

where the parameter ε is a measure of the anisotropy. The average Raman tensor is

$$\begin{aligned} \langle \alpha'_{ij} \alpha'_{kl} \rangle &= \alpha'^2_0 (\delta_{ij}\delta_{kl} + \varepsilon (\delta_{ij} \langle u_k u_l \rangle + \delta_{kl} \langle u_i u_j \rangle) + \varepsilon^2 \langle u_i u_j u_k u_l \rangle) \\ &= \alpha'^2_0 \left[\left(1 + \frac{2}{3}\varepsilon \right) \delta_{ij}\delta_{kl} + \frac{\varepsilon^2}{15} (\delta_{ij}\delta_{kl} + \delta_{ik}\delta_{jl} + \delta_{il}\delta_{jk}) \right], \end{aligned} \tag{5.33}$$

for an isotropic orientation distribution.

Using these results, the depolarization ratio is

$$\rho = \frac{\langle \alpha'^2_{yz} \rangle}{\langle \alpha'^2_{yy} \rangle} = \frac{\varepsilon^2}{15 + 10\varepsilon + 3\varepsilon^2} . \tag{5.34}$$

Measuring the depolarization ratio, therefore, provides a direct, spectroscopic determination of the structure and dynamics of vibrational modes. Furthermore, this experiment can be used to determine the parameter ε, which must be known for quantitative measurements of the order parameters in oriented systems.

5.3 General Form of the Raman Tensor for Transversely Isotropic Systems

In general, the derived polarizability tensor will have a biaxial form,

$$\alpha'_{ij} = \begin{bmatrix} \alpha'_3 & 0 & 0 \\ 0 & \alpha'_2 & 0 \\ 0 & 0 & \alpha'_1 \end{bmatrix} , \tag{5.35}$$

in the frame coaxial with its principal axes. As in the case of the absorption cross section tensor, pictured in Figure 5.4, these axes will not necessarily be aligned with the associated molecular segment. This arrangement was shown in Figure 4.4 for the polarizability tensor. The general form of α'_{ij} is

$$\alpha'_{ij} = \alpha'_2 \delta_{ij} + \tilde{\Delta}_{12} u_i u_j + \tilde{\Delta}_{32} p_i p_j , \tag{5.36}$$

where $\tilde{\Delta}_{ij} = \alpha'_i - \alpha'_j$. As in the case of the polarizability tensor, derived in section 4.1.2, the vectors u_i and p_i are defined in Figure 4.4. In terms of the unit vectors r_i and n_i defining the segment orientation, this tensor is

$$\alpha'_{ij} = \alpha'_2\delta_{ij} + Ar_{ij} + Bn_{ij} + C(n_ir_j + r_in_j), \tag{5.37}$$

where

$$\begin{aligned} A &= \tilde{\Delta}_{12}c_\theta^2 + \tilde{\Delta}_{32}s_\theta^2, \\ B &= \tilde{\Delta}_{12}s_\theta^2 + \tilde{\Delta}_{32}c_\theta^2, \\ C &= \tilde{\Delta}_{13}s_\theta c_\theta, \\ r_{ij} &= r_ir_j, \\ n_{ij} &= n_in_j. \end{aligned} \tag{5.38}$$

This expression is identical in form to equation (5.20). In the case of Raman scattering, however, it is necessary to compute the average Raman tensor, $\langle\alpha'_{ij}\alpha'_{kl}\rangle$. For a transversely isotropic system, the segment is free to spin about the r_i axis, and the vector n_i is averaged over the unit circle normal to r_i. In addition to $\langle n_i\rangle = 0$ and equation (5.22), we require the result,

$$\begin{aligned} \langle n_{ijkl}\rangle = \langle n_in_jn_kn_l\rangle &= \frac{1}{8}(\delta_{ij}\delta_{kl} + \delta_{ik}\delta_{jl} + \delta_{il}\delta_{jk}) \\ &- \frac{1}{8}(\delta_{ij}r_{kl} + \delta_{kl}r_{ij} + \delta_{ik}r_{jl} + \delta_{jl}r_{ik} + \delta_{il}r_{jk} + \delta_{jk}r_{il}) + \frac{3}{8}r_{ijkl}. \end{aligned} \tag{5.39}$$

The average Raman tensor is then found to be

$$\begin{aligned} \langle\alpha'_{ij}\alpha'_{kl}\rangle &= D_1\delta_{ij}\delta_{kl} + D_2(\delta_{ik}\delta_{jl} + \delta_{il}\delta_{jk}) + D_3(\delta_{ij}\langle r_{kl}\rangle + \delta_{kl}\langle r_{ij}\rangle) \\ &+ D_4(\delta_{ik}\langle r_{jl}\rangle + \delta_{jl}\langle r_{ik}\rangle + \delta_{il}\langle r_{jk}\rangle + \delta_{jk}\langle r_{il}\rangle) + D_5\langle r_{ijkl}\rangle, \end{aligned} \tag{5.40}$$

where

$$\begin{aligned} D_1 &= \alpha'^2_2 + \alpha'_2B + \frac{B^2}{8}, \\ D_2 &= \frac{B^2}{8}, \\ D_3 &= \alpha'_2A - \frac{1}{2}\alpha'_2B - \frac{B^2}{8} + \frac{AB}{2}, \\ D_4 &= -\frac{B^2}{8} + \frac{C^2}{8}, \\ D_5 &= A^2 - AB + \frac{3}{8}B^2 + 2C^2. \end{aligned} \tag{5.41}$$

5.4 Raman Scattering Jones Matrix for Oriented Systems

Raman scattering, like infrared dichroism, is a vibrational spectroscopic method capable of measuring orientation. As an example, consider the simple forward Raman scattering process where both the incident and the Raman scattered light propagate along the z axis. The Jones matrix describing the interaction of the derived polarizability tensor with the incident light is simply [22]

$$\mathbf{J}_R = \begin{bmatrix} \alpha'_{xx} & \alpha'_{xy} \\ \alpha'_{yx} & \alpha'_{yy} \end{bmatrix}. \tag{5.42}$$

The Mueller matrix can then be calculated using equation (2.3). Because the Mueller matrix represents observable intensities, these elements must be averaged over probability distributions functions that represent the system. This matrix is

$$\begin{bmatrix} (\mathbf{M}_R)_{11} \\ (\mathbf{M}_R)_{22} \end{bmatrix} = \frac{1}{2}(\langle\alpha'^2_{xx}\rangle + \langle\alpha'^2_{yy}\rangle \pm 2\langle\alpha'^2_{xy}\rangle),$$

$$(\mathbf{M}_R)_{12} = (\mathbf{M}_R)_{21} = \frac{1}{2}(\langle\alpha'^2_{xx}\rangle - \langle\alpha'^2_{yy}\rangle),$$

$$\begin{bmatrix} (\mathbf{M}_R)_{13} \\ (\mathbf{M}_R)_{23} \end{bmatrix} = \begin{bmatrix} (\mathbf{M}_R)_{31} \\ (\mathbf{M}_R)_{32} \end{bmatrix} = \langle\alpha'_{xy}(\alpha'_{xx} \pm \alpha'_{yy})\rangle, \tag{5.43}$$

$$\begin{bmatrix} (\mathbf{M}_R)_{33} \\ (\mathbf{M}_R)_{44} \end{bmatrix} = \langle\alpha'_{xx}\alpha'_{yy} \pm \alpha'^2_{xy}\rangle,$$

$$(\mathbf{M}_R)_{14} = (\mathbf{M}_R)_{24} = (\mathbf{M}_R)_{34} = (\mathbf{M}_R)_{41} = (\mathbf{M}_R)_{42} = (\mathbf{M}_R)_{43} = 0,$$

where it has been assumed that the Raman tensor is real. This assumption is linked to the fact that the Raman tensor is derived from gradients in the real, polarizability tensor. It is instructive at this point to adopt a simple model of the Raman tensor and demonstrate the utility of this measurement in measuring properties of the orientation distribution of a system. If the sample can be represented by a field of orientation unit vectors, $\mathbf{u}$, then the Raman tensor suggested by equation (5.32) is appropriate. Considering the Mueller matrix element, $(\mathbf{M}_R)_{12}$, one can show that

$$(\mathbf{M}_R)_{12} = \frac{1}{2}\alpha'^2_0(2\varepsilon\langle u_x^2 - u_y^2\rangle + \varepsilon^2\langle u_x^4 - u_y^4\rangle). \tag{5.44}$$

Evidently, Raman scattered light contains information about both the *second* and the *fourth* moments of the orientation distribution function. This is in contrast to birefringence and dichroism measurements, which respond only to anisotropies in the second moments.

If the orientation distribution function is uniaxial, and symmetric about the x axis, the averages in equation (5.44) take on the following simple forms,

$$\langle u_x^2 - u_y^2 \rangle = \frac{1}{2}(3\langle \cos^2\theta \rangle - 1) = P_2, \tag{5.45}$$

$$\langle u_x^4 - u_y^4 \rangle = \frac{1}{18}(5\langle \cos^4\theta \rangle + 6\langle \cos^2\theta \rangle - 3) = \frac{1}{7}(P_4 + 6P_2). \tag{5.46}$$

The functions P_2 and P_4 are Legendre polynomials and are functions of the angle θ, which defines the orientation of a representative unit vector $\mathbf{u}$ relative to the symmetry axis of the orientation distribution function for this uniaxial example. As will be shown in Chapter 10, experiments can be designed that can extract sufficient information to determine both Legendre polynomials, and thereby both the second and fourth moments of the orientation distribution function.

The example presented above concerned the generation of Raman scattered light by an oriented sample but did not account for the fact that such a sample will also be birefringent. Since the scattering process is sensitive to the polarization of the incident light, however, it is important to properly account for any birefringence that might be present since this will continuously alter the electric vector as the light propagates through the material. The schematic diagram in Figure 5.8 describes this process [22].

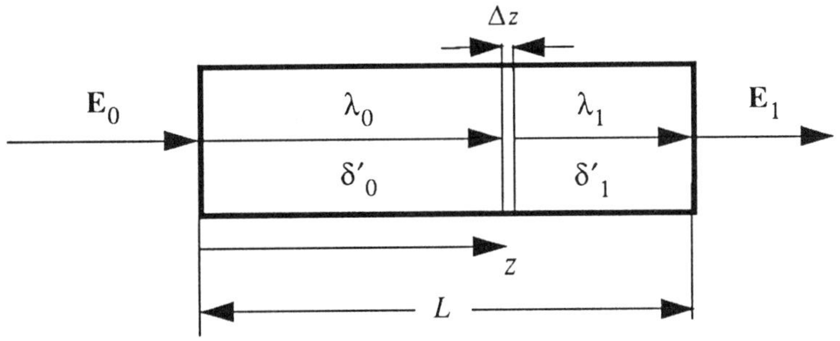

Figure 5.8 The Raman scattering process in a birefringent material.

The particular example described in Figure 5.8 concerns zero-angle Raman scattered light generated with an incident beam traveling along the z axis. For that reason, the electric vectors of the incident and scattered light will lie in the (x, y) plane. As illustrated in the figure, the sequence of events involved with this process begins with the transmission of the incident beam with wavelength λ_0 through a length z of material with retardation $\delta'_0 = (2\pi\Delta n' z)/\lambda_0$. The birefringence, $\Delta n'$, is assumed to be independent of wavelength. The Raman scattering event occurs uniformly throughout the sample and transforms the wavelength to λ_1. In this particular illustration, the Raman scattering takes place within the differential element of thickness Δz. Assuming no multiple Raman scattering occurs, the light then exits the sample through a material of retardation

$\delta'_1 = [2\pi\Delta n'(L-z)]/\lambda_1$. The Jones matrix describing this sequence of events is simply the following result:

$$\mathbf{J}_R = \begin{bmatrix} e^{-i\frac{\delta'\gamma(L-z)}{2L}} & 0 \\ 0 & e^{i\frac{\delta'\gamma(L-z)}{2L}} \end{bmatrix} \cdot \begin{bmatrix} \alpha'_{xx} & \alpha'_{xy} \\ \alpha'_{xy} & \alpha'_{yy} \end{bmatrix} \cdot \begin{bmatrix} e^{-i\frac{\delta' z}{2L}} & 0 \\ 0 & e^{i\frac{\delta' z}{2L}} \end{bmatrix}, \tag{5.47}$$

where

$$\delta' = \frac{2\pi\Delta n' L}{\lambda_0}, \tag{5.48}$$

$$\gamma = \lambda_0/\lambda_1. \tag{5.49}$$

In writing this result, it has been assumed that the birefringence is oriented parallel to either the x or y axes. Completing the multiplication in equation (5.47) yields the following Jones matrix elements

$$\begin{aligned} J_{11} &= \alpha'_{xx} e^{-i\frac{\delta'}{2}[1-\beta(1-s)]}, \\ J_{22} &= \alpha'_{yy} e^{i\frac{\delta'}{2}[1-\beta(1-s)]}, \\ J_{21} &= \alpha'_{xy} e^{i\frac{\delta'}{2}[(1-2s)-\beta(1-s)]}, \\ J_{12} &= \alpha'_{xy} e^{-i\frac{\delta'}{2}[(1-2s)-\beta(1-s)]}. \end{aligned} \tag{5.50}$$

The parameter $\beta = (\lambda_1 - \lambda_0)/\lambda_1$ is normally a small number and $s = z/L$.

The electric field generated by scattering elements within the differential slab is simply $E_{s,i} = J_{ij}E_{0,j}$, and the total intensity from the entire sample is

$$I = \int_0^1 ds J_{ij} J^*_{ik} E_{0,j} E^*_{0,k}. \tag{5.51}$$

It is required to calculate the Mueller elements by first calculating the differential elements for the slab using equation (2.3) and integrating along the path of light propagation as indicated in equation (5.51). The Mueller matrix elements are

$$M_{11} = \frac{1}{2}\langle \alpha'^2_{xx} + \alpha'^2_{yy} + 2\alpha'^2_{xy} \rangle,$$

$$M_{12} = M_{21} = \frac{1}{2}\langle \alpha'^2_{xx} - \alpha'^2_{yy} \rangle,$$

$$M_{22} = \frac{1}{2}\langle \alpha'^2_{xx} + \alpha'^2_{yy} - 2\alpha'^2_{xy} \rangle,$$

$$M_{13} = M_{31} = M_{14} = M_{41} = M_{23} = M_{32} = 0, \tag{5.52}$$

$$M_{33} = S\left(\frac{\beta\delta'}{2}\right)\cos\left[\delta'\left(1-\frac{\beta}{2}\right)\right]\langle \alpha'_{xx}\alpha'_{yy} \rangle + S\left(\delta'\left[1-\frac{\beta}{2}\right]\right)\cos\left(\frac{\beta\delta'}{2}\right)\langle \alpha'^2_{xy} \rangle,$$

$$M_{34} = -M_{43} = S\left(\delta'\left[1-\frac{\beta}{2}\right]\right)\left[S\left(\frac{\beta\delta'}{2}\right)\langle \alpha'_{xx}\alpha'_{yy} \rangle + \frac{\sin(\beta\delta'/2)}{1-\beta/2}\langle \alpha'^2_{xy} \rangle\right],$$

$$M_{44} = S\left(\frac{\beta\delta'}{2}\right)\cos\left[\delta'\left(1-\frac{\beta}{2}\right)\right]\langle \alpha'_{xx}\alpha'_{yy} \rangle - S\left(\delta'\left[1-\frac{\beta}{2}\right]\right)\cos\left(\frac{\beta\delta'}{2}\right)\langle \alpha'^2_{xy} \rangle,$$

where the function $S(\alpha) = \sin\alpha/\alpha$. It is evident that the birefringence of a sample will directly affect the Raman scattered light from an oriented sample and it must be measured. The design of instrumentation to simultaneously measure birefringence and Raman scattered light is discussed in section 8.6.

5.5 Polarized Fluorescence

Although polarized fluorescence arises from a very different physical phenomenon, it is similar to Raman scattering since this interaction produces emitted radiation of a different wavelength than the incident light. In addition, this measurement can also provide both the second and fourth moments of the orientation distribution function. The basic sequence giving rise to fluorescence is described in Figure 5.9 [18,23]. The process proceeds with the absorption of light and the transition of a molecule from its ground state ($\mathbf{S}_0$) to an excited state ($\mathbf{S}_1$ or $\mathbf{S}_2$). This occurs very quickly over a duration on the order of 10^{-15} sec. This is usually followed by a relaxation of the state of the molecule by "internal conversion" processes to the lowest vibrational state of the first excited state, $\mathbf{S}_1$. Alternatively, relaxation to an available triplet state, $\mathbf{T}_1$, can take place by an "inter-system crossing" from $\mathbf{S}_1$. Relative to the subsequent fluorescence process, these are also fast events, occuring in 10^{-12} sec. From the states $\mathbf{S}_1$ and $\mathbf{T}_1$, two types of luminescence processes can occur, depending on the relative spin states of electrons in $\mathbf{S}_1$ and $\mathbf{T}_1$ to the ground state. For the excited singlet state, $\mathbf{S}_1$, the spin of the electron will be the same as found in the ground state. This allows the decay from $\mathbf{S}_1$ to $\mathbf{S}_0$ to occur rapidly at a time of approximately 10^{-8} sec. The energy released by this decay is emitted as fluorescence. For the triplet state, however, the spin is opposite relative to the ground state, and this is "forbidden" and occurs very slowly, over a time span of 0.1 to 10 seconds. The emitted radiation is termed phosphorescence. This section is only concerned with fluorescence.

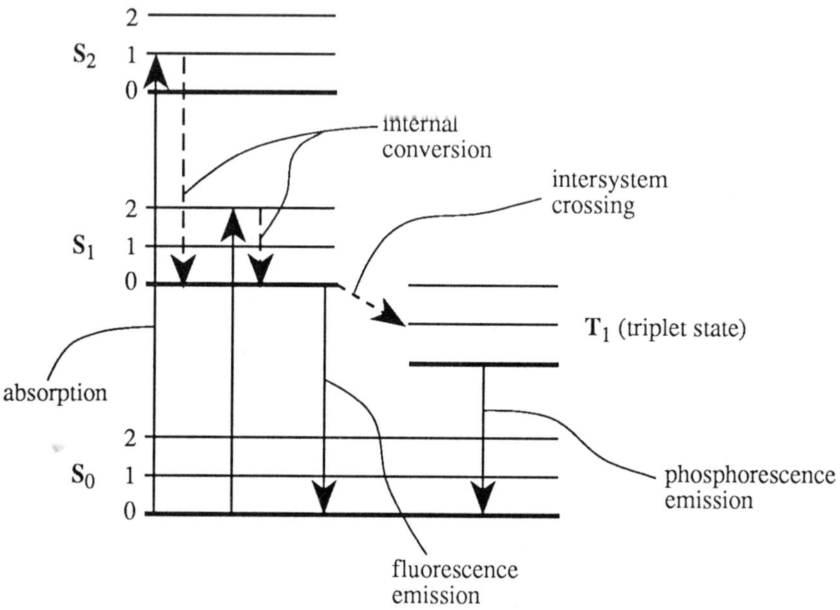

Figure 5.9 The Jablonski diagram describing absorption and emission processes.

Both the initial absorption event, and the subsequent fluorescence emission are directional processes. The strength of a particular fluorescence process will then be proportional to $\langle \cos^2\Omega_a \cos^2\Omega_e \rangle$, where Ω_a is the angle defining the relative orientation of the polarization of the incident light and the absorption axis. Similarly, Ω_e is the angle between the emission axis and the polarization direction of the measured radiation. The difference between these two angles is normally very small and is often taken as zero. One reason for a finite difference between Ω_a and Ω_e is that the absorption and emission processes are not simultaneous events, and if the target molecule has an opportunity to rotate, these two axes will not be coincident [17,24]. If the direction cosines of the absorption and emission axes are $\mathbf{f}^a$ and $\mathbf{f}^e$, the intensity of the light measured in an experiment with an incident and analyzing polarizations aligned parallel to the unit vectors $\mathbf{n}^i$ and $\mathbf{n}^f$ is

$$I = A\sum\langle(\mathbf{n}^i\cdot\mathbf{f}^a)^2(\mathbf{n}^f\cdot\mathbf{f}^e)^2\rangle = A\sum\langle[\mathbf{n}^i\cdot(\mathbf{f}^a\mathbf{f}^e)\cdot\mathbf{n}^e]^2\rangle. \quad (5.53)$$

Here A is a constant that depends on the efficiencies of the absorption and emission processes, and the summation is over all fluorescent groups in the system contributing to the observed signal. This result is identical in form to equation (5.29), and the tensor $f_i^a f_j^e$ is analogous to the Raman scattering tensor, α'_{ij}. Indeed, if the angles Ω_a and Ω_e are identical, then this tensor will assume a form similar to equation (5.32).

6 Laser Doppler Velocimetry and Dynamic Light Scattering

The methods described in this book are primarily concerned with the measurement of the microstructure of complex fluids subject to the application of external, orienting fields. In the case of flow, it is also of interest to measure the kinematics of the fluid motion. This chapter describes two experimental techniques that can be used for this purpose: laser Doppler velocimetry for the measurement of fluid velocities, and dynamic light scattering (or photon correlation spectroscopy) for the determination of velocity gradients.

6.1 Laser Doppler Velocimetry

Laser Doppler velocimetry is a powerful technique for the *in situ* measurement of fluid velocities. The basic optical configuration for the measurement is shown in Figure 6.1. The velocity measurement is made at the intersection of two laser beams that are focused to a point in the flow. The use of laser radiation is essential since the light must be monochromatic and coherent. This is required since the intersection of the two beams must create an interference pattern within the fluid. Such a pattern is shown in Figure 6.2, where two plane waves intersect at an angle 2ϕ. The two waves will have the following form [55]:

$$E_1 = A_1 e^{i\{\omega t - [k(\cos\phi z + \sin\phi y) + \alpha_1]\}},$$
$$E_2 = A_2 e^{i\{\omega t - [k(\cos\phi z - \sin\phi y) + \alpha_2]\}}, \tag{6.1}$$

where α_i is a phase factor and A_i is an amplitude.

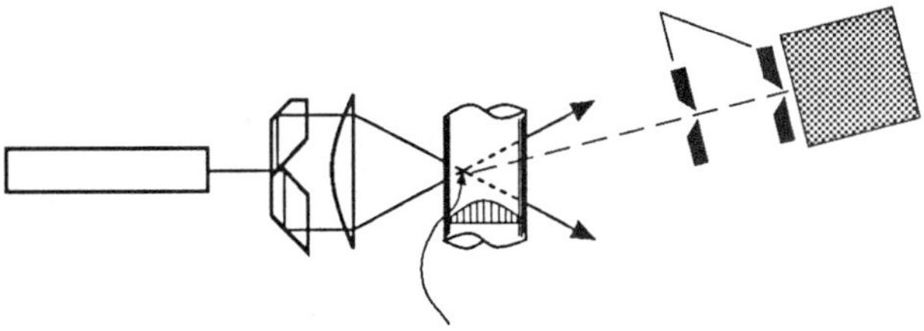

Figure 6.1 Optical configuration for a laser Doppler velocimeter.

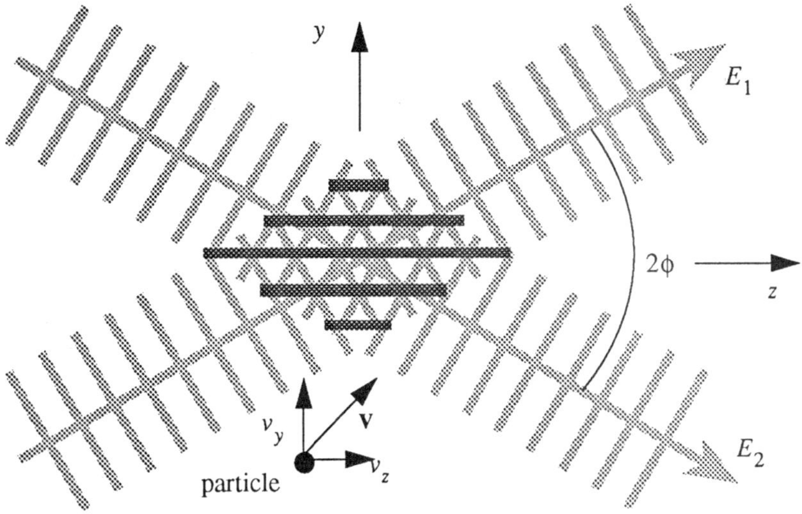

Figure 6.2 Fringe pattern formed by two intersecting plane waves.

The intensity of the combined beams in the intersection region is simply

$$I = (E_1 + E_2)(E_1 + E_2)^* = A_1^2 + A_2^2 + 2A_1A_2\cos\left[\left(\frac{4\pi n}{\lambda}\right)(\sin\phi)\,y\right], \tag{6.2}$$

if the two beams have the same phase ($\alpha_1 = \alpha_2$). This simple expression does not account for the radial intensity distribution (normally Gaussian) of the laser beams, but this complication can be easily included. The result are interference fringes (shown as the dark, horizontal lines in Fig. 6.2) where the intensity oscillates from bright to dark with a spacing of $d = \lambda/(2\sin\phi)$.

Also shown in Figure 6.2 is a particle moving with a velocity $\mathbf{v} = (v_x, v_y, v_z)$. If such a particle traverses the intersection region, light scattered by the particle will have an intensity that is modulated at a temporal frequency,

$$\nu = v_y/d = \frac{2v_y \sin\phi}{\lambda}. \quad (6.3)$$

The signal produced by the photomultiplier tube in Figure 6.1 will have an "ac" component with a frequency ν that is proportional to the velocity component of particles parallel to the fringe pattern. This frequency can be retrieved by a Fourier transformation of the intensity signal, which can be conveniently accomplished using digital signal processing methods.

A successful application of laser Doppler velocimetry requires a proper seeding of the flow with scattering particles. Frequently, such particles are naturally present in fluid systems in the form of contaminants such as dust. Their size should be chosen to produce an adequate scattering signal, and this can depend on the location of the photomultiplier tube. When possible, forward scattering should be used since the scattering by large particles will be greater at small scattering angles. Geometrical constraints, however, may require placing the photomultiplier tube in a back scattering location. Clearly, the particle size should be smaller than the fringe spacing. The concentration of the particles should be chosen so that it is in a range that avoids multiple particles residing simultaneously in the measurement volume, and yet is high enough to provide frequent "bursts" of particle scattering. The minimum number concentration of the particles can be simply calculated as $c_{min} = \sin 2\phi/(\pi R^3)$, which is the reciprocal of the volume of the intersection region and R is the radius of the focused laser beams.

The simple optical configuration in Figure 6.1 is limited by only being able to measure a single velocity component (the component v_y in the example above) and cannot distinguish between positive or negative values of that velocity. Measurement of two orthogonal velocity components simultaneously can be accomplished by using two pairs of intersecting beams, all targeting the same measurement volume. Such an arrangement is shown in Figure 6.3. Two separate photomultiplier tubes are used to measure the scattered light from the separate fringe patterns from the two pairs of beams. Two methods are normally used to distinguish the two patterns: polarization discrimination and two-color techniques. With polarization, each pair of beams is given an orthogonal polarization [for example, the pair in the (x, z) plane could be polarized along the y axis, and the pair lying in the (y, z) plane would be polarized in the x direction]. By placing a polarizer in front of each photomultiplier tube, and by aligning the polarizers with the polarization of one or the other pairs of beams, the measured intensity signals will follow orthogonal projections of the particle velocities. The two-color technique follows a similar strategy, except that two separate wavelengths are used to identify the two pairs of beams. As in the case of two color flow birefringence, discussed in Chapter 8, an argon ion laser is used to generate the two colors. Such a laser can be operated in a manner that simultaneously generates intense beams at wavelengths of 488 and 514 nm. Each color is split into pairs of converging beams with orthogonal planes of intersection. Color separation is achieved at the photomultiplier tubes by placing appropriate laser line filters in front of each tube.

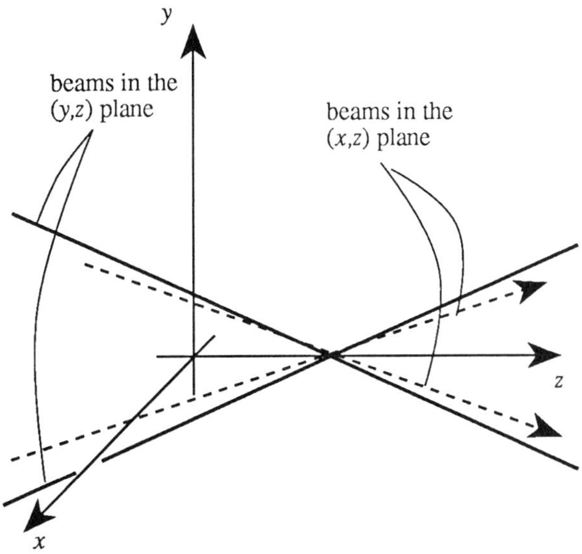

Figure 6.3 Two-component laser Doppler velocimetry beam arrangement.

The problem of distinguishing between positive and negative velocities is linked to the fact that in the configuration shown in Figure 6.1, the Doppler shifted frequencies are shifted from zero frequency. Velocities of opposite direction, but with the same magnitude, will produce the same frequency shift and will be indistinguishable. This deficiency can be overcome by frequency shifting one of the beams making up the fringe pattern using a Bragg cell. This is an acousto-optic component that will modify one of the beams (beam "1" in equation (6.1), for example) so that the relative phase of the two beams is $\alpha_1 - \alpha_2 = \omega_B t$, where ω_B is the Bragg shifted frequency. The intensity of the fringe pattern is then,

$$I = (E_1 + E_2)(E_1 + E_2)^* = A_1^2 + A_2^2 + 2A_1 A_2 \cos\left[\left(\frac{4\pi n}{\lambda}\right)(\sin\phi)\,y + \omega_B t\right] . \quad (6.4)$$

The fringe pattern is predicted to "translate" in space with a speed proportional to ω_B^{-1}. In this case, particles at rest within the fringe pattern will scatter light with an intensity that oscillates in time. Particles in motion in the same direction as the fringe pattern translation will produce a scattered light intensity with a decreased frequency, whereas particles traveling in the opposite direction will produce an increased frequency.

6.2 Dynamic Light Scattering

In equation (4.51) of Chapter 4, the "equal time structure factor," $C(\mathbf{x})$, was defined. For light scattered from fluctuations in the dielectric properties of a material, it was shown that the light intensity was proportional to this quantity. In the problem of total intensity light scattering discussed in that chapter, the measurement is integrated over time and time-dependent fluctuations are not directly observed. When time-dependent fluctuations are taken into account, however, the intensity of scattered light will have the following, more general form, [44]

$$I(\mathbf{q},t) = \frac{1}{(4\pi r)^2}\left|\mathbf{k}_s\times\mathbf{k}_s\times\mathbf{n}_i\right|^2\left(\frac{\partial\varepsilon_0}{\partial c}\right)^2\left(\iint d\mathbf{x}'d\mathbf{x}''e^{-i\mathbf{q}\cdot(\mathbf{x}'-\mathbf{x}'')}\right) \times \langle E_0(\mathbf{x}',t)\,\delta c(\mathbf{x}',t)\,E_0^*(\mathbf{x}'',0)\,\delta c(\mathbf{x}'',0)\rangle, \tag{6.5}$$

where the fluctuations in the dielectric properties are assumed to arise from concentration fluctuations. In addition, spatial variation of the incident beam within the scattering volume has been accounted for by allowing it to depend on the position of the scatterer.

The number density of the scattering elements is taken to be

$$c(\mathbf{x},t) = \sum_{j=1}^{N}\delta(\mathbf{x}-\mathbf{r}_j(t)), \tag{6.6}$$

where $\mathbf{r}_j(t)$ is the position of element j at time t and there are N such elements. Using this expression, and carrying out the spatial integrations in equation (6.5), the intensity is

$$I(\mathbf{q},t) = \frac{1}{(4\pi r)^2}\left|\mathbf{k}_s\times\mathbf{k}_s\times\mathbf{n}_i\right|^2\left(\frac{\partial\varepsilon_0}{\partial c}\right)^2\sum_{j=1}^{N}\langle E_{0j}(t)\,E_{0j}^*(0)\,e^{-i\mathbf{q}\cdot[\mathbf{r}_j(t)-\mathbf{r}_j(0)]}\rangle. \tag{6.7}$$

The intensity is proportional to the defined function,

$$F_1(\mathbf{q},t) = \sum_{j=1}^{N}\langle E_{0j}(t)\,E_{0j}^*(0)\,e^{-i\mathbf{q}\cdot[\mathbf{r}_j(t)-\mathbf{r}_j(0)]}\rangle, \tag{6.8}$$

which is named the heterodyne correlation function. In dynamic light scattering, the intensity signal is sent to an autocorrelator, which calculates the autocorrelation function,

$$C(\mathbf{q},t) = \langle I(\mathbf{q},t)\,I(\mathbf{q},0)\rangle = B\left|F_1(\mathbf{q},t)\right|^2 = BF_2(\mathbf{q},t), \tag{6.9}$$

where B is a constant and $F_2(\mathbf{q},t)$ is the homodyne autocorrelation function. The distinction between homodyne and heterodyne light scattering refers to whether a portion of the incident light beam is mixed with the scattered light. That is the case for heterodyne scattering. In a homodyne experiment, only the scattered light is measured.

The function F_1 can be simplified if the timescales associated for variation of the amplitude, $E_{0j}(t)$ and the phase factor, $\mathbf{q}\cdot[\mathbf{r}_j(t)-\mathbf{r}_j(0)]$ are sufficiently different. The amplitude factor will vary on a timescale comparable to the transit time of a scattering element as it convects through the scattering volume. If this volume is characterized by a length scale, L, that timescale is

$$\tau_t = L/U, \tag{6.10}$$

where U is the mean velocity of the particle. The relevant length scale for the phase factor, on the other hand, is q^{-1} and the associated timescale will be shown to be

$$\tau_s = (q\dot{\gamma}L)^{-1}, \tag{6.11}$$

where $\dot{\gamma}$ is the average velocity gradient within the volume. The ratio of these timescales is

$$\tau_t/\tau_s = q\dot{\gamma}L^2/U \gg 1, \tag{6.12}$$

under normal experimental conditions. Since this ratio of timescales can be safely assumed to be very large, the amplitude factor can be taken to be independent of time during the measurement of the autocorrelation function. In this case we have

$$F_1(\mathbf{q}, t) = \sum_{j=1}^{N} I_j \langle \exp\{i\mathbf{q} \cdot [\mathbf{r}_j(t) - \mathbf{r}_j(0)]\}\rangle = \sum_{j=1}^{N} I_j F_{sj}(\mathbf{q}, t), \tag{6.13}$$

where $I_j = E_{0j}(t) E_{0j}^*(0)$.

Calculation of the autocorrelation function proceeds by noting that the single particle scattering function, $F_{sj}(\mathbf{q}, t)$, is simply the Fourier transform of $G_{sj}(\mathbf{x}_j)$, the Van-Hove self space-time correlation function. [8] In other words,

$$\begin{aligned} G_{sj} &= \int d^3\mathbf{q}\, e^{-i\mathbf{q} \cdot \mathbf{x}_j} \langle e^{i\mathbf{q} \cdot [\mathbf{r}_j(t) - \mathbf{r}_j(0)]} \rangle \\ &= \langle \delta(\mathbf{x}_j - [\mathbf{r}_j(t) - \mathbf{r}_j(0)]) \rangle \end{aligned} . \tag{6.14}$$

Examination of equation (6.14) reveals that G_{sj} is the probability of finding a particle at a position $\mathbf{r}_j(t)$, given that it started at $\mathbf{r}_j(0)$ at time $t = 0$. This is the solution to the classical convection diffusion equation for a scatterer subject to the combined influence of diffusion and convection. This equation is

$$\frac{\partial G_{sj}}{\partial t} + \nabla \cdot (\mathbf{v}_j G_{sj}) - D_t \nabla^2 G_{sj} = 0, \tag{6.15}$$

subject to

$$G_{sj}(\mathbf{x}_j, t) = \delta(\mathbf{x}_j), \tag{6.16}$$

where D_t is the diffusion coefficient and $\mathbf{v}_j$ is the velocity of the jth scatterer. Since the scattering volume dimensions are normally small relative to length scales associated with the velocity field, the velocity can be expanded as

$$\mathbf{v}_j = \mathbf{V}_j + \mathbf{G} \cdot \mathbf{x}_j, \tag{6.17}$$

where $\mathbf{G} = \nabla \mathbf{v}$ is the velocity gradient tensor. Substituting equation (6.17) into (6.15) and taking the Fourier transform of the result yields

$$\frac{\partial F_{sj}}{\partial t} - i\mathbf{V}_j \cdot \mathbf{q} F_{sj} - \mathbf{q} \cdot \mathbf{G} \cdot \nabla_q F_{sj} + D_t q^2 F_{sj} = 0,$$
$$F_{sj}(\mathbf{q}, 0) = \frac{1}{(2\pi)^{3/2}}. \tag{6.18}$$

Equation (6.18) can be solved using the method of characteristics to give

$$F_{sj}(\mathbf{q}, t) = \frac{1}{(2\pi)^{3/2}} \exp\left[\int_0^t dt' \left[D_t q'^2(t') + i\mathbf{V}_j \cdot \mathbf{q}'(t')\right]\right],$$
$$\frac{d}{dt}\mathbf{q}' = -\mathbf{G}^T \cdot \mathbf{q}',$$
$$\mathbf{q}'(0) = \mathbf{q}. \tag{6.19}$$

Combining these results, and converting the summation over particles in equation (6.13) to a volume integral we have

$$F_1(\mathbf{q}, t) = \iiint d^3\mathbf{x} I(\mathbf{x}) \exp\left[-\int_0^t dt' \left[D_t q'^2(t') + i\mathbf{V}(\mathbf{x}) \cdot \mathbf{q}'(t')\right]\right], \tag{6.20}$$

where $I(\mathbf{X})$ is the intensity of the incident beam within the scattering volume, and $\mathbf{V}(\mathbf{x})$ is the velocity of the particles, which is taken as the following approximation,

$$\mathbf{V}(\mathbf{x}) = \mathbf{U} + \mathbf{G} \cdot \mathbf{x}. \tag{6.21}$$

Here $\mathbf{U}$ is the mean velocity within the scattering volume and not the velocity of a specific scattering element. One finally obtains the following result:

$$F_1(\mathbf{q}, t) = \tilde{I}\left[\int_0^{t'} dt' \mathbf{q}'(t') \cdot \mathbf{G}\right] \exp\left[-\int_0^t dt' \left[D_t q'^2(t') + i\mathbf{U} \cdot \mathbf{q}'(t')\right]\right], \tag{6.22}$$

where

$$\tilde{I}\left[\int_0^{t'} dt' \mathbf{q}'(t') \cdot \mathbf{G}\right] = \iiint d^3\mathbf{x} I(\mathbf{x}) e^{-i\int_0^{t'} dt' q'(t') \cdot \mathbf{G} \cdot \mathbf{x}} \tag{6.23}$$

is the Fourier transform of the intensity profile within the scattering volume. The corresponding homodyne autocorrelation function is then

$$F_2(\mathbf{q}, t) = \exp\left[-2\int_0^t dt' D_t q'^2(t')\right] \left| \tilde{I}\left[\int_0^{t'} dt' \mathbf{q}'(t') \cdot \mathbf{G}\right]\right|^2 . \tag{6.24}$$

It is important to note that the homodyne autocorrelation function is independent of the mean velocity within the scattering volume and is influenced only by the diffusion and velocity gradient induced motions of the particles. Furthermore, if there is no velocity gradient, the result is that the correlation function is the following, familiar result,

$$F_2(\mathbf{q}, t) = e^{-2D_t q^2 t} , \tag{6.25}$$

from which the diffusion coefficient can be determined.

In the presence of flow, the correlation function will be affected by gradients in the velocity. For example, in the case of simple shear flow,

$$\mathbf{G} = \dot{\gamma}\begin{bmatrix} 0 & 1 & 0 \\ 0 & 0 & 0 \\ 0 & 0 & 0 \end{bmatrix} , \tag{6.26}$$

the homodyne autocorrelation function is computed to be

$$F_2(\mathbf{q}, t) = \exp\left[-2D_t\left(q_x^2\left(1 + \frac{1}{3}(\dot{\gamma}t)^2\right) + \dot{\gamma}t q_x q_y + q_y^2 + q_z^2\right)t\right] \left|\tilde{I}\left(\dot{\gamma}t(0, q_x, 0)^T\right)\right|^2 \tag{6.27}$$

There are three timescales controlling the shape of $F_2(\mathbf{q}, t)$: $\tau_s = (q\dot{\gamma}L)^{-1}$, $\tau_f = \dot{\gamma}^{-1}$ and $\tau_D = (q^2 D_t)^{-1}$. In most cases, experimental conditions will present the condition that $\tau_s \ll (\tau_f, \tau_D)$, so that the correlation function will assume the simpler form,

$$F_2(\mathbf{q}, t) \approx \left|\iiint d^3\mathbf{x} I(\mathbf{x}) e^{-iq_x\dot{\gamma}yt}\right|^2 \tag{6.28}$$

for simple shear flow. In the case of a more general form of the velocity gradient tensor, this limiting condition will produce the following autocorrelation function,

$$F_2(\mathbf{q}, t) \approx \left|\iiint d^3\mathbf{x} I(\mathbf{x}) e^{-i\mathbf{q} \cdot \mathbf{G} \cdot \mathbf{x}t}\right|^2 . \tag{6.29}$$

Equation (6.29) is the governing equation for the use of homodyne, dynamic light scattering for measurement of velocity gradients. What is left is to specify the shape of the intensity distribution of the incident light within the scattering volume. As an example, if a simple Gaussian distribution is used so that $I(\mathbf{x}) = \exp[-(x^2 + y^2 + z^2)/L^2]$, the autocorrelation function becomes

$$F_2(\mathbf{q}, t) = e^{-\frac{1}{2}q^2\dot{\gamma}L^2t^2} , \tag{6.30}$$

for a simple shear flow where $\mathbf{q} \cdot \mathbf{G} \cdot \mathbf{x}t = q_x \dot{\gamma} y t$. Measurement of this function, and its timescale, $(q\dot{\gamma}L)^{-1}$, will produce a result directly proportional to the inverse of the velocity gradient. These results indicate that the correlation function measured for diffusing particles in the presence of flow will usually be dominated by the effects of velocity gradients, and it normally will be very difficult to extract diffusion information.

7 Microstructural Theories of Optical Properties

The optical measurements presented in the previous chapters can be used to either characterize local, microstructural properties or as probes of bulk responses to orientation processes. In either case, it is normally desirable to make the connection between experimental observables and their molecular or microstructural origins. The particular molecular properties that are probed will naturally depend on the physical interaction between the light and the material. This chapter explores molecular models and theories that describe these interactions and identifies the properties of complex materials that can be extracted from measurements of optical anisotropies. The presentation begins with a discussion of molecular models that are applied to polymeric materials. Using these models, optical phenomena such as birefringence, dichroism, and Rayleigh and Raman scattering are predicted. Models appropriate for particulate systems are also developed.

7.1 Molecular and Polymeric Systems

Orientation of molecules can normally be detected using polarized light in either transmission or scattering experiments. The prediction of the observed effect (birefringence or dichroism, for example) requires a description of the basic "light-matter" interaction combined with a calculation of the distribution of orientations associated with the individual molecules or segments of molecules.

7.1.1 The Lorentz-Lorenz Equation

In this subsection, the connection is made between the molecular polarizability, α, and the macroscopic dielectric constant, ε, or refractive index, n. This relationship, referred to as the Lorentz-Lorenz equation, is derived by considering the immersion of a dielectric material within an electric field, and calculating the resulting polarization from both a macroscopic and molecular point of view. Figure 7.1 shows the two equivalent problems that are analyzed.

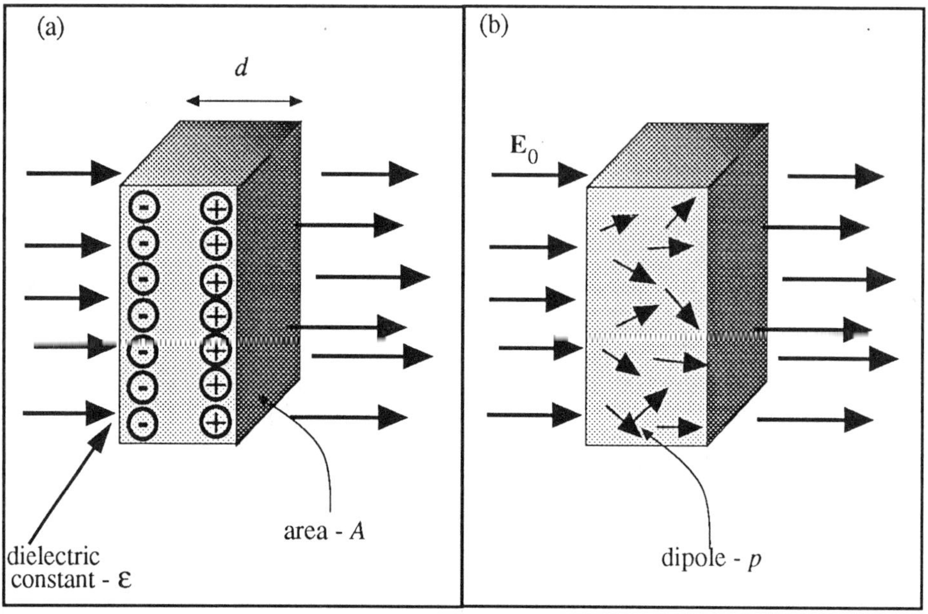

Figure 7.1 Insertion of a polarizable material into a uniform electric field, $\mathbf{E}_0$. (a) A slab of material with macroscopic dielectric constant, ε. (b) A slab of material containing a distribution of dipoles of strength p.

It is desired to calculate the polarization, $\boldsymbol{P}$, which is defined as the total dipole moment of the system per unit volume. For the macroscopic model in Figure 7.1a, the slab becomes polarized upon insertion within the field and charges collect on both surfaces. The result is the presence of a surface charge density, σ, and a total charge σA, where A is the area of each surface. The total dipole moment, $\mathbf{p}$, is simply the total charge multiplied by the distance separating the two surfaces. The polarization, $\boldsymbol{P}$, is then

$$\boldsymbol{P} = \frac{\mathbf{p}}{Ad} = \frac{(\sigma A)\, d}{Ad}\mathbf{e}_z = \sigma \mathbf{e}_z, \tag{7.1}$$

where $\mathbf{e}_z$ is the unit vector parallel to the electric field. The surface charge density can be determined by applying the first boundary condition in (1.69) so that

$$\sigma = \frac{1}{4\pi}(\varepsilon - 1)\, E_0 . \tag{7.2}$$

To complete this analysis, a result for the polarization based on the microscopic picture of the problem in Figure 7.1b must be developed. If the number density of dipoles in the slab is ν, the polarization is simply

$$\boldsymbol{P} = \nu \mathbf{p} = \nu \langle \alpha_M \rangle \mathbf{E}, \tag{7.3}$$

where $\langle\alpha_M\rangle$ is the effective molecular polarizability averaged over the distribution of orientations of polarizable elements. In general, the effective molecular polarizability can be a function of the electric field and could involve permanent as well as induced dipole moments.

The electric field, $\mathbf{E}$, in equation (7.3) is the total field experienced by each dipole and is the sum of the external field, $\mathbf{E}_0$, and an internal field, $\mathbf{E}_i$. This latter field accounts for contributions to the electric field felt by each dipole from interactions with neighboring dipoles. In a dilute system of dipoles, this augmented field can be estimated in a mean field sense by calculating the increase of an electric field in the vicinity of a dipole immersed in a uniform field. An expression for such an increase can be obtained by examining the case of a sphere immersed in a uniform field, developed in section 4.1.3. From equation (4.29), the polarization due to a single sphere is

$$P = -\frac{3}{4\pi}\frac{(\varepsilon_1 - \varepsilon_2)}{\varepsilon_1 + 2\varepsilon_2}\mathbf{E}_0, \tag{7.4}$$

and from the second equation in (4.24) the external field experienced by the sphere on its surface is

$$\mathbf{E} = \mathbf{E}_0 - \frac{(\varepsilon_1 - \varepsilon_2)}{(\varepsilon_1 + 2\varepsilon_2)}\mathbf{E}_0 = \mathbf{E}_0 + \frac{4\pi}{3}P. \tag{7.5}$$

Although this result was derived from the special case of an isotropic sphere, the above relationship between the total field and the polarization also applies to anisotropic systems. Furthermore, the volume of the sphere can be made arbitrarily small to envelop a single dipole.

Combining equations (7.1), (7.2), (7.3), and (7.5) the results for the polarization using both the macroscopic and microscopic models can be equated to yield

$$\langle\alpha_M\rangle = \frac{3}{4\pi\nu}\left(\frac{\varepsilon - 1}{\varepsilon + 2}\right) = \frac{3}{4\pi\nu}\left(\frac{n^2 - 1}{n^2 + 2}\right), \tag{7.6}$$

which is the Lorentz-Lorenz equation. This simple relation connects the molecular polarizability to the macroscopic refractive index.

7.1.2 Birefringence of a Rigid Rod Polymer

The Lorentz-Lorenz equation can be used directly to model the birefringence of a solution of rigid rod molecules subject to an orienting, external field. Figure 7.2 shows a representative molecule, which is modeled as having a uniaxial polarizability of the form

$$\underset{\sim}{\alpha} = \begin{bmatrix} \alpha_2 & 0 & 0 \\ 0 & \alpha_2 & 0 \\ 0 & 0 & \alpha_1 \end{bmatrix}, \tag{7.7}$$

where α_1 and α_2 are the principal values of the polarizability in the coordinate system that is coincident with the axes of the rod.

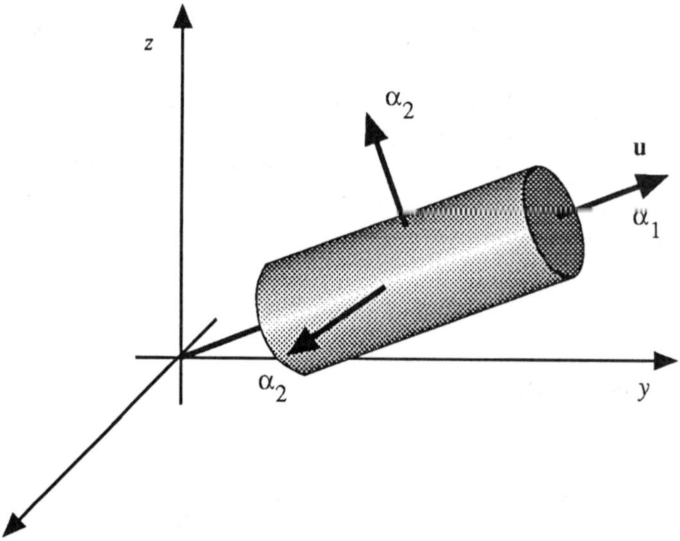

Figure 7.2 Rigid rod polymer with uniaxial polarizability and orientation vector **u**

The Lorentz-Lorenz equation can be used to express the components of the refractive index tensor in terms of the polarizability tensor. Recognizing that the birefringence normalized by the mean refractive index is normally very small, ($|\Delta n'|/n' \ll 1$), it is assumed that $|\Delta\alpha|/\bar{\alpha} \ll 1$, where the mean polarizability is $\bar{\alpha} = (\alpha_1 + 2\alpha_2)/3$ and the polarizability anisotropy is $\Delta\alpha = \alpha_1 - \alpha_2$. It is expected that the macroscopic refractive index tensor will be coaxial with the averaged polarizability tensor, and the Lorentz-Lorenz equation can be used to obtain the coefficient of proportionality. This is carried out by expressing the Lorentz-Lorenz equation for any of the principal values of these tensors. Along the "1" direction we have

$$\frac{4\pi\nu}{3}\langle\alpha_1\rangle = \frac{n_1^2 - 1}{n_1^2 + 2}, \tag{7.8}$$

or, in terms of the average polarizability and refractive index, and their anisotropies,

$$\frac{4\pi\nu}{3}\langle\bar{\alpha} + \frac{2}{3}\Delta\alpha\rangle = \frac{\left(\bar{n} + \frac{2}{3}\Delta n\right)^2 - 1}{\left(\bar{n} + \frac{2}{3}\Delta n\right)^2 + 2}. \tag{7.9}$$

Expanding equation (7.9) for small birefringences, we have

$$\Delta n = \frac{4\pi\nu}{18}\frac{(\bar{n}^2+2)^2}{\bar{n}}\langle\Delta\alpha\rangle . \tag{7.10}$$

This expression can be combined with equation (4.12) to obtain the following expression for the refractive index tensor as a function of the orientation distribution of the rods making up the bulk sample:

$$\mathbf{n} = \frac{4\pi\nu}{18}\frac{(\bar{n}^2+2)^2}{\bar{n}}[\alpha_2\mathbf{I} + (\alpha_1-\alpha_2)\langle\mathbf{uu}\rangle] . \tag{7.11}$$

Using this model, the birefringence and orientation angle, χ, of the principal directions of the refractive index tensor can be determined. If light is propagating along the y axis, we find that

$$\Delta n = \frac{4\pi\nu}{18}\frac{(\bar{n}^2+2)^2}{\bar{n}}(\alpha_1-\alpha_2)\sqrt{\langle u_z^2-u_x^2\rangle^2+4\langle u_z u_x\rangle^2} \tag{7.12}$$

$$\tan 2\chi = \frac{2\langle u_z u_x\rangle}{\langle u_z^2-u_x^2\rangle} . \tag{7.13}$$

If the sample is uniaxial and aligned along the z axis, then

$$\Delta n \approx \langle u_z^2-u_x^2\rangle = \frac{1}{2}\langle 3\cos^2\theta - 1\rangle = P_2(\langle\cos^2\theta\rangle) , \tag{7.14}$$

which agrees with the result obtained for dichroism in equation (5.11).

7.1.3 The Kuhn and Grun Model of a Flexible Chain

The rigid rod model discussed above is simplified by the fact that the conformation of such a molecule is characterized by a single unit vector defining its orientation. In the case of a flexible polymer chain, the conformation will be a function of distortion and orientation of the chain segments. Figure 7.3 pictures such a chain, which is modeled as a sequence of N_K links that are attached through freely rotating joints. The subscript "K" identifies these links as being "Kuhn" segments making up the chain. The end-to-end vector that separates the two ends of the chain is taken to be $\mathbf{R}$. Each segment is taken to have a polarizability of the form given in equation (7.7) and a length a.

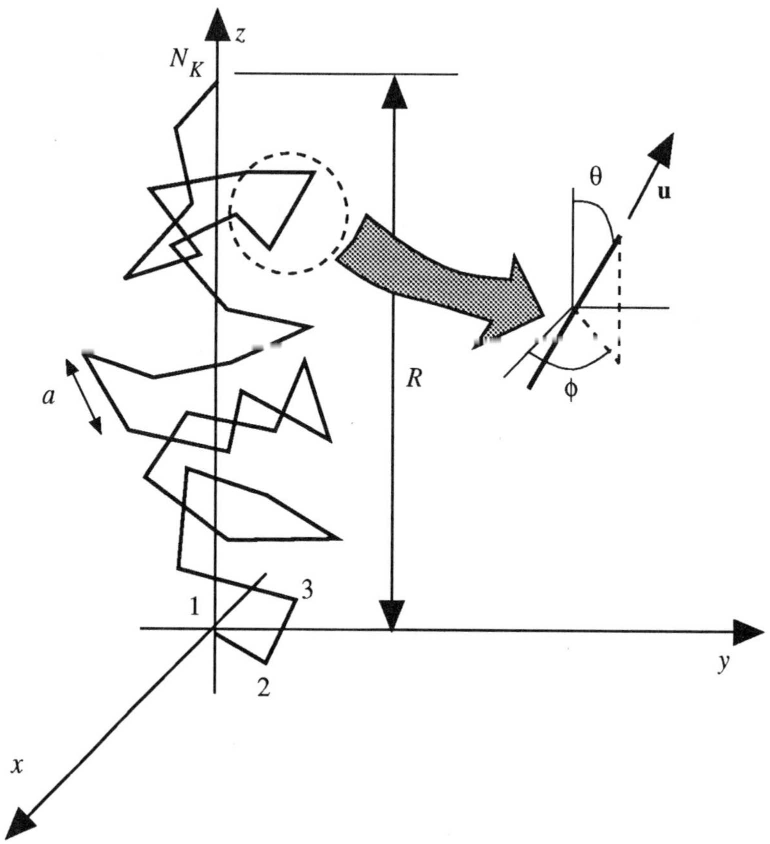

Figure 7.3 Kuhn-Grun model of a flexible polymer chain.

As before, the polarizability of the ith individual segment is $\underline{\alpha}_i = \alpha_2\mathbf{I} + (\alpha_1 - \alpha_2)(\mathbf{u}_i\mathbf{u}_i)$. The polarizability of the entire chain is the summation of this polarizability over all the segments and its average over the conformation of the macromolecule. In other words,

$$\underline{\alpha} = N_K\alpha_2\mathbf{I} + N_K(\alpha_1 - \alpha_2)\langle\mathbf{uu}\rangle, \tag{7.15}$$

where it has been assumed that the segments are identical and independent in orientation. The average in equation (7.15) is over the distribution function, $P(\mathbf{u};\mathbf{R})$, which is defined as the probability that a segment has an orientation $\mathbf{u}$ given that the ends of the polymer chain are separated by the vector $\mathbf{R}$ (see Figure 7.3). In 1942, Kuhn and Grun [57,56] showed that this distribution is

$$P(\mathbf{u};\mathbf{R}) = \frac{1}{4\pi}\frac{\beta}{\sinh\beta}e^{\frac{\beta}{R}(\mathbf{R}\cdot\mathbf{u})}, \tag{7.16}$$

where the function β is related to the fractional extension of the chain, $r = R/(N_K a)$, by

$$r = \frac{R}{N_K a} = \coth\beta - \frac{1}{\beta} = L(\beta), \tag{7.17}$$

where $L(\beta)$ is the Langevin function. The solution for β is the inverse Langevin function, which has the following expansion for small values of extension [56]:

$$\beta = L^{-1}(r) \approx 3r + \frac{9}{5}r^3 + \ldots \quad . \tag{7.18}$$

The average in equation (7.15) will have the general form,

$$\langle \mathbf{uu} \rangle = A\mathbf{I} + B\langle R^2 \rangle \mathbf{I} + C\langle \mathbf{RR} \rangle, \tag{7.19}$$

where A, B, and C are constants. Taking the trace of both sides of this equation, evidently, $A = 1/3$ and $B = -C/3$. The specific value of C can be determined by taking the end-to-end vector along the z axis and evaluating the "3,3" component of $\langle \mathbf{uu} \rangle$. This leads to

$$\langle \cos^2\theta \rangle = \frac{1}{3} + \frac{2}{3}C\langle R^2 \rangle, \tag{7.20}$$

where

$$\langle \cos^2\theta \rangle = 1 - 2\langle \frac{r}{\beta} \rangle \approx \frac{1}{3} + \frac{2}{5}\langle r^2 \rangle, \tag{7.21}$$

for small extensions. Using this result to evaluate the constant C, we have

$$\langle \mathbf{uu} \rangle = \frac{1}{3}\mathbf{I} - \frac{1}{5}\frac{1}{(N_K a)^2}\langle R^2 \rangle \mathbf{I} + \frac{3}{5}\frac{1}{(N_K a)^2}\langle \mathbf{RR} \rangle. \tag{7.22}$$

Combining these results and dropping the isotropic terms, the anisotropic part of the polarizability of the polymer chain is

$$\boldsymbol{\alpha} = \frac{3}{5}(\alpha_1 - \alpha_2)\frac{1}{N_K a^2}\langle \mathbf{RR} \rangle, \tag{7.23}$$

which, when applied to the expansion of the Lorentz-Lorenz equation, leads to the following expression for the refractive index tensor

$$\mathbf{n} = \frac{2\pi\nu}{15}\frac{(\bar{n}^2 + 2)^2}{\bar{n}}(\alpha_1 - \alpha_2)\frac{1}{N_K a^2}\langle \mathbf{RR} \rangle. \tag{7.24}$$

Equation (7.24) predicts that the refractive index tensor is proportional to the second-moment tensor of the orientation distribution of the end-to-end vector. This expression was developed using a number of assumptions, however, and is strictly valid only for small chain extensions. By avoiding the expansion of the inverse Langevin function in equation (7.21), this restriction can be removed. Other approximations include the use of the uniaxial segmental orientation distribution, $P(\mathbf{u};\mathbf{R})$, whereas nonuniaxial fields (simple shear flow, for example) would lead to an anisotropic dependence on the angle ϕ shown in Figure 7.3. Furthermore, the Kuhn and Grun model neglects the effects of excluded volume.

7.1.4 Molecular Theories of the Raman Tensor

The Raman tensor for a chemical bond oriented along the unit vector $\mathbf{u}$, but attached to a segment oriented along a vector $\mathbf{r}$, can be computed using the analysis given in section 5.3. For a flexible polymer chain, a procedure similar to the Kuhn and Grun analysis of section 7.1.3 can be used to provide a connection between the Raman tensor and the orientation of the end-to-end vector, $\mathbf{R}$. We first express the Raman tensor of an individual segment,

$$\begin{aligned}\langle \alpha'_{ij}\alpha'_{kl}\rangle &= D_1\delta_{ij}\delta_{kl} + D_2(\delta_{ik}\delta_{jl} + \delta_{il}\delta_{jk}) + D_3(\delta_{ij}\langle r_{kl}\rangle + \delta_{kl}\langle r_{ij}\rangle) \\ &\quad + D_4(\delta_{ik}\langle r_{jl}\rangle + \delta_{jl}\langle r_{ik}\rangle + \delta_{il}\langle r_{jk}\rangle + \delta_{jk}\langle r_{il}\rangle) + D_5\langle r_{ijkl}\rangle,\end{aligned} \tag{7.25}$$

where the constants D_i are defined in equation (5.41). It is assumed that all of the Raman scattering elements distributed along the chain are identical so that the total Raman tensor for the polymer is simply $N_K\langle \alpha'_{ij}\alpha'_{kl}\rangle$. It is now left to determine the averages $\langle r_{ij}\rangle$ and $\langle r_{ijkl}\rangle$ using the Kuhn and Grun distribution. These are found as in section 7.1.3, except that the following higher order moments are required:

$$\begin{aligned}\langle \cos^2\theta\rangle &= \frac{1}{3} + \frac{2}{45}\beta^2 - \frac{14}{945}\beta^4 \approx \frac{1}{3} + \frac{2}{5}\rho^2 + \frac{24}{175}\rho^4 + O(\rho^6) \\ \langle \cos^4\theta\rangle &= \frac{1}{5} + \frac{4}{105}\beta^2 - \frac{16}{4725}\beta^4 \approx \frac{1}{5} + \frac{12}{35}\rho^2 + \frac{24}{175}\rho^4 + O(\rho^6)\end{aligned} \tag{7.26}$$

so that

$$\begin{aligned}\langle r_{ij}\rangle &= \left(\frac{1}{3} - \frac{1}{5}\langle\rho^2\rangle\right)\delta_{ij} + \frac{3}{5}\langle\rho_{ij}\rangle + \frac{36}{175}\left(\langle\rho^2\rho_{ij}\rangle - \frac{1}{3}\langle\rho^4\rangle\delta_{ij}\right) \\ \langle r_{ijkl}\rangle &= \left(\frac{1}{15} - \frac{2}{35}\langle\rho^2\rangle - \frac{3}{175}\langle\rho^4\rangle\right)(\delta_{ij}\delta_{kl} + \delta_{ik}\delta_{jl} + \delta_{il}\delta_{jk}) \\ &\quad + \frac{3}{35}(\langle\rho_{kl}\rangle\delta_{ij} + \langle\rho_{ij}\rangle\delta_{kl} + \langle\rho_{ik}\rangle\delta_{jl} + \langle\rho_{jk}\rangle\delta_{il} + \langle\rho_{il}\rangle\delta_{jk} + \langle\rho_{jl}\rangle\delta_{ik}) \\ &\quad + \frac{3}{175}(\langle\rho^2\rho_{kl}\rangle\delta_{ij} + \langle\rho^2\rho_{ij}\rangle\delta_{kl} + \langle\rho^2\rho_{ik}\rangle\delta_{jl} \\ &\quad + \langle\rho^2\rho_{jk}\rangle\delta_{il} + \langle\rho^2\rho_{il}\rangle\delta_{jk} + \langle\rho^2\rho_{jl}\rangle\delta_{ik}) + \frac{3}{35}\langle\rho_{ijkl}\rangle,\end{aligned} \tag{7.27}$$

where $\rho_i = R_i/N_K a$ is the fractional end to end vector. Combining these results, the Raman tensor of a flexible chain in the Kuhn-Grun limit is

$$\langle \alpha'_{ij}\alpha'_{kl}\rangle/N_K = \langle A_1(\rho^2,\rho^4)\rangle\delta_{ij}\delta_{kl} + \langle A_2(\rho^2,\rho^4)\rangle(\delta_{ik}\delta_{jl}+\delta_{il}\delta_{jk})$$
$$+ \langle A_3(\rho^2)(\delta_{ij}\rho_{kl}+\delta_{kl}\rho_{ij})\rangle \tag{7.28}$$
$$+ \langle A_4(\rho^2)(\delta_{ik}\rho_{jl}+\delta_{kj}\rho_{il}+\delta_{il}\rho_{jk}+\delta_{jl}\rho_{ik})\rangle + A_5\langle \rho_{ijkl}\rangle$$

where

$$A_1 = D_1 + 2D_3\left(\frac{1}{3}-\frac{1}{5}\rho^2-\frac{12}{175}\rho^4\right)+D_5\left(\frac{1}{15}-\frac{2}{35}\rho^2-\frac{3}{175}\rho^4\right),$$
$$A_2 = D_2 + 2D_4\left(\frac{1}{3}-\frac{1}{5}\rho^2-\frac{12}{175}\rho^4\right)+D_5\left(\frac{1}{15}-\frac{2}{35}\rho^2-\frac{3}{175}\rho^4\right),$$
$$A_3 = D_3\left(\frac{3}{5}+\frac{36}{175}\rho^2\right)+D_5\left(\frac{3}{35}+\frac{3}{175}\rho^2\right), \tag{7.29}$$
$$A_4 = D_4\left(\frac{3}{5}+\frac{36}{175}\rho^2\right)+D_5\left(\frac{3}{35}+\frac{3}{175}\rho^2\right),$$
$$A_5 = \frac{3}{35}D_5,$$

and the constants D_i are found in equation (5.41).

7.1.5 Form Contributions of Birefringence and Dichroism

The Kuhn and Grun model considers only the intrinsic birefringence arising from anisotropy in the segmental polarizability. It neglects the additional contribution to optical anisotropy that can arise whenever there is a sufficient difference between the refractive indices of the solute and the solvent. Such a difference, accompanied by an anisotropic shape of the particle or macromolecule in solution, can give rise to both birefringence and dichroism. When this occurs, the refractive index can be complex even though the polarizability may be real and the particles or macromolecules nonabsorbing.

Among the first theories of form birefringence was the calculation of Peterlin and Stuart [58] who solved for the anisotropy in the dielectric tensor of a spheroidal particle with different dielectric constants ε_1 and ε_2 parallel and perpendicular to its symmetry axis, respectively. If the spheroid is aligned along the z axis, and resides in a fluid of dielectric constant ε_s, the contribution of a single particle to the difference between the principal values of the macroscopic dielectric tensor of the fluid is

$$\varepsilon_{zz}-\varepsilon_{xx} = \frac{(\varepsilon_1-\varepsilon_2)-\varepsilon_s\left(1-\frac{\varepsilon_1}{\varepsilon_2}\right)(L_1-L_2)}{\left[1-\left(1-\frac{\varepsilon_1}{\varepsilon_s}\right)L_1\right]\left[1-\left(1-\frac{\varepsilon_2}{\varepsilon_s}\right)L_2\right]}, \tag{7.30}$$

where L_1 and L_2 are shape factors that depend only on the aspect ratio of the spheroid, and satisfy the relationship, $L_1 + 2L_2 = 1$. In the limit that $|\varepsilon_1 - \varepsilon_2| \ll |\varepsilon_s - (\varepsilon_1 + 2\varepsilon_2)/3| \ll 1$, equation (7.30) becomes

$$\varepsilon_{zz} - \varepsilon_{xx} = (\varepsilon_1 - \varepsilon_2) - \frac{(\Delta\varepsilon)^2}{\varepsilon_s}(L_1 - L_2) , \tag{7.31}$$

where $\Delta\varepsilon = \varepsilon_s - (\varepsilon_1 + 2\varepsilon_2)/3$. The first term in equation (7.31) represents the intrinsic contribution to the birefringence, whereas the second term is the form effect that arises from anisotropy in shape as well as a finite difference between the particle and solvent refractive indices. The form anisotropy, therefore, can be present even for a particle with isotropic internal properties $(\varepsilon_1 = \varepsilon_2)$. This treatment cannot, however, describe form dichroism and the model predicts only a complex anisotropy in the refractive index if the particle is absorbing. This is due to the restriction in the original analysis that the particle is much smaller in dimension than the wavelength of the light. The theory of Onuki and Doi [13], however, can be used to calculate form dichroism as well as form birefringence.

It is important to note that the form effect is proportional to the square of the dielectric contrast, $\Delta\varepsilon$, and will always be positive for prolate particles ($L_1 > L_2$), and negative for oblate shapes ($L_2 > L_1$). The intrinsic contribution can change sign depending on the relative magnitudes of the principal values of the polarizability tensor of the particle.

7.1.5.1 Theory of Copic for Form Birefringence of a Flexible Chain

The original theory of Kuhn and Grun for the polarizability of a flexible chain accounted only for the intrinsic anisotropy of segmental polarizabilities and did not include augmentation of the polarizability through local dipolar interactions within the chain. Copic has treated this problem by calculating the electric field sensed by a chain segment i, which will be different from the incident field, $\mathbf{E}_0$, according to

$$\mathbf{E} = \mathbf{E}_0 + \sum_{k=1}^{N_K} \mathbf{E}_{ik} , \tag{7.32}$$

where the perturbation, $\mathbf{E}_{ik}$, is the electric field generated by a polarized segment k and sensed at segment i. These contributions are then summed over all N_K segments of the macromolecule. The perturbation field is given by

$$\mathbf{E}_{ik} = \mathbf{T}_{ik} \cdot \mathbf{p}_k , \tag{7.33}$$

where $\mathbf{p}_k$ is the induced dipole moment associated with segment k and $\mathbf{T}_{ik}$ is the interaction tensor,

$$\mathbf{T}_{ik} = \frac{1}{n_s r_{ik}^3}\left(3\frac{\mathbf{r}_{ik}\mathbf{r}_{ik}}{r_{ik}^2} - \mathbf{I}\right), \tag{7.34}$$

and $\mathbf{r}_{ik} = \mathbf{r}_i - \mathbf{r}_k$.

Equation (7.34) is the leading term in a near field expansion of the electric field generated from a radiating dipole. Substituting equations (7.34) and (7.33) into equation (7.32), and multiplying the result by the mean polarizability α_i of segment i, the following equation for the dipole moment $\mathbf{p}_i$ is found:

$$\mathbf{p}_i = \alpha_i \mathbf{E}_0 + \alpha_i \sum_{k=1}^{N_K} \mathbf{T}_{ik} \cdot \mathbf{p}_k. \tag{7.35}$$

Equation (7.35) is a set of $3N_K$ coupled linear equations that must be solved to determine the dipole moments existing at each segment of the polymer chain. The analysis proceeds by averaging over the distribution of segmental orientations given by equation (7.16) to obtain the following approximate form for the anisotropy in the polarizability for the chain pictured in Figure 7.3:

$$\langle \alpha_{zz} - \alpha_{xx} \rangle = \Theta_i \left(\frac{R}{\sqrt{N_k} a}\right)^2 + \Theta_f \left(\frac{R}{\sqrt{N_k} a}\right) + \dots, \tag{7.36}$$

where the first term is the intrinsic contribution and is identical to the Kuhn and Grun calculation. The second term arises from the form effect. The coefficients appearing in equation (7.36) are

$$\Theta_i = \frac{3}{5}(\alpha_1 - \alpha_2) \tag{7.37}$$

and

$$\Theta_f = \left(\frac{9M}{4\pi\rho N_A}\right)^2 (\sqrt{N_K} a)^{-3} \left[n_s \frac{(n^2 - n_s^2)}{(n^2 - 2n_s^2)} \right]^2, \tag{7.38}$$

where M is the molecular weight, ρ is the density of the polymer, and N_A is Avogadro's number.

As will be discussed in section 7.3, since the form contribution is proportional to $r = R/\sqrt{N_K} a$ instead of r^2 (as is the case for the intrinsic contribution), this effect will have significant consequences for the "stress-optical rule." Indeed, whenever form effects for the birefringence are significant, this rule cannot be used in its simplest form. The form effect is also seen to be positive, whereas the intrinsic contribution can be positive or negative. Therefore, depending on the relative importance of each contribution, the birefringence of certain polymer systems (polystyrene, for example), can change sign if Θ_i is negative. The form factor varies with molecular weight as $M^{2-3\zeta}$, where ζ is in the range of $0.5-0.6$, depending on the quality of the solvent. Therefore, the form effect will generally be more pronounced for higher molecular weight materials.

Form birefringence can offer important insights into chain conformations induced by flow and is more closely related to the overall shape of the coil than intrinsic birefringence, which is primarily associated with the degree of segmental orientation. Unfortunately, it is very difficult to separate the two effects experimentally. In principal, however, form dichroism can also be present which, for most macromolecules in the visible wavelength range, would not be complicated by the presence of any intrinsic dichroism effects since absorption is normally absent. The Onuki-Doi theory, discussed in the following section, can be used to predict this phenomena in polymer systems.

7.1.5.2 Onuki-Doi Theory

In section 4.7, the Onuki-Doi theory for form birefringence and dichroism was developed and presented in equations (4.91) and (4.92). It is left to calculate the structure factor, $S(\mathbf{q})$, as a function of flow. This was done in the limit of weak oscillatory flow for the dilute case where the single chain structure factor was calculated by Onuki and Doi.[13] Those authors showed that their calculation of form birefringence is equivalent to the result obtained by Copic in the limit of small wavenumbers.

More recently, the Onuki-Doi theory has been experimentally verified for the problem of flow-induced enhancement of concentration fluctuations in polymer solutions [60]. This problem is explored in greater detail in section 7.1.7. A similar verification has been achieved in the problem of electric field induced enhancement of fluctuations [61].

7.1.6 The Dynamics of Polymer Molecules

In the previous sections, theories were reviewed where the optical properties of polymer liquids were cast in terms of the microscopic properties of the constituent chains. The dynamics of polymer chains subject to external fields that orient and distort these complex liquids are considered in this section for a variety of systems ranging from dilute solutions to melts. Detailed descriptions of theories for the dynamics and structure of polymer fluids subject to flow are found in a number of books, including those by Bird *et al.* [62], Doi and Edwards [63], and Larson [64].

7.1.6.1 Dilute Solutions of Flexible Chains

A flexible macromolecule subject to flow will distort and orient in response to the opposing forces of hydrodynamic drag on its segments and the restoring action of entropy. The latter force will drive the chain to assume its most random configuration. The simplest model of such a molecule is the "dumbbell" model shown in Figure 7.4. The polymer chain is represented as a single vector, $\mathbf{R}$, that defines the orientation and magnitude of the end-to-end vector of the macromolecule. The frictional resistance of the segments of the chain are embedded within two beads characterized by a friction factor, ζ. Neglecting hydrodynamic interaction between the beads, the frictional resistance will give rise to a hydrodynamic force on the beads that is proportional to the relative velocity of the beads and the solvent. For bead i, this force is

$$F_i^H = \zeta [\mathbf{v}_i - \mathbf{v}_s(\mathbf{R}_i)] . \tag{7.39}$$

Since the length scale of a polymer chain is much smaller than the distances over which a macroscopic velocity field will change, the velocity of the solvent is normally expanded as

$$\mathbf{v}_s(\mathbf{R}) = \mathbf{v}_s(0) + \boldsymbol{G} \cdot \mathbf{R} , \tag{7.40}$$

where $\boldsymbol{G} = (\nabla \mathbf{v})$ is the velocity gradient tensor.

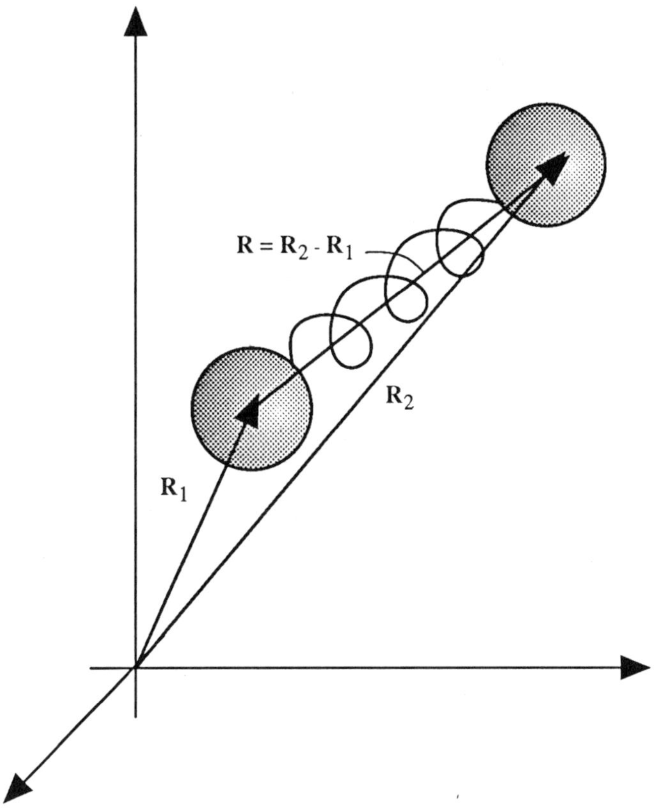

Figure 7.4 The dumbbell model of a flexible polymer chain.

An elastic spring tethers the two beads together and exerts a force on the beads that tends to restore the system to its equilibrium coil-like shape. If the spring is taken to have a force constant, $K(|\mathbf{R}|)$, the spring force is

$$F_i^S = (-1)^i K(|\mathbf{R}|) \mathbf{R} , \tag{7.41}$$

where $i = 1, 2$.

The force constant can be determined by considering the force required to stretch the ends of a flexible chain away from their equilibrium separation. If the distribution of conformations of the dumbbell is given by $\Psi(\mathbf{R})$, the free energy is given by

$$W = -TS = k_B T \log[\Psi(\mathbf{R})], \tag{7.42}$$

where S is the entropy and T is the absolute temperature. If the end-to-end vector, $\mathbf{R}$, is stretched or compressed, the required force is

$$F^S = \frac{\partial W}{\partial \mathbf{R}}. \tag{7.43}$$

The form of the distribution function will depend on the approximations that have been incorporated into the model. In its simplest form, where finite extensibility, hydrodynamic interaction and excluded volume have been neglected, the following Gaussian function describes the distribution of conformations,

$$\Psi(\mathbf{R}) = \left(\frac{3N_K}{2\pi}\right)^{\frac{3}{2}} e^{-\frac{3N_K}{2}\left(\frac{R}{N_K a}\right)^2}. \tag{7.44}$$

Using this expression in (7.43) leads to

$$F^S = \left(\frac{3kT}{N_K a^2}\right)\mathbf{R}, \tag{7.45}$$

which gives the form of a linear spring force with force constant as $K = 3kT/N_K a^2$.

The Gaussian distribution is deficient in a number of ways. In particular, it suggests that the dumbbell can be stretched without limit, since there is a finite probability for any value of the end-to-end distance. The constraint that the end-to-end distance should not exceed the contour length $N_K a$ can be incorporated into the model and a discussion of that procedure is provided in reference 65 in section 5a(i). The probability distribution function then has the form:

$$\Psi(\mathbf{R}) = \frac{(2\pi N_K a^2)^{-3/2}\,[L^{-1}(r)]^2}{r\{1-[L^{-1}(r)\,\mathrm{cosech}L^{-1}(r)]^2\}^{1/2}}\left[\frac{\sinh L^{-1}(r)}{L^{-1}(r)\exp(rL^{-1}(r))}\right]^{N_K}, \tag{7.46}$$

where $r = R/(N_K a)$ and $L^{-1}(r)$ is the inverse Langevin function defined in (7.17). Using this expression, the spring force becomes [65]

$$F^S = \frac{k_B T}{a} L^{-1}(r)\frac{\mathbf{R}}{|\mathbf{R}|}. \tag{7.47}$$

This force has the property of tending to an infinite value as the end-to-end distance approaches the contour length ($r \to 1$). An alternative, simpler form that captures the quantitative features of the Langevin spring law is the "Warner spring" [62], where

$$F^S = \left(\frac{3kT}{Na^2}\right)\frac{\mathbf{R}}{1-r^2}.$$

The remaining force that must be considered is that due to Brownian interactions between the solvent molecules and the segments of the polymer. This force is taken as

$$F_i^B = -k_B T \frac{\partial}{\partial \mathbf{R}_i} \Psi(\mathbf{R}) . \tag{7.48}$$

The three forces discussed above can be combined in a force balance on each bead to supply an expression for the rate of change of the end-to-end vector, $\dot{\mathbf{R}}$. Neglecting inertial forces, we find that

$$\frac{\partial \mathbf{R}}{\partial t} = G \cdot \mathbf{R} - \frac{2k_B T}{\zeta} \frac{\partial}{\partial \mathbf{R}} \log \Psi - \frac{2}{\zeta} K(\mathbf{R})\,\mathbf{R} . \tag{7.49}$$

This result, when multiplied by the distribution function, gives the "flux" of probability at a point in the space of conformations. Since probability is conserved, the following continuity equation applies:

$$\frac{\partial \Psi}{\partial t} + \frac{\partial}{\partial \mathbf{R}}\left(\frac{\partial \mathbf{R}}{\partial t}\right)\Psi = 0 , \tag{7.50}$$

which, when combined with (7.49) leads to

$$\frac{\partial \Psi}{\partial t} = -G \cdot \mathbf{R} \cdot \frac{\partial}{\partial \mathbf{R}} \Psi + \frac{2}{\zeta} \frac{\partial}{\partial \mathbf{R}} \cdot K(\mathbf{R})\,\mathbf{R}\Psi + \frac{2k_B T}{\zeta} \nabla^2 \Psi . \tag{7.51}$$

In principle, once the probability distribution function is available, bulk solution properties can be evaluated by averaging appropriate functions of conformation space and time. From the Kuhn and Grun analysis leading to equation (7.24) for the refractive index tensor, we are particularly interested in the second moment tensor, $\langle \mathbf{RR} \rangle = \int d\mathbf{R}\ \mathbf{RR}\Psi(\mathbf{R})$. In this case, it is more convenient to generate an expression for the rate of change of $\langle \mathbf{RR} \rangle$ which will be solved instead of obtaining an explicit expression for Ψ itself. This relation is obtained by multiplying equation (7.51) by $\mathbf{RR}$ and averaging over the distribution function. The rate of change of the second-moment tensor is

$$\frac{d}{dt}\langle \mathbf{RR} \rangle = G \cdot \langle \mathbf{RR} \rangle + \langle \mathbf{RR} \rangle \cdot G^+ - \frac{4}{\zeta} \langle K(\mathbf{R})\,\mathbf{RR} \rangle + \frac{4k_B T}{\zeta} \mathbf{I} . \tag{7.52}$$

For the case of the linear spring given in equation (7.45), equation (7.52) is a linear, ordinary differential equation that can be solved exactly. If the initial condition is a quiescent solution at equilibrium, then $\langle \mathbf{RR} \rangle_{t=0} = \frac{k_B T}{K} \mathbf{I}$. The use of a nonlinear spring function that incorporates the effect of finite extensibility, such as the Warner spring, leads to an unclosed form of equation (7.52). One approximate method of solution is to use a preaveraging approximation where $K(\mathbf{R})$ is replaced by $K(\sqrt{tr\langle \mathbf{RR} \rangle})$. The result is a set of closed, nonlinear ordinary differential equations that can be solved numerically.

Other modifications to the elastic dumbbell have been considered, such as the concept of internal viscosity, where an additional spring force proportional to the *rate* of change of the end-to-end vector, $\frac{\partial \mathbf{R}}{\partial t}$, is included [67,68,69]. Another refinement is the use of a conformational dependent friction factor, $\zeta(\mathbf{R})$, which recognizes the fact that hydrodynamic interactions among the segments of a chain will cause the frictional resistance of a polymer chain to change as it is deformed. The simplest form of such a model is a friction factor that is linear in the end-to-end distance [70,71,72].

The dumbbell model offers a coarse grained picture of a macromolecule and only considers its largest length scale, the end-to-end distance. The many-bead-and-spring model shown in Figure 7.5 recognizes the fact that a polymer chain will have multiple modes of relaxation by constructing the polymer from a sequence of beads linked together by springs. There are $N_K + 1$ beads, where N_K is the number of statistical Kuhn segments of the macromolecule. Although N_K is proportional to the number of monomers making up the chain, each segment is made of a collection of monomers that is sufficiently large to ensure that they are statistically independent of one another. As in the case of the dumbbell model, the frictional resistance of the molecule is concentrated on the beads and a force balance will be simply

$$F_i^H + F_i^S + F_i^B = 0, \tag{7.53}$$

where the subscript i identifies the bead in question.

The force balance on bead i has a form similar to equation (7.49) and is

$$\zeta\left[\frac{\partial \mathbf{R}}{\partial t}_i - (G \cdot \mathbf{R}_i)\right] + K_B \sum_{j=1}^{N_K} A_{ij}\mathbf{R}_j + 2k_B T \frac{\partial \Psi}{\partial \mathbf{R}_i} = 0, \tag{7.54}$$

where

$$A_{ij} = \begin{cases} 2 & \text{if } i = j \\ -1 & \text{if } i = j \pm 1 \\ 0 & \text{otherwise} \end{cases} \tag{7.55}$$

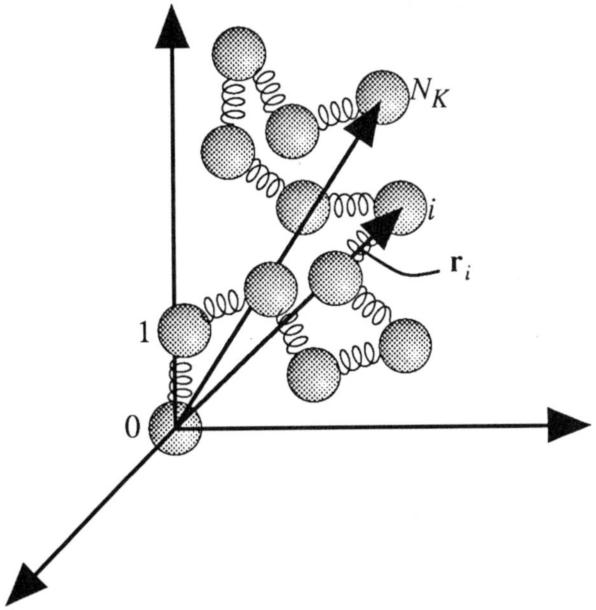

Figure 7.5 Bead-and-spring model of a polymer chain.

$\mathbf{R}_i = \mathbf{r}_i - \mathbf{r}_{i-1}$ is the "end-to-end vector" of Kuhn segment i and A_{ij} is the "Rouse matrix" that couples the beads together through the spring force interactions. The spring force coefficient for a single spring in the sequence will be simply N_K times the force constant for the whole chain. In other words, $K_B = N_K K = 3k_B T/a^2$, if the springs are linear. The distribution function, $\Psi(\mathbf{R}_1, \mathbf{R}_2, \mathbf{R}_3, \ldots, \mathbf{R}_{N_K}; t)$, specifies the probability that the conformation of the chain at time t is given by the sequence of end-to-end vectors, $(\mathbf{R}_1, \mathbf{R}_2, \mathbf{R}_3, \ldots, \mathbf{R}_{N_K})$. Following equation (7.50), the continuity equation for this distribution is

$$\frac{\partial \Psi}{\partial t} + \sum_{i=1}^{N_K} \frac{\partial}{\partial \mathbf{R}_i} \cdot \left(\frac{\partial \mathbf{R}_i}{\partial t}\right) \Psi = 0. \tag{7.56}$$

Solving equation (7.55) for $\frac{\partial \mathbf{R}_i}{\partial t}$, the convection-diffusion equation for Ψ becomes

$$\frac{\partial \Psi}{\partial t} = \sum_{i=1}^{N_K} \left[-G \cdot \mathbf{R}_i \cdot \frac{\partial}{\partial \mathbf{R}_i} \Psi + \frac{2}{\zeta} K_B \left(\frac{\partial}{\partial \mathbf{R}_i} \cdot \sum_{j=1}^{N_K} A_{ij} \mathbf{R}_j \Psi \right) + \frac{2k_B T}{\zeta} \nabla_i^2 \Psi \right]. \tag{7.57}$$

The procedure to solve this equation is to seek a transformation of the end-to-end vectors, $\mathbf{Y}_i = \sum_{j=1}^{N_K} Q_{ij}\mathbf{R}_j$, that diagonalizes the Rouse matrix. This normal mode analysis was first carried out by Rouse [73]. The result of such a transformation is

$$\frac{\partial \Psi}{\partial t} = \sum_{i=1}^{N_K} \left[-G \cdot \mathbf{Y}_i \cdot \frac{\partial}{\partial \mathbf{Y}_i} \Psi + \frac{2}{\zeta} K_B \left(\lambda_i \frac{\partial}{\partial \mathbf{Y}_i} \cdot \mathbf{Y}_j \Psi \right) + \frac{2kT}{\zeta} \nabla^2_{Y_i} \Psi \right], \tag{7.58}$$

where $\lambda_i = 4 \sin^2 [i\pi/2(N_K + 1)]$ are the eigenvalues of the Rouse matrix. Since the "Rouse" modes are independent and uncoupled, the distribution function can be factored as $\Psi(\mathbf{Y}_1, \mathbf{Y}_2, \ldots, \mathbf{Y}_{N_K}) = \prod_{i=1}^{N_K} \Psi_i(\mathbf{Y}_i)$, where

$$\frac{\partial \Psi_i}{\partial t} = -\left(G \cdot \mathbf{Y}_i \cdot \frac{\partial}{\partial \mathbf{Y}_i} \Psi_i \right) + \frac{2}{\zeta} K_B \left(\lambda_i \frac{\partial}{\partial \mathbf{Y}_i} \cdot \mathbf{Y}_j \Psi_i \right) + \frac{2k_B T}{\zeta} \nabla^2_{Y_i} \Psi_i. \tag{7.59}$$

Because this equation has the same form as equation (7.51), the second-moment tensor of the modes of relaxation can be solved using identical procedures to those discussed previously.

7.1.6.2 Dilute Solutions of Rigid Rods

The simplest model of a rigid chain is the rigid dumbbell. This is constructed by simply replacing the elastic spring in Figure 7.4 by a rigid link of length L, as shown in Figure 7.6. This link is taken to have zero mass, and its contribution to hydrodynamic resistance is neglected.

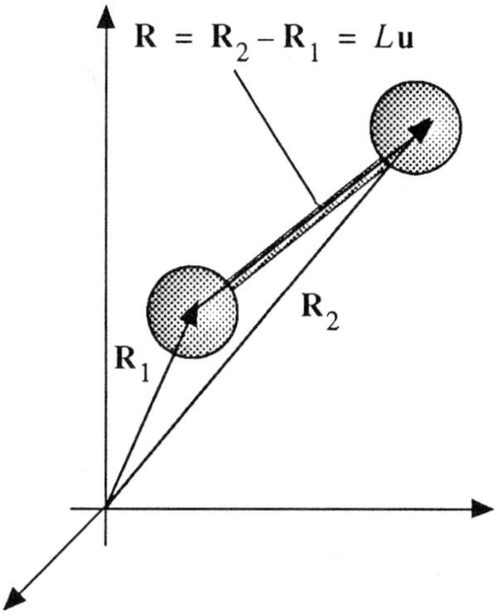

Figure 7.6 The rigid dumbbell model.

The probability distribution function describing the configuration of a collection of rigid dumbbells will have the form, $\Psi(\mathbf{u}, t)$, and will only depend on the unit vector $\mathbf{u}$ and time. The conservation of this probability is expressed as

$$\frac{\partial \Psi}{\partial t} + \frac{\partial}{\partial \mathbf{u}} \cdot (\dot{\mathbf{u}}\Psi) = 0. \tag{7.60}$$

The two contributions to the rate of rotation, $\dot{\mathbf{u}}$, of the rod are convection and Brownian diffusion. Unlike the elastic dumbbell, where the springs were allowed to deform by the flow, the fixed separation of the beads in the rigid dumbbell must be maintained. For that reason, the vector $\mathbf{u}$ can rotate, but it cannot stretch. This constraint is satisfied by ensuring that $\frac{du^2}{dt} = 0$. For that reason, the convective term is $\dot{\mathbf{u}}_H = \boldsymbol{G} \cdot \mathbf{u} - (\mathbf{uu}{:}\boldsymbol{D})\,\mathbf{u}$, where $\boldsymbol{D} = (\boldsymbol{G} + \boldsymbol{G}^+)/2$ is the symmetric, rate of strain tensor. The term $(\mathbf{uu}{:}\boldsymbol{D})\,\mathbf{u}$ accounts for the stretching of the vector $\mathbf{u}$ so that the remaining term represents rotation of the end-to-end unit vector by the flow. Adding Brownian rotations, we have [62,64]

$$\dot{\mathbf{u}} = \boldsymbol{G} \cdot \mathbf{u} - (\mathbf{uu}{:}\boldsymbol{D})\,\mathbf{u} - D_r^0 \frac{\partial}{\partial \mathbf{u}} (\log \Psi), \tag{7.61}$$

where $D_r^0 = 2k_B T/\zeta L^2$ is the rotational diffusivity of a single rod. Substituting this result into the continuity equation gives

$$\frac{\partial \Psi}{\partial t} + \frac{\partial}{\partial \mathbf{u}}\left[\boldsymbol{G} \cdot \mathbf{u}\Psi - (\mathbf{uu}{:}\boldsymbol{D})\,\mathbf{u}\Psi - D_r^0 \frac{\partial \Psi}{\partial \mathbf{u}}\right] = 0. \tag{7.62}$$

As before, it is of interest to evaluate the second-moment tensor, $\langle \mathbf{RR} \rangle = L^2 \langle \mathbf{uu} \rangle$. Multiplying equation (7.62) by $(\mathbf{uu})$ and integrating over the unit sphere gives the following equation of motion:

$$\frac{d}{dt}\langle \mathbf{RR} \rangle = (\boldsymbol{G} \cdot \langle \mathbf{RR} \rangle + \langle \mathbf{RR} \rangle \cdot \boldsymbol{G}^+) - \frac{1}{\tau_R^0}\langle \mathbf{RR} \rangle + \frac{L^2}{3\tau_R^0}\mathbf{I} - \frac{2}{L^2}\boldsymbol{D}{:}\langle \mathbf{RRRR} \rangle. \tag{7.63}$$

where $\tau_R^0 = 1/6D_r^0$ is the rotational relaxation time of the dumbbells.

This equation is not closed in the unknown, second-moment tensor due to the presence of the $\langle \mathbf{RRRR} \rangle$ term. One solution procedure often used, is to invoke a closure approximation, of which the form $\langle \mathbf{RRRR} \rangle \approx \langle \mathbf{RR} \rangle \langle \mathbf{RR} \rangle$ is the simplest. This approximation, however, is only quantitatively accurate in the limit of nearly perfect orientation of the dumbbells, but is able to offer correct, qualitative responses for many purposes.

7.1.6.3 Models of Concentrated Flexible Chains

When polymer chains become sufficiently concentrated, entanglement interactions lead to cooperative relaxation phenomena that can dominate the dynamical response. The relaxation times dramatically slow down with concentration and nonlinear flow phenomena become accessible at much lower flow strengths. The first models of concentrated systems are classified as transient network models and combined elastic, entropic spring interactions with phenomenological descriptions of temporary network junctions. These models recognize that for sufficiently small times, polymeric liquids respond much like an elastic solid, and then commence to flow as the chains begin to slip through their mutual entanglements. The transient network model, originating with the papers of Yamamoto [75] and Lodge [76], describes the state of the network through a distribution function, $f(\mathbf{r}, N, t)$, defined such that $f\mathrm{d}^3\mathbf{r}$ is the number of elastic segments with end-to-end vector $\mathbf{r}$ at time t. The segments contain N subunits. The distribution function obeys the continuity equation,

$$\frac{\partial f}{\partial t} + \nabla \cdot (G \cdot \mathbf{r}) f = H(\mathbf{r}, N) - \beta(\mathbf{r}, N) , \tag{7.64}$$

where the function $H(\mathbf{r}, N)$ governs the rate of entanglement creation and is defined such that $H(\mathbf{r}, N)\,\mathrm{d}^3\mathbf{r}\mathrm{d}t$ is the number of segments formed with N subunits and end-to-end vector $\mathbf{r}$ in the time interval $\mathrm{d}t$. The destruction of entanglement junctions is assumed to occur at a rate $\beta(\mathbf{r}, N)$, which is a function of the segment conformation. Since equation (7.64) is a first-order partial differential equation, the method of characteristics can be used to obtain an exact solution for any choice of the functions H and β. Once this solution is available, any number of bulk solution properties (the stress tensor and the refractive index tensor, for example) can be evaluated by averaging over the distribution function. This approach was used by Fuller and Leal [77] for various of choices for the creation and destruction functions. It is worth pointing out that if the destruction function, β, is taken to be a constant that is independent of $\mathbf{r}$, the predictions of this model are similar to the linear elastic dumbbell model for dilute solutions.

The transient network model, in its various forms, is able to predict a wide variety of nonlinear, viscoelastic phenomena. These models, however, do not offer a physical description of an entanglement, and in particular the molecular weight and concentration dependence of such an interaction. There is experimental evidence for the molecular weight dependence of the longest timescale associated with the relaxation of entanglements that gives insight to the molecular basis of this phenomena. It is observed for melts that a critical molecular weight exists, M_c, where a transition occurs in the dependence of the zero shear viscosity on molecular weight. It is found that [78]

$$\eta \sim \left(\begin{array}{ll} M & \text{if } M < M_c \\ M^{3.4} & \text{if } M > M_c \end{array} \right) . \tag{7.65}$$

The critical molecular weight, M_c is related to the *entanglement molecular weight*, M_e, as $M_e = M_c/2$. The entanglement molecular weight specifies the length of a polymer that is required to cause interpenetration of the chains. If solvent is added to the melt, the effective molecular weight required to cause entanglement is increased according to $M_e(c) = M_e(0)\rho/c$, where ρ is the density of the melt and c is the solution concentration. Because the longest relaxation timescale is proportional to the zero shear viscosity, this result implies that highly entangled chains relax their conformations on a timescale proportional to $M^{3.4}$.

Another important observation on the nature of entanglements concerns the frequency dependence of the storage modulus, $G'(\omega)$. This property corresponds to that part of the stress displayed by a material that is in-phase with an applied, oscillatory strain. A perfectly elastic material, with permanent entanglements will respond completely in-phase with the strain and $G'(\omega)$ is constant, with no frequency dependence. This reflects the absence of relaxation mechanisms in a perfect elastic. A purely viscous liquid, however, produces a stress that is completely out-of-phase with the applied strain and shows no storage modulus. Such a material relaxes instantaneously. A viscoelastic fluid, such as a polymeric liquid, is characterized by a modulus that increases with frequency. At low frequencies, the fluid can quickly relax and behaves as a viscous liquid and the storage modulus is small. As the imposed frequency is increased, relaxation processes in the material do not have sufficient time to respond and the material behaves increasingly elastic. Concentrated polymer liquids are characterized by a plateau in the storage modulus [78], as shown schematically in Figure 7.7. As molecular weight is increased, the plateau becomes more pronounced and can cover many orders of magnitude in the applied frequency. The plateau is only present if the molecular weight is sufficiently large, and first appears when the molecular weight surpasses M_e.

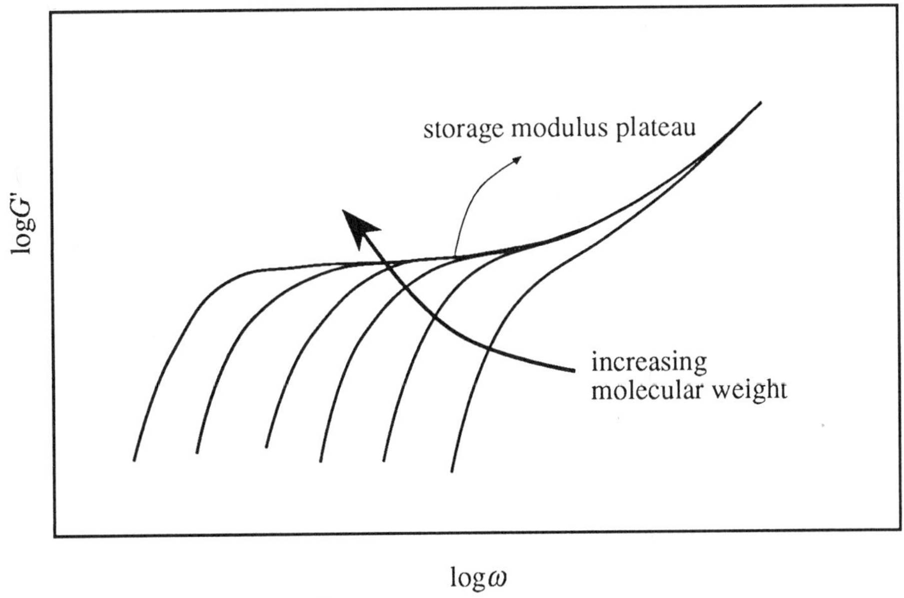

Figure 7.7 Storage modulus as a function of frequency for various molecular weights.

Evidently, in the plateau region, polymer fluids of sufficient molecular weight and concentration behave dynamically much like a cross-linked material. These observations suggest that the motion of high molecular weight chains is constrained about the entanglements. In a classical paper, de Gennes [79] offered the model of *reptation* as a description of chain dynamics. The process of relaxation of a chain from its entanglements using the reptation model is pictured in Figure 7.8. The entanglements are assumed to envelop the chain within a tube that constrains its motion to occur only along its axis. Transverse motion is forbidden. Relaxation of the chain's conformation is accomplished by a random walk along its backbone and is completed once the chain escapes the tube. The escape process proceeds as the chain ends exit out the end of the tube and become entangled with other segments in the liquid. Portions of the original tube that no longer contain the chain do not constrain it and "disappear." The relaxation time calculated using this model is found by analyzing the diffusion process associated with the disengagement phenomena. Since the original tube has a contour length of $L = N_K a$, this time is

$$\tau_R = \frac{L^2}{D_R} = \frac{N_K^2 a^2}{D_R} = \frac{N_K^2 a^2}{k_B T/\zeta}, \tag{7.66}$$

where D_R is the diffusion coefficient for reptation and ζ is the chain's total friction factor. This choice of the diffusion coefficient recognizes an important simplification of a highly concentrated, entangled system of chains; hydrodynamic interactions between segments are effectively screened. Unlike an isolated chain, where such interactions lead to a friction factor that is proportional to its radius of gyration, each segment of an entangled chain will contribute equally its friction. The chain friction factor is then proportional to the contour length, L. Therefore, the relaxation time is

$$\tau_R = \frac{N_K^2 a^2}{k_B T/(N_K \zeta_0)} = \frac{a^2 \zeta_0}{k_B T} N_K^3, \tag{7.67}$$

where ζ_0 is the friction factor of a single segment. This simple model predicts that the relaxation timescales with the molecular weight to the power of 3 (instead of the experimentally observed value of 3.4) and offers a simple physical picture of the dynamics of entanglements.

In a celebrated series of papers, Doi and Edwards [80] used the reptation, or tube model, to develop a theory for the rheological properties of entangled chains. In this model, the dynamics of a chain is described by the tube of entanglements. The configuration of the chain is defined by a set of unit vectors, $\{\mathbf{u}\}$, that are parallel to the segments spanning the entanglements. Following a deformation, the chain will deform and orient in a manner that increases the length of those segments and that reduces the number of entanglements. In the original Doi-Edwards theory, however, it was assumed that a rapid relaxation process occurs, which returns the segments of the chain to their original length. This is justified formally in the context of the "independent alignment approximation."

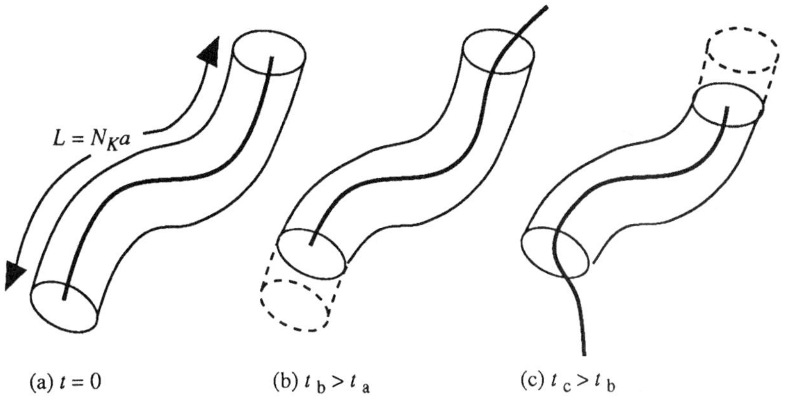

Figure 7.8 Escape of a chain from its entanglements (modeled as a tube) according to the reptation model. At time zero (a) the chain is fully enveloped within a tube. As time proceeds, the chain ends escape the original tube and portions of the tube progressively disappear.

It is desired to determine the distribution of segment orientations and we follow here the derivation offered by Marrucci and Grizzuti [87]. At equilibrium, this distribution function is uniform and equal to

$$\psi_0(\mathbf{u})\,d\mathbf{u} = \frac{1}{4\pi}d\mathbf{u}, \tag{7.68}$$

where $d\mathbf{u} = d\theta \sin\theta d\phi$. Since probability is conserved, the following relationship holds between the distribution of the deformed state, $\psi(\mathbf{u}')$, and the equilibrium state,

$$\psi(\mathbf{u}')\,d\mathbf{u}' = \psi_0(\mathbf{u})\,d\mathbf{u}. \tag{7.69}$$

Following a deformation specified by the strain tensor, E, the unit vector in the deformed state is related to its initial value by

$$\mathbf{u}' = \frac{E \cdot \mathbf{u}}{|E \cdot \mathbf{u}|}. \tag{7.70}$$

Bulk material properties can be determined quite simply using this model. For example, consider the calculation of the second-moment tensor, $\mathbf{Q} = \langle \mathbf{u}'\mathbf{u}' \rangle$, which is required for the stress and refractive index tensors. Using the independent alignment approximation, we have

$$\mathbf{Q} = \langle \mathbf{u}'\mathbf{u}' \rangle = \langle \frac{E \cdot \mathbf{u}\, E \cdot \mathbf{u}}{|E \cdot \mathbf{u}|^2} \rangle_0, \tag{7.71}$$

where the subscript "0" means that the final average is evaluated over the *equilibrium* distribution. For a shear deformation, the deformation tensor is

$$E = \begin{bmatrix} 1 & \gamma & 0 \\ 0 & 1 & 0 \\ 0 & 0 & 1 \end{bmatrix}, \tag{7.72}$$

where γ is the strain. The " (x, y) " component of $\mathbf{Q}$ is

$$Q_{xy} = \left\langle \frac{(u_x + \gamma u_y) u_y}{(u_x + \gamma u_y)^2 + u_y^2 + u_z^2} \right\rangle_0 . \tag{7.73}$$

In the limit of small deformations $(\gamma \ll 1)$, this becomes

$$\begin{aligned} Q_{xy} &\approx \gamma \langle u_y^2 - 2u_x^2 u_y^2 \rangle + O(\gamma^2) \\ &= \frac{1}{5}\gamma. \end{aligned} \tag{7.74}$$

It is important to note that the averaging process only requires the equilibrium distribution. The above calculation, however, was concerned only with the instantaneous configuration of the melt, immediately following the imposition of deformation. The relaxation of the deformation proceeds by the reptation of the chain away from the original, deformed tube. Once a portion of the chain escapes the tube, it is able to relax. De Gennes calculated the fraction of a chain remaining within an original tube and found it to be the following function of time:

$$P(t) = \sum_{i \text{ odd}} \frac{8}{\pi^2 i^2} e^{-i^2 \frac{t}{\lambda^2}}, \tag{7.75}$$

where $\lambda \approx M^{3.4}$ is the longest relaxation time. In the original Doi-Edwards model, the relaxation of the $\mathbf{Q}$ tensor is simply $\mathbf{Q}P(t)$.

7.1.6.4 Models of Semidilute Rigid Rods

The dynamics of semidilute, rigid rods are strongly affected by interactions that restrict their free rotation. The analysis described here is by Doi and Edwards [88] and involves the following range in the number concentration, ν ,

$$1/L^3 \ll \nu \ll 1/(L^2 d), \tag{7.76}$$

where L is the length of the rods and d is their diameter. The lower limit ensures that the rods are at a sufficiently high concentration to restrict in their rotation by nearest neighbors. The upper limit avoids the possibility of being close to the order/disorder transition to a nematic state, characteristic of lyotropic liquid crystals.

For a thin rod in dilute solution, the translational diffusivity is anisotropic, with different values for diffusion parallel to the rod axis, D_{tp}^0, and perpendicular to the axis, D_{tr}^0. These diffusivities are

$$D_{tp}^{0} = \frac{k_B T \log(L/d)}{2\pi\eta_s L},$$
$$D_{tr}^{0} = D_{tp}^{0}/2. \tag{7.77}$$

The rotational diffusivity, D_r^0, is defined as

$$D_r^0 = D_t^0/L^2, \tag{7.78}$$

where $D_t^0 = (D_{tp}^0 + 2D_{tr}^0)/3$. The rotational diffusivity is a measure of the frequency of an isolated rod to execute a complete rotation by Brownian interactions.

The superscript "0" on the diffusivities listed above refers to the fact that these are for dilute solutions. In a concentrated system the rate of rotation will be slowed down considerably because of steric hinderance from nearest neighbors. The nature of the entanglements from other rods onto a test rod is such that the translational motion perpendicular to the rod axis becomes highly constrained. The translation along the chain axis, on the other hand, is for the most part unaffected. The steric interactions imposed by the neighboring rods on a single test rod can be modeled by placing such a rod within a tube of radius a_c and length L. This tube constrains the rotational movement of the test rod to small angular motions of magnitude $a_c/L \ll 1$. The translational motion of the rod along its axis is not restricted, however, and the rod can exit the tube by translational Brownian motions. By a combined action of small angular motions of magnitude a_c/L and translational movements, the rod can still undergo an end-over-end rotation about its axis, but the timescale for one such orbit will be extremely long in the concentrated system. An estimate of the hindered rotational diffusivity can be obtained by considering the mechanism shown in Figure 7.9. In this caricature, one possible mode by which the test rod can rotate about its axis is described. Here the rod starts out in a tube at time $t = 0$. It can only rotate slightly through a small angle, $\theta \approx a_c/L$, within that tube. It can then exit the tube by translational Brownian diffusion and enter another tube. If it repeats this action of small angular motions followed by translation into new tubes, it can work its way around a large circle of radius R. Through simple geometric arguments, this radius is $R = 2(L/a_c)L$. Since the motion is controlled by the translational diffusivity, D_{tp}^0, the time required for the rod to completely travel the circumference of the circle of radius R is

$$\tau \approx \frac{R^2}{D_{tp}^0}. \tag{7.79}$$

The rotational diffusivity in the concentrated system is simply the inverse of this time,

$$D_r = \frac{D^0_{tp}}{R^2} \approx \left(\frac{a_c}{L}\right)^2 \frac{D^0_{tp}}{L^2} = \beta\left(\frac{a_c}{L}\right)^2 D^0_r, \qquad (7.80)$$

where β is a parameter that is thought to be of order unity.

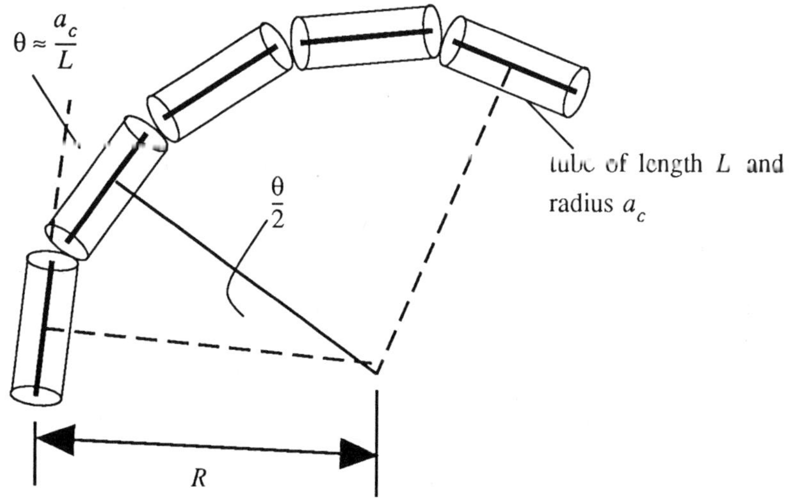

Figure 7.9 The rotational diffusion process of a rigid rod in semidilute solution.

The next step in the calculation of the rotational diffusivity is to evaluate the radius, a_c. The subscript c on this parameter is meant to remind us that it is a function of concentration. Clearly, as the concentration is increased, this radius must decrease. In addition, one can anticipate that a_c will depend on the orientation of the rods. As the rods become increasingly aligned parallel to one another, the hinderance to rotation will drop off. To obtain a_c, one must determine the number of rods, $N(a_c)$, which can cut through the imaginary cylinder constraining the test rod. To determine the probability of such an intersection, consider the diagram below.

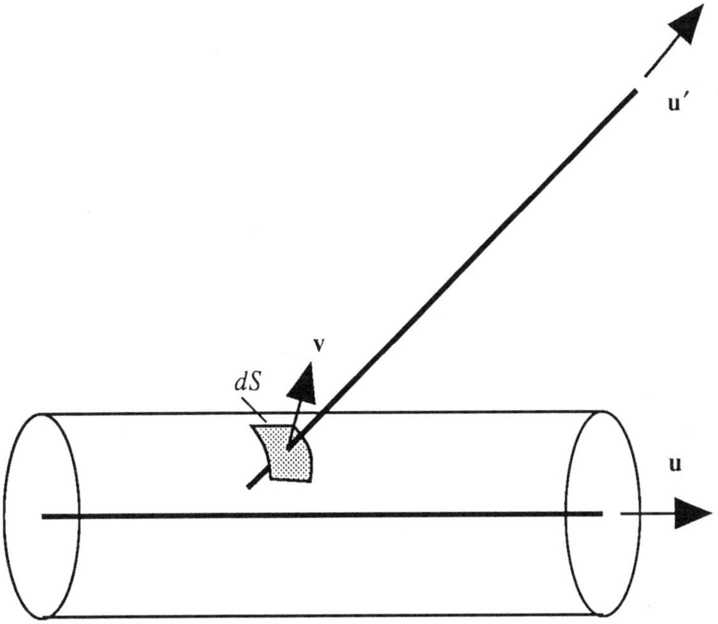

Figure 7.10 The probability of the intersection of a neighboring rod of orientation $\mathbf{u}'$ intersecting the tube enveloping a test rod of orientation $\mathbf{u}$.

In Figure 7.10, an elemental surface area, dS, on the surface of the tube is shown. It is defined by a normal vector, $\mathbf{v}$. For a neighboring rod to penetrate the surface element, the rod must lie within the element of the fluid given by $dS\ L\mathbf{v}\cdot\mathbf{u}'$, where $\mathbf{u}'$ is the orientation of the rod penetrating the surface. If ν is the number concentration of rods in the solution per unit volume, the number of rods penetrating the surface element is

$$dN = \nu dS\ L\int d\mathbf{u}\ \psi(\mathbf{u}')\,\mathbf{v}\cdot\mathbf{u}', \tag{7.81}$$

where we have integrated over all possible orientations of the penetrating rod using the orientational probability distribution function, $\psi(\mathbf{u}')$. To obtain the total number of rods penetrating the cylinder, an integration over all surface elements must be performed. This integral must be divided by two due to the fact that the vast majority of rods will penetrate two surfaces (one on the top and one on the bottom). The number of penetrating rods is

$$N = \frac{1}{2}\nu L\int dS\int d\mathbf{u}'\psi(\mathbf{u}')\,\mathbf{u}'\cdot\mathbf{v}. \tag{7.82}$$

The coordinate system shown in Figure 7.11 is used to carry out the integration over the surface elements. Here a test rod inside the tube is given an orientation defined by the unit vector $\mathbf{u}$ and is taken to be parallel to the z axis. The rod penetrating the surface is given an orientation $\mathbf{u}'$ such that it is at an angle of θ with respect to the z axis, and its projection onto the (x, y) plane is coincident with the x axis. When integrating over all surface elements, we shall be sampling over all orientations of the normal vector $\mathbf{v}$ defining the surface elements. If the angle between the vectors $\mathbf{u}'$ and $\mathbf{v}$ is α, as shown in Figure 7.11, then $\cos\alpha = \mathbf{u}' \cdot \mathbf{v}$. If the vector $\mathbf{v}$ is at a azimuthal angle of ϕ in the (x, y) plane, then the relationship between α, θ, and ϕ is $\cos\alpha = \sin\theta\cos\phi$, which can be demonstrated by straightforward trigonometry. The integration over all surface elements is

$$\int dS = a_c L \int d\phi . \tag{7.83}$$

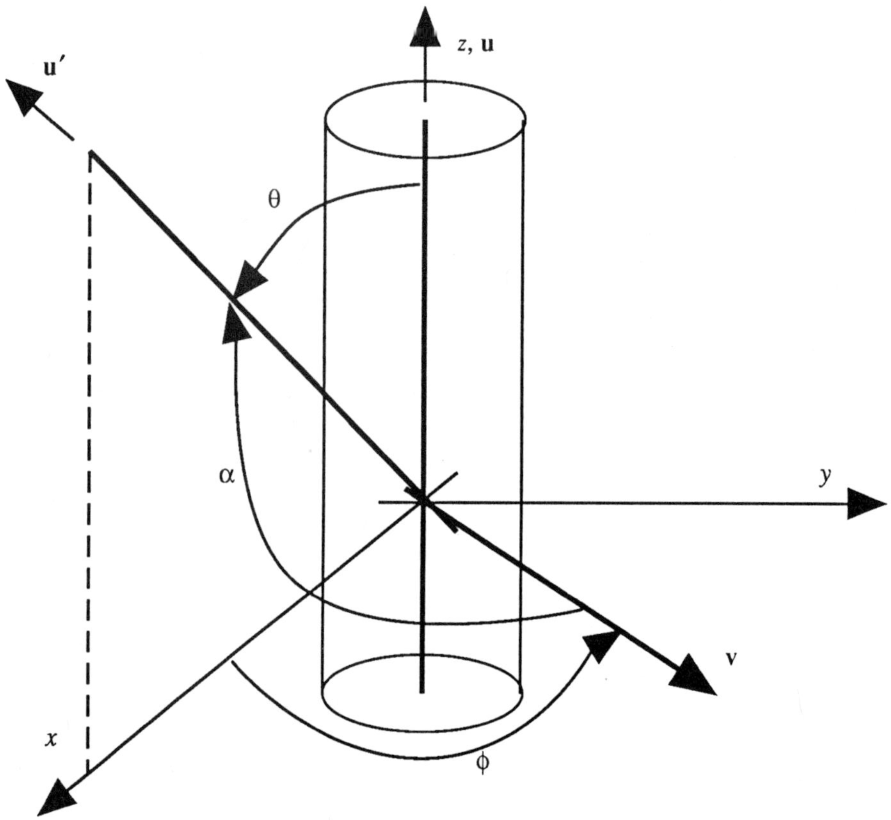

Figure 7.11 Coordinate system used to evaluate the integral describing the number of rods penetrating the tube of a test rod.

When integrating equation (7.82), it is important to note that in averaging $(\mathbf{u}' \cdot \mathbf{v})$, the absolute value of this quantity is required. This is because it does not matter which direction the rods are pointing when passing through the tube. The number of penetrating rods is

$$\begin{aligned} N &= \frac{1}{2}\nu a_c L^2 \int d\mathbf{u}' \psi(\mathbf{u}') \sin\theta \int_0^{2\pi} d\phi |\cos\phi| \\ &= 2\nu a_c L^2 \int d\mathbf{u}' \psi(\mathbf{u}') \sin\theta \\ &= 2\nu a_c L^2 \int d\mathbf{u}' \psi(\mathbf{u}') \sin(\widehat{\mathbf{u}\mathbf{u}'}), \end{aligned} \tag{7.84}$$

where $(\widehat{\mathbf{u}\mathbf{u}'})$ is the angle between the test rod and the penetrating rod. At equilibrium, $\psi(\mathbf{u}') = 1/(4\pi)$, and

$$N = \frac{1}{2}\pi\nu a_c L^2. \tag{7.85}$$

To determine a_c, N is set equal to unity since the tube represents the restrictions placed on the test rod by its nearest neighbors. At equilibrium, a_c is

$$a_c/L = 1/\left(\frac{1}{2}\pi\nu L^3\right), \tag{7.86}$$

from which the following expression for the rotational diffusivity for a semidilute solution of rigid rods at rest is calculated to be

$$\hat{D}_r = \kappa(\nu L^3)^{-2} D_r^0, \tag{7.87}$$

where κ is a constant of order unity.

When subject to an external field, the rods will become aligned and the steric hindrance on the rotation will decrease. Using the general result in equation (7.84) to calculate a_c, the rotational diffusivity of a rod of orientation $\mathbf{u}$ in a concentrated solution described by a distribution function $\psi(\mathbf{u})$ is

$$D_r(\mathbf{u}, \psi) = \hat{D}_r\left[\frac{4}{\pi}\int d\mathbf{u}' \psi(\mathbf{u}') \sin(\widehat{\mathbf{u}\mathbf{u}'})\right]^{-2}. \tag{7.88}$$

Inspection of equation (7.88) shows that as the rods become increasingly aligned, the average of $\sin(\widehat{\mathbf{u}\mathbf{u}'})$ will decease and the rotational diffusivity will increase, as it should.

The convection-diffusion equation for $\psi(\mathbf{u}, t)$ will be of the same form as the rigid dumbbell model of section 7.1.6.2 except that the diffusivity must be replaced by $D_r(\mathbf{u}, \psi)$ to give

$$\frac{\partial\psi}{\partial t} + \frac{\partial}{\partial\mathbf{u}}\left[G\cdot\mathbf{u}\psi - (\mathbf{u}\mathbf{u}{:}D)\mathbf{u}\psi - D_r(\mathbf{u}, \psi)\frac{\partial\psi}{\partial\mathbf{u}}\right] = 0. \tag{7.89}$$

Equation (7.62) is a highly nonlinear partial differential equation because of the appearance of ψ within the diffusivity. For this reason, bulk solution properties are normally calculated using this model by replacing $D_r(\mathbf{u}, \psi)$ by the following preaveraged result:

$$\langle D_r \rangle = \hat{D}_r \left[\frac{4}{\pi} \int d\mathbf{u}\ \psi(\mathbf{u}) \int d\mathbf{u}'\ \psi(\mathbf{u}') \sin(\widehat{\mathbf{u}\mathbf{u}'}) \right]^{-2}, \tag{7.90}$$

and the diffusion equation can be solved using standard methods.

7.1.7 The Structure Factor of Flowing Complex Liquid Mixtures

The polymer dynamics presented above have all been concerned with the orientation and distortion of the liquids at the level of the segments making up the polymer. For uniform, homogoneous flows, no attention has been given to the fact that the density of segments will fluctuate spatially. Such fluctuations will invariably cause light to scatter and will lead to the possibility of form contributions to the birefringence and dichroism. In certain polymer systems, these effects can be very pronounced due to a coupling of segmental concentration fluctuations to flow-induced stresses. This is the basis for a wide range of phenomena in polymer mixtures and solutions that are frequently on the edge of miscibility. These effects include flow-induced turbidity, mixing, demixing, gelation, crystallization, and molecular weight fractionation. These phenomena can be found reviewed recently by Rangel-Nafaile *et al.* [89] and Larson [90].

We derive here the governing equations necessary to describe the structure factor, $S(\mathbf{q}, t)$, for a complex liquid mixture subject to flow. Since this observable is the Fourier transformation of the spatial correlation of concentration functions, it is first required to develop an equation of motion for $\delta c(\mathbf{r}, t)$. The approach described here employs a modified *Cahn-Hilliard* equation and is described in greater detail in the book by Goldenfeld [91]. To describe the physical system, an "order parameter," $\psi(\mathbf{r}, t)$, is introduced. In a complex mixture, this parameter would simply be $\psi(\mathbf{r}, t) = c(\mathbf{r}, t) - \langle c \rangle$, where $c(\mathbf{r}, t)$ is the local concentration of one of the constituents and $\langle c \rangle$ is its mean concentration. The order parameter has the property of being zero in a disordered, or on phase region, and non-zero in the ordered or two-phase region. The observed structure factor, which is the object of this calculation, is simply

$$S(\mathbf{k}, t) = \langle \psi(\mathbf{k}, t) \psi^*(\mathbf{k}, t) \rangle = \langle \psi_{\mathbf{k}} \psi_{-\mathbf{k}} \rangle. \tag{7.91}$$

An expression for the rate of change of the order parameter is required. Since the concentration is conserved, the following equation of continuity applies,

$$\frac{\partial \psi}{\partial t} + \nabla \cdot \mathbf{j} = 0, \tag{7.92}$$

where the flux, $\mathbf{j}$, is assumed to arise from diffusion down a gradient in the chemical potential. Since the chemical potential is $\mu = \frac{\partial f}{\partial \psi} \approx \delta F / \delta \psi$,

$$\mathbf{j} = -\Xi a^3 \nabla \frac{\delta F}{\delta \psi(\mathbf{r})}, \tag{7.93}$$

where F is the Cahn-Hilliard free energy, a is the length scale of the diffusing species, and Ξ is the phenomenological mobility coefficient. A common form for F is the following expansion in the order parameter and its derivatives,

$$F = \int d\mathbf{r}\left[\frac{1}{2}K(\nabla\psi(\mathbf{r}))^2 + f(\psi(\mathbf{r}))\right], \tag{7.94}$$

where K is a parameter that will be shown to be related to the correlation length of the fluctuations, and $f(\psi)$ is the free energy density. This form for the free energy function is motivated by the Landau free energy functional used in the theory of phase transitions. The operator, $\delta/\delta\psi(\mathbf{r})$, in equation (7.93) is the functional derivative and is computed as

$$\frac{\delta F[\psi(\mathbf{r})]}{\delta\psi(\mathbf{r})} = \lim_{\varepsilon\to 0}\frac{F[\psi(\mathbf{r}') + \varepsilon\delta(\mathbf{r}'-\mathbf{r})] - F[\psi(\mathbf{r}')]}{\varepsilon}. \tag{7.95}$$

Applying this operator in equation (7.93), the dynamical expression for the order parameter becomes

$$\frac{\partial\psi}{\partial t} = \Xi a^3\nabla^2\left(\frac{\partial f}{\partial\psi} - K\nabla^2\psi\right) + \vartheta(\mathbf{r}, t), \tag{7.96}$$

where the function $\vartheta(\mathbf{r}, t)$ is a noise term that has been included to account for random, microscopic thermal interactions that induce fluctuations. It is assumed to be a Gaussian random function with the following moments,

$$\langle\vartheta(\mathbf{r}, t)\rangle = 0; \qquad \langle\vartheta(\mathbf{r}, t)\vartheta(\mathbf{r}', t)\rangle = k_B T\Xi\delta(\mathbf{r}-\mathbf{r}')\delta(t). \tag{7.97}$$

Equation (7.96) is commonly referred to as a Langevin equation and has been widely used to describe spinodal decomposition of binary mixtures where concentration fluctuations grow to develop complex, phase-separated patterns. Since the free energy density, $f\{\psi(\mathbf{r})\}$, can be a nonlinear function of the order parameter, numerical techniques are normally used to solve the problem. We seek here, however, a solution to the problem when the system is in the single phase region. If the phase diagram is characterized by a *upper critical solution temperature*, for example, the temperature is taken to be slightly above this point. In this case, the fluctuations are small and the free energy density is expanded as $f(\psi) \approx f_0 + f'\psi + \dots$. To generate the dynamical expression for the structure factor, $S(\mathbf{k}, t)$, the Langevin equation is Fourier transformed to give

$$\frac{\partial S}{\partial t} = -2\Xi a^3 k^2(f'' + Kk^2)S + 2D_t k^2, \tag{7.98}$$

where $D_t = k_B T\Xi$ is the diffusion coefficient. At steady state, the structure factor is predicted to have the following "Orstein-Zernike" form [8]:

$$S(\mathbf{k}, t\to\infty) = \frac{k_B T/a^3}{f'' + Kk^2} = \left(\frac{k_B T}{f''a^3}\right)\frac{1}{1+\xi^2 k^2}, \tag{7.99}$$

where the correlation length, $\xi = \sqrt{K/f''}$ has been introduced. This result leads to the important observation that the structure factor diverges at the critical point where f'' tends to zero.

In the presence of flow, the time derivative in equation (7.98) is modified to include a convective term so that

$$\frac{\partial S}{\partial t} - \mathbf{k} \cdot \boldsymbol{G} \cdot \nabla_{\mathbf{k}} S = -2\Xi a^3 k^2 (f'' + Kk^2) S + 2D_t k^2 , \tag{7.100}$$

where $\boldsymbol{G}$ is the velocity gradient tensor. This first-order partial differential equation can be solved using the method of characteristics.

The application of flow will generate mechanical stresses. If these stresses are not equally shared by the components in a mixed system, an addition flux will result. This problem was considered for the case of polymeric liquids by Helfand and Fredrickson [92]. The existence of a mechanical stress tensor, $\underline{\tau}$, acting on the fluid will lead to a force, $\mathbf{F}_M = -\nabla \cdot \underline{\tau}$, and a corresponding flux, $\mathbf{j}_M = -\Xi \nabla \cdot \underline{\tau}$. Expanding this flux for small values of the order parameter, the equation of motion for the structure factor becomes

$$\frac{\partial S}{\partial t} - \mathbf{k} \cdot \boldsymbol{G} \cdot \nabla_{\mathbf{k}} S = -2\Xi a^3 k^2 \left(f'' + Kk^2 - \hat{\mathbf{k}}\hat{\mathbf{k}} : \frac{\partial}{\partial \psi} \underline{\tau} \right) S + 2D_t k^2 , \tag{7.101}$$

where $\hat{\mathbf{k}} = \mathbf{k}/|\mathbf{k}|$. Again, this equation can be solved directly using the method of characteristics. At steady state, and in the limit where the convection term can be neglected, the following result can be used to estimate the shape of the structure factor,

$$S(\mathbf{k}) \approx \left(\frac{k_B T}{a^3} \right) \frac{1}{f'' + Kk^2 - \hat{\mathbf{k}}\hat{\mathbf{k}} : \frac{\partial}{\partial \psi} \underline{\tau}} . \tag{7.102}$$

For example, if the stress tensor is dominated by the shear stress, the term $\hat{\mathbf{k}}\hat{\mathbf{k}} : \frac{\partial}{\partial \psi} \underline{\tau}$ is simply

$$\hat{\mathbf{k}}\hat{\mathbf{k}} : \frac{\partial}{\partial \psi} \underline{\tau} = \hat{\mathbf{k}}^{\mathrm{T}} \cdot \begin{bmatrix} 0 & \frac{\partial \tau_{xy}}{\partial \psi} \\ \frac{\partial \tau_{xy}}{\partial \psi} & 0 \end{bmatrix} \cdot \hat{\mathbf{k}} = 2\hat{k}_x \hat{k}_y \left(\frac{\partial \tau_{xy}}{\partial \psi} \right) = 2\hat{k}_x \hat{k}_y \eta' G , \tag{7.103}$$

where η' is the differential viscosity, and G is the shear rate. The structure factor is then

$$S(\mathbf{k}) \approx \left(\frac{k_B T}{f'' a^3} \right) \frac{1}{1 + \xi^2 k^2 - \left(\frac{2\eta' G}{K} \right) \xi^2 \hat{k}_x \hat{k}_y} . \tag{7.104}$$

This function is plotted in Figure 7.12 for the case of $(2\eta' \xi^2 G/K) = 0.8$.

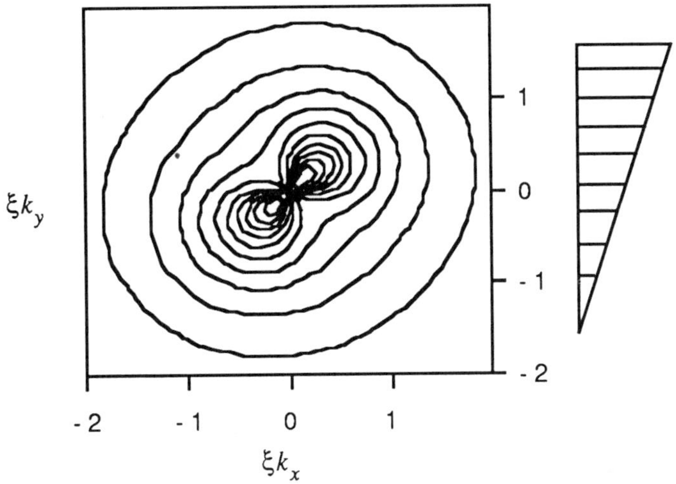

Figure 7.12 Structure factor of a fluid subject to simple shear flow using the Helfand-Fredrickson model. Only shear stresses have been considered.

This simple model predicts that the structure factor will develop a "butterfly pattern" and grow along an axis that is at 45° with respect to the flow direction, which is parallel to the principal axis of strain in this flow. Since the structure factor is the Fourier transform of the pattern of concentration fluctuations causing the scattering, the model predicts an enhancement of fluctuations *perpendicular* to the principal axis of strain.

In a complex, polymeric liquid, normal stresses as well as the shear stress can be present, and these contributions will influence the shape of the structure factor. The simplest rheological constitutive model that can account for normal stresses is the "second-order fluid model" [64], where the first and second normal stress differences are quadratic functions of the shear rate. Calculations using this model [92,93,94,90,60], indicate that the appearance of normal stresses can rotate the structure factor towards the direction of flow in the case of simple shear flow and can induce a four-fold symmetry in the case of extensional flow.

7.2 Particulate Suspensions and Dispersions

Suspensions and dispersions of particles are characterized by distributions describing their orientation and spatial distribution. Knowledge of these distributions then allow the calculation of bulk mechanical and optical properties. In the following sections, theories are presented for both dilute and semidilute systems.

7.2.1 Dynamics of Particulates

7.2.1.1 Dilute Suspensions

The dynamics of rigid, isolated spheroids was first analyzed for the case of shear flow by Jeffery[95]. When subject to a general linear flow with velocity gradient tensor G, the time rate of change of the unit vector defining the orientation of the symmetry axis of such a particle will have the following general form,

$$\frac{d\mathbf{u}}{dt} = A\mathbf{u} \cdot \boldsymbol{W} + B\mathbf{u} \cdot \boldsymbol{D} + C\mathbf{u}\,(\mathbf{uu}{:}\boldsymbol{D})\,, \tag{7.105}$$

where the antisymmetric vorticity tensor, $\boldsymbol{W} = (\boldsymbol{G} - \boldsymbol{G}^{+})/2$, measures the rotational tendency of the flow and the rate of strain tensor is $\boldsymbol{D} = (\boldsymbol{G} + \boldsymbol{G}^{+})/2$. The constants A, B, and C depend only on the aspect ratio of the particles, $\mu = \frac{a}{b}$, where a is the length of the symmetry axis of the particle. A prolate, rodlike particle is characterized by $\mu > 1$ and an oblate, disklike particle has $\mu < 1$. For a purely rotational flow ($\boldsymbol{D} = 0$), the particle must undergo a simple, rigid body rotation with the flow so that $A = 1$. Likewise, in purely extensional flow ($\boldsymbol{W} = 0$), a particle aligned along the principal axis of the rate of strain tensor cannot rotate and $B = -C$. From the original analysis by Jeffery for simple shear flow $(\boldsymbol{D} = \boldsymbol{W})$, the constant B is identified as $B = (\mu^2 - 1)/(\mu^2 + 1)$ and

$$\frac{d\mathbf{u}}{dt} = \mathbf{u} \cdot \boldsymbol{W} + \left(\frac{\mu^2 - 1}{\mu^2 + 1}\right)[\mathbf{u} \cdot \boldsymbol{D} - \mathbf{u}\,(\mathbf{uu}{:}\boldsymbol{D})]\,. \tag{7.106}$$

In the limit of an infinite aspect ratio $(\mu \to \infty)$, this result tends to the dynamics of the rigid dumbbell shown in equation (7.61).

It is instructive to present the solution to equation (7.106) for the case of simple shear flow. For a spheroid oriented with its symmetry axis defined by the polar angle θ relative to the z axis, and azimuthal axis ϕ measured in the (x, y) plane relative to the x axis, equation (7.106) produces the following two equations for a simple shear flow of the form, $\mathbf{v} = G\,(0, x, 0)$:

$$\begin{aligned} \frac{d\theta}{dt} &= \frac{G}{4}\left(\frac{\mu^2 - 1}{\mu^2 + 1}\right)\sin 2\theta \sin 2\phi, \\ \frac{d\phi}{dt} &= \frac{G}{\mu^2 + 1}\,(\mu^2 \cos^2\phi + \sin^2\phi)\,. \end{aligned} \tag{7.107}$$

These two equations can be integrated with the initial conditions $\theta\,(t = 0) = \theta_0$ and $\phi\,(t = 0) = \phi_0$ to give

$$\begin{aligned} \tan\theta &= \frac{C\mu}{\sqrt{\mu^2 \cos^2\phi + \sin^2\phi}}, \\ \tan\phi &= \mu \tan\left(\frac{2\pi t}{T} + \kappa\right), \end{aligned} \tag{7.108}$$

where $C = C(\theta_0, \phi_0)$ and $\kappa = \kappa(\phi_0)$ are the "orbit constants." The limits on these constants are $-\infty < C < \infty$ and $0 \le \kappa \le 2\pi$. The parameter T is the period of oscillation and is the time required for the spheroid to complete one full orbit. It is

$$T = \frac{2\pi}{G}(\mu + 1/\mu) . \tag{7.109}$$

The period of oscillation is shortest for a sphere and increases as the particles become either oblate or prolate. In either case, the motion of the particle will be a periodic orbit where the symmetry axis "tumbles" about the vorticity axis of the flow and is referred to as a "Jeffery orbit."

The Jeffery orbits are deterministic, and the particles will precess indefinitely in the flow. The following effects can perturb a particle and deflect it from an orbit and send it into a new one:

1. Brownian motion resulting from thermal fluctuations of solvent molecules onto the particle.
2. Colloidal, interparticle interactions (such as electrostatic forces) can apply additional torques to the particles and influence their motion.
3. Hydrodynamic interactions between particles.
4. External fields such as gravity or electric fields.
5. Inhomogeneity of temperature and velocity fields.

To account for Brownian motion, the equation of motion for $\mathbf{u}$ given equation (7.106) must be augmented by the term $-D_r^0 \frac{\partial}{\partial \mathbf{u}}(\log \Psi)$. This result can be combined with the continuity equation for the orientation distribution function given in equation (7.60). The result is a convective diffusion equation that is very similar to equation (7.62). Exact solutions for the distribution function cannot be found and approximate methods or numerical procedures are employed to evaluate bulk solution properties. Asymptotic solutions have been obtained in the limit of large and small aspect ratio, as well as large and small Peclet numbers [96].

An exact solution for the orientation distribution function can be found in the absence of Brownian motion since the motion of the particles is deterministic and given by equation (7.107). In this case the orbit of each particle is uniquely determined from its initial conditions specified by the orbit constants, C and κ. Indeed, the orientation distribution in the angular coordinates, $\psi(\theta, \phi)$, can be directly mapped onto a second distribution, $\bar{\psi}(C, \kappa)$, that specifies the initial distribution of particle orientations. This mapping means that the following equality holds:

$$\psi(\theta, \phi)\, d\theta d\phi = \bar{\psi}(C, \kappa)\, dC d\kappa . \tag{7.110}$$

The Jacobian matrix specifying the transformation from (θ, ϕ) to (C, κ) is immediately derived from equation (7.107) and is

$$\begin{aligned} dC d\kappa &= J\begin{bmatrix} C, \kappa \\ \theta, \phi \end{bmatrix} d\theta d\phi \\ &= \left(\frac{\partial C}{\partial \theta}\frac{\partial \kappa}{\partial \phi} - \frac{\partial C}{\partial \phi}\frac{\partial \kappa}{\partial \theta}\right) d\theta d\phi \qquad (7.111) \\ &= \frac{\sec^2\theta}{\sqrt{\mu^2\cos^2\phi + \sin^2\phi}} d\theta d\phi. \end{aligned}$$

Therefore, the relationship between ψ and $\overline{\psi}$ is

$$\psi(\theta, \phi) = \frac{\sec^2\theta}{\sqrt{\mu^2\cos^2\phi + \sin^2\phi}} \overline{\psi}(C, \kappa). \qquad (7.112)$$

Equation (7.112) must be valid for all times, including the initial condition, $t = 0$. If the initial distribution function is $\psi(\theta, \phi; t = 0) = \psi_0(\theta_0, \phi_0)$, then

$$\overline{\psi}(C, \kappa) = \cos^2\theta_0 \sqrt{\mu^2\cos^2\phi + \sin^2\phi}\, \psi_0(\theta_0, \phi_0), \qquad (7.113)$$

and the distribution function at any time is

$$\psi(\theta, \phi, t) = \cos^2\theta_0 \sec^2\theta \sqrt{\frac{\mu^2\cos^2\phi_0 + \sin^2\phi_0}{\mu^2\cos^2\phi + \sin^2\phi}}\, \psi_0(\theta_0, \phi_0). \qquad (7.114)$$

It is now necessary to use the Jeffery orbit equations (7.107) to express the current angles, θ and ϕ, existing of time t, in terms of the initial angles, θ_0 and ϕ_0. After some algebra, one obtains

$$\psi(\theta, \phi, t) = \frac{\psi_0\left[\tan^{-1}(\Lambda\tan\theta), \tan^{-1}\left(\dfrac{\tan\phi - \mu\tan\left(\frac{2\pi t}{T}\right)}{1 + \frac{1}{\mu}\tan\left(\frac{2\pi t}{T}\right)}\right)\right]}{\Lambda(\cos^2\theta + \Lambda^2\sin^2\theta)}, \qquad (7.115)$$

where

$$\begin{aligned} \Lambda^2 &= \Lambda_1\sin^2\phi + \Lambda_2\sin\phi\cos\phi + \Lambda_3\cos^2\phi, \\ \Lambda_1 &= \frac{1}{2}\left(1 + \mu^{-2} + (1 - \mu^{-2})\cos\frac{4\pi t}{T}\right), \\ \Lambda_2 &= (\mu^{-1} - \mu)\sin\frac{4\pi t}{T}, \qquad (7.116) \\ \Lambda_3 &= \frac{1}{2}\left(1 + \mu^2 + (1 - \mu^2)\cos\frac{4\pi t}{T}\right). \end{aligned}$$

This result was originally derived by Mason and coworkers [97] for the motion of noncolloidal suspensions of axisymmetric particles. It predicts that the distribution function at any time is directly related to its initial distribution in contrast to the case of Brownian motion where the initial conditions are eventually forgotten. Of particular interest is when the flow is started from equilibrium so that initially the distribution of orientations in random. Then $\psi_0(\theta_0, \phi_0) = (1/4\pi)\sin\theta_0$, and the distribution function is

$$\psi(\theta, \phi, t) = \frac{1}{4\pi}\frac{\sin[\tan^{-1}(\Lambda\tan\theta)]}{\Lambda(\cos^2\theta + \Lambda^2\sin^2\theta)} = \frac{1}{4\pi}\frac{\sin\theta}{(\cos^2\theta + \Lambda^2\sin^2\theta)^{3/2}}, \tag{7.117}$$

where the identity $\sin[\tan^{-1}(\Lambda\tan\theta)] = \Lambda\sin\theta/\sqrt{\cos^2\theta + \Lambda^2\sin^2\theta}$ has been used.

The distribution function is predicted to oscillate indefinitely in time with a period of $T/2$. The reason the period is half the Jeffery period is due to the fore-aft symmetry of the particles. The distribution starts out as a random distribution at $t = 0$ and returns to that state whenever $t = mT/2$, where m is an integer. At intermediate times, the equation predicts a net orientation of the particles along the flow direction (the y axis in this example).

7.2.1.2 Semidilute Suspensions of Hard Spheres

The microstructure of a semidilute suspension of hard spheres will become distorted with the application of flow. This is manifested by anisotropy in the spatial position of the centers of mass of the spheres, which is described the pair distribution function, $P_2(\mathbf{r})$. This is defined as the probability of finding a second sphere at a position $\mathbf{r}$ measured from a sphere at an arbitrary origin. A detailed account of the dynamics of colloidal dispersions can be found in the book by Russel *et al.* [98] This problem was first considered by Batchelor and Green [99], who calculated the effect of particle-particle interactions to the particle stress to second order in the particle concentration. The equation of motion for $P_2(\mathbf{r})$ is a balance between convection, particle interaction, and Brownian diffusion forces. In the limit of weak Peclet numbers ($Pe = \dot{\gamma}/D_r^0 \ll 1$), the pair distribution function has the following, approximate solution for a shear flow with a rate of strain tensor, $\mathbf{E}$,

$$P_2(\mathbf{r}) = n^2 g(r)\left[1 - \frac{3\pi\eta a^3}{k_B T}\mathbf{r}\cdot\mathbf{E}\cdot\mathbf{r} f(r)\right], \tag{7.118}$$

where n is the number of particles per unit volume, $g(r)$ is the equilibrium radial distribution function, η is the solvent viscosity, and a is the radius of the spheres. The equilibrium radial distribution function is found following classical statistical thermodynamic considerations[100]. $f(r)$ is a scalar function of the interparticle distance that is found from solving an ordinary differential equation that results from inserting equation (7.118) in the equation of motion for $P_2(\mathbf{r})$.

7.3 The Stress Tensor and the Stress-Optical Rule

The following derivation of the stress tensor of a polymeric liquid follows the description offered in the book by Bird *et al.* [62] The relationship between the stress tensor of a polymeric liquid and its constituent polymer chains is easily derived by considering the forces generated by a polymer segment on a plane of fluid, as shown in Figure 7.13. The plane is one surface of a cubic volume element, the volume of which is $1/\nu$, where ν is the number concentration of polymer segments. On average, such a cube will contain a single segment. There is a probability of $(\mathbf{R}\cdot\mathbf{n})/\nu^{-1/3}$ that the segment will straddle the plane (this is simply the ratio of the projected length of the segment to the length of the cube). If the segment exerts a force $\mathbf{F}_S = K(\mathbf{R})\mathbf{R}$ on the surface, the stress will be [62]

$$d\underline{\tau}\cdot\mathbf{n} = \frac{K(\mathbf{R})\mathbf{R}\dfrac{(\mathbf{R}\cdot\mathbf{n})}{\nu^{-1/3}}}{\nu^{-2/3}}\Psi(\mathbf{R})\,d\mathbf{R} = [\nu K(\mathbf{R})\mathbf{R}\mathbf{R}\Psi(\mathbf{R})\,d\mathbf{R}]\cdot\mathbf{n}, \tag{7.119}$$

where the result has been multiplied by the probability, $\Psi(\mathbf{R})\,d\mathbf{R}$, that the segment has a given end-to-end vector. Averaging over all possible conformations yields the following result for the polymer contribution to the stress tensor,

$$\underline{\tau}_p = \nu\langle K(\mathbf{R})\mathbf{R}\mathbf{R}\rangle. \tag{7.120}$$

If the segments are dissolved within a solvent of viscosity η_s, the total stress tensor is

$$\underline{\tau} = \nu\langle K(\mathbf{R})\mathbf{R}\mathbf{R}\rangle + 2\eta_s D, \tag{7.121}$$

where D is the rate of strain tensor. For a linear spring with constant $K = 3kT/N_K a^2$, then $\underline{\tau}_p = 3\nu kT/N_K a^2\langle\mathbf{R}\mathbf{R}\rangle$ and the polymer contribution to the stress tensor has the same form as the refractive index tensor given in equation (7.24). Since both tensors are proportional to the second moment tensor, $\langle\mathbf{R}\mathbf{R}\rangle$, this suggests that the stress and refractive index are related according to

$$\mathbf{n} = C\underline{\tau} + A\mathbf{I}, \tag{7.122}$$

where the constant A depends on the isotropic portions of these two tensors. The stress optical coefficient C is

$$C = \frac{2\pi}{45kT}\frac{(\bar{n}^2+2)^2}{\bar{n}}(\alpha_1 - \alpha_2). \tag{7.123}$$

In terms of the shear stress, τ_{xy}, and the first normal stress difference, $N_1 = \tau_{xx} - \tau_{yy}$, the stress optical rule is

$$\tau_{xy} = \frac{1}{2C}\Delta n' \sin 2\chi \tag{7.124}$$

and

$$N_1 = \frac{1}{C}\Delta n' \cos 2\chi . \tag{7.125}$$

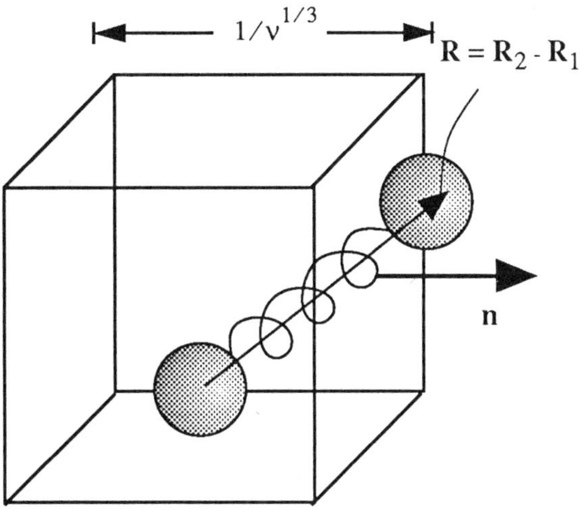

Figure 7.13 A polymer segment exerting a force on a plane embedded with a fluid element. The plane is identified by the unit vector, **n**. The cube has a volume of 1/ν, where ν is the number concentration of polymer segments.

The stress optical rule enjoys wide validity for melts and concentrated solutions of flexible chains [29]. From the simple derivation given above, however, it is evident that it is expected to fail under a number of conditions. Large chain extensions, for example, where the spring constant becomes nonlinear will lead to a deviation from this rule. In polymer melts, however, extremely large stresses are required to achieve such nonlinearity, and the stress optical rule is found to hold well into the region of non-Newtonian rheological properties. In polymer solutions, if the solvent contribution is correctly subtracted from both the stress and the birefringence, the rule is found to hold in simple shear flow [81]. In extensional flow, however, strong flow conditions will cause the rule to fail [82,83]. In such flows, the orientation of chain segments can saturate, giving a birefringence which is independent of strain rate. The stresses, however, continue to rise with increasing velocity gradient.

The stress optical coefficient is normally determined by plotting the ratio of $2\Delta n' \sin\chi/\tau_{xy}$ as a function of shear rate. The independence of this ratio with shear rate verifies the stress optical rule. Alternatively, the ratio of $2\tau_{xy}/N_1$ can be compared with $\tan 2\chi$ determined from birefringence measurements.

As indicated by equation (7.123), the stress optical coefficient is found to be independent of molecular weight. For polymer solutions, however, it is a function of the solvent due to local solvent-segment interactions [85]. These interactions cause solvent molecules to couple in their orientation with the polymer chains. However, because these interactions are very local, the stress optical coefficient is only weakly dependent on solvent concentration as long as the solvent is in sufficient quantity to fully solvate the chains.

Scattering or form birefringence contributions will cause a deviation in the stress optical rule. As seen in equation (7.36), these effects do not depend on the second-moment tensor, but increase linearly with chain extension.

The term $\langle K(\mathbf{R})\,\mathbf{RR}\rangle$ is common to both equations (7.52) and (7.121) and can be eliminated to provide the following alternative form for the stress tensor:

$$\underset{=}{\tau} = -\frac{\nu\zeta}{2}\left[\frac{d}{dt}\langle\mathbf{RR}\rangle - (\boldsymbol{G}\cdot\langle\mathbf{RR}\rangle + \langle\mathbf{RR}\rangle\cdot\boldsymbol{G}^{+})\right] + 2\eta_s\boldsymbol{D}. \tag{7.126}$$

This result is referred to as the Giesekus expression [62,86] and can be used to develop the form of the stress tensor for the rigid dumbbell model. Equation (7.63) for the rate of change of the second-moment tensor for this model is used to give the following result:

$$\underset{=}{\tau} = \frac{3\nu kT}{L^2}\langle\mathbf{RR}\rangle + \frac{kT\nu}{D_r^0L^4}\boldsymbol{D}:\langle\mathbf{RRRR}\rangle - \nu kT\mathbf{I} + 2\eta_s\boldsymbol{D}. \tag{7.127}$$

The presence of the fourth-rank tensor in (7.127) and its absence in (7.11) suggests that the stress optical rule should not apply for dilute solutions of rigid rods. Unfortunately, because of the difficulty of acquiring truly rigid rods, and the problems of making measurements of stress in dilute systems, there are no data available on dilute rigid rod solutions where the stress optical rule can be investigated on this class of polymer liquids.

When rigid rods become concentrated, and enter the semidilute region discussed in section 7.1.6.4, equation (7.127) is still appropriate, except that $1/D_r^0$ is no longer the appropriate timescale and should be replaced by $1/D_r^0 = (\nu L^3)^{-2}/\bar{D}_r$. When the Peclet number is moderate or weak, $\dot{\gamma}/\bar{D}_r \le O(1)$, the term involving the fourth-rank tensor can be neglected since $\nu L^3 \gg 1$. The anisotropic part of the stress tensor for concentrated rigid rods becomes

$$\underset{=}{\tau} = \frac{3\nu kT}{L^2}\langle\mathbf{RR}\rangle, \tag{7.128}$$

where the solvent contribution is also assumed to be negligible. Evidently, the stress optical rule applies to solutions of rigid rods in the semidilute region.

8 Design of Optical Instruments

The design of an optical instrument begins with a consideration of the structural and dynamic information that is desired from a sample, and their relationship to its optical properties. This information must be combined with the particular flow or sample geometry that will be probed and the timescales of the dynamics to be monitored.

The optical arrangements presented in this chapter are presented according to whether the light is measured in transmission, or as scattered light. The latter category is further subdivided into small angle, wide angle, dynamic, and Raman scattering. Throughout this chapter, schematic diagrams will be offered that describe the various designs of instruments. The notation outlined in the following table is used to represent different optical elements in these figures. As an example, a typical element would be described as depicted in the following figure, where a birefringent sample of retardation δ' and orientation θ is shown. The symbols above the element refer to the type of device that is represented. Below the element are symbols describing its properties (the type of anisotropy and its angle of orientation in the optical train).

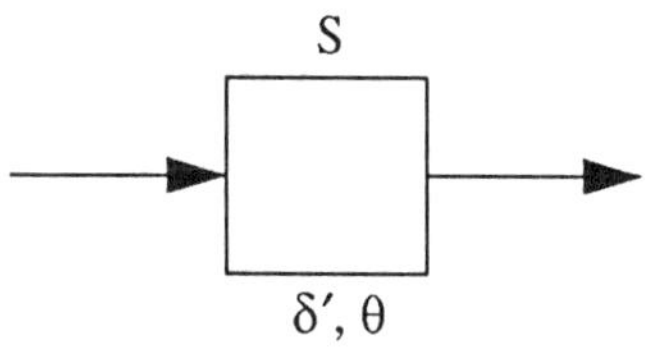

Figure 8.1 Description of a typical element in the schematic of an optical train.

TABLE 8.1 Notation for Instrument Schematics

Optical Element	Symbol
analyzing polarizer	A
beam splitter	B
circular polarizer	CP
detector	D
color filter	F
half-wave plate	H
light source	L
lens	LS
monochromator	M
notch filter	NF
polarization state analyzer	PSA
polarization state generator	PSG
polarizer	P
quarter-wave plate	Q
rotating half-wave plate	RH
rotating polarizer	RP
rotating quarter-wave plate	RQ
photoelastic modulator	PEM
sample	S

8.1 Transmission Experiments: Polarimeters

In transmission, it is the polarization of the light that is normally analyzed. The basic scheme of an optical polarimeter is shown below. The elements of the apparatus are (1) a light source, L; (2) a polarization state generator, PSG, that defines the polarization of the light prior to transmission through the sample; (3) the sample, S; (4) a polarization state analyzer, PSA; and (5) a detector. This organization follows the general scheme devised by Hauge [25]. Each element is described by an appropriate Mueller matrix so that the Stokes vector of the light received at the detector is

$$\mathbf{S}_3 = \mathbf{M}_{PSA} \cdot \mathbf{M}_S \cdot \mathbf{M}_{PSG} \cdot \mathbf{S}_0 . \tag{8.1}$$

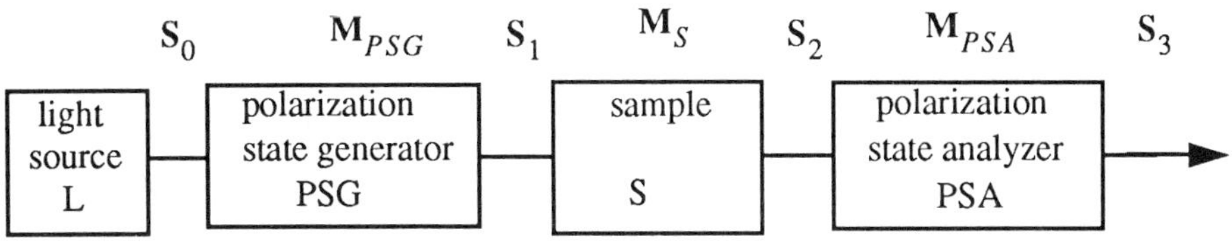

Figure 8.2 Basic polarimeter design.

The strategy in the design of any optical polarimeter is the use of the appropriate optical elements comprising the PSG and PSA to retrieve the desired information from the sample. Symbolically, this can be understood by considering the following expression for the intensity measured at the detector,

$$I = I_0 (\mathbf{S}_{PSA}^T \cdot \mathbf{M}_S \cdot \mathbf{S}_{PSG}),$$

$$I = I_0 (m_{11}^{PSA}, m_{12}^{PSA}, m_{13}^{PSA}, m_{14}^{PSA})^T \cdot \begin{bmatrix} m_{11}^S & m_{12}^S & m_{13}^S & m_{14}^S \\ m_{21}^S & m_{22}^S & m_{23}^S & m_{24}^S \\ m_{31}^S & m_{32}^S & m_{33}^S & m_{34}^S \\ m_{41}^S & m_{42}^S & m_{43}^S & m_{44}^S \end{bmatrix} \cdot \begin{bmatrix} m_{11}^{PSG} \\ m_{21}^{PSG} \\ m_{31}^{PSG} \\ m_{41}^{PSG} \end{bmatrix} . \quad (8.2)$$

Here, m_{ij}^k, k= *PSA*, *S*, and *PSG* are elements of the Mueller matrices for the PSA, S and PSG, respectively. This equation was written for the arbitrary choice of unpolarized, incident light, $\mathbf{S}_0 = (I_0, 0, 0, 0)^T$. Since the first element in the PSG is a polarizer, the specific choice for $\mathbf{S}_0$ is arbitrary.

If particular elements of $\mathbf{M}_S$ are required, then the design of the PSG and the PSA must serve to highlight specific components of $\mathbf{M}_{PSG}$ and $\mathbf{M}_{PSA}$. For example, if linear dichroism and its associated orientation angle are sought, examination of equation (I.10) in Appendix I suggests that the PSG must highlight m_{21}^{PSG} and m_{31}^{PSG}. This must be combined with a PSA that highlights m_{11}^{PSA}. For example, if $m_{12}^{PSA} = m_{13}^{PSA} = m_{14}^{PSA} = m_{41}^{PSG} - 0$,

$$I = I_0 (m_{11}^{PSG} m_{11}^S + m_{21}^{PSG} m_{12}^S + m_{31}^{PSG} m_{13}^S) m_{11}^{PSA}, \quad (8.3)$$

and the desired coefficients, m_{12}^S and m_{13}^S can be extracted assuming that the elements m_{21}^{PSG} and m_{31}^{PSG} can be isolated in the measured signal.

Below, a variety of designs of optical analyzers are presented. The designs are grouped according to the types of PSGs and PSAs that are employed. In the following tables, designs of PSGs and PSAs are presented along with the respective Stokes vectors, $\mathbf{S}_{PSG}$ and $\mathbf{S}_{PSA}$.

TABLE 8.2 PSG Designs

NAME	Design	S_{PSG}
$(P_\theta)_{PSG}$	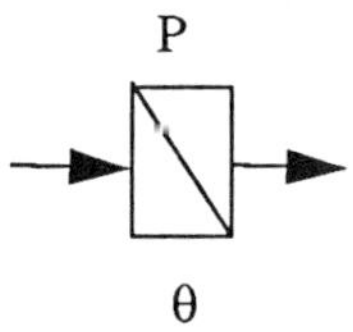	$\frac{1}{2}\begin{bmatrix} 1 \\ c_{2\theta} \\ s_{2\theta} \\ 0 \end{bmatrix}$
$(P/RP)_{PSG}$	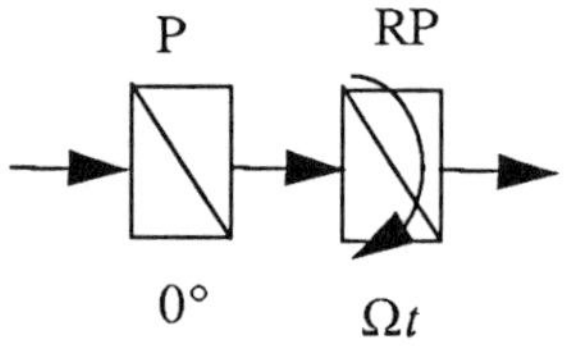	$\frac{1}{4}\begin{bmatrix} 1 + c_{2\Omega t} \\ c_{2\Omega t} + \frac{1}{2}(1 + c_{4\Omega t}) \\ s_{2\Omega t} + \frac{1}{2}s_{4\Omega t} \\ 0 \end{bmatrix}$
$(P/RH)_{PSG}$	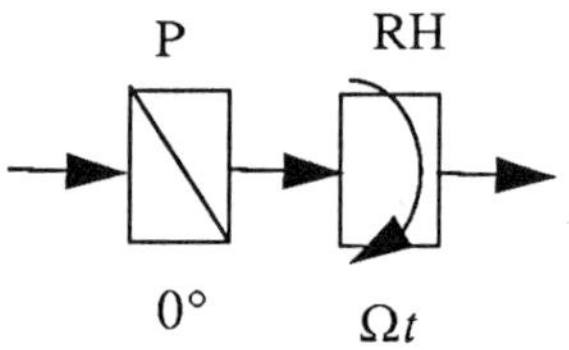	$\frac{1}{2}\begin{bmatrix} 1 \\ c_{4\Omega t} \\ s_{4\Omega t} \\ 0 \end{bmatrix}$
$(P/Q/RP)_{PSG}$	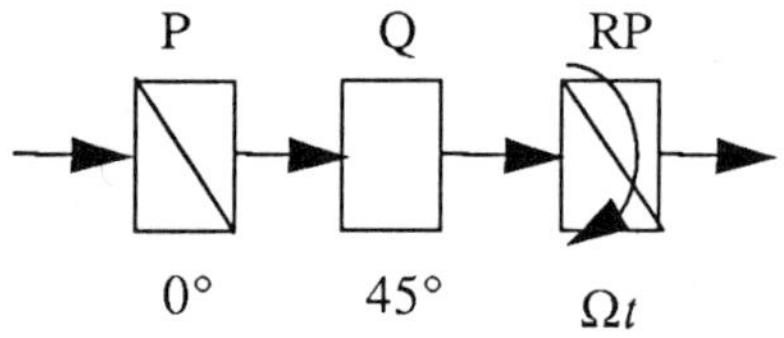	$\frac{1}{4}\begin{bmatrix} 1 \\ c_{2\Omega t} \\ s_{2\Omega t} \\ 0 \end{bmatrix}$
$(P/RH/Q)_{PSG}$	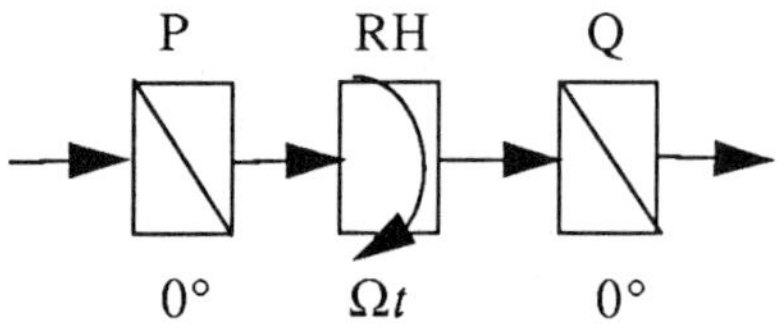	$\frac{1}{2}\begin{bmatrix} 1 \\ c_{4\Omega t} \\ 0 \\ -s_{4\Omega t} \end{bmatrix}$

$(P/RQ)_{PSG}$	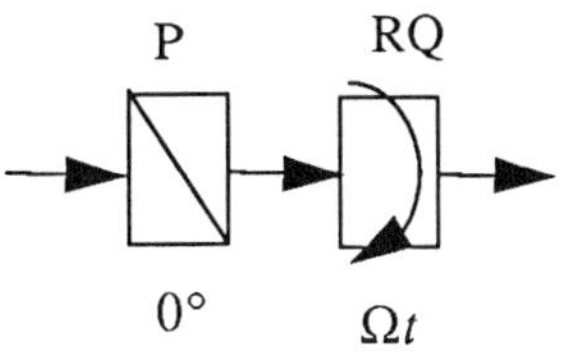	$\frac{1}{2}\begin{bmatrix} 1 \\ \frac{1}{2}(1+c_{4\Omega t}) \\ \frac{1}{2}s_{4\Omega t} \\ s_{2\Omega t} \end{bmatrix}$
$(P/PEM)_{PSG}$	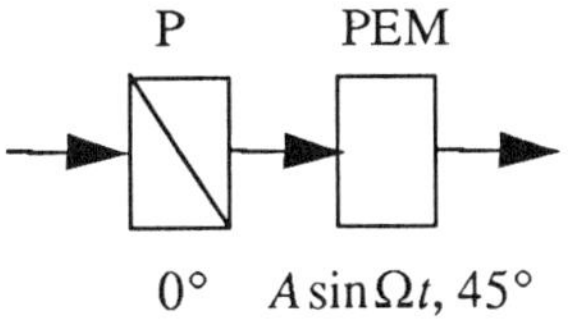	$\frac{1}{2}\begin{bmatrix} 1 \\ c_{A\sin\Omega t} \\ 0 \\ s_{A\sin\Omega t} \end{bmatrix}$
$(P/PEM/Q)_{PSG}$	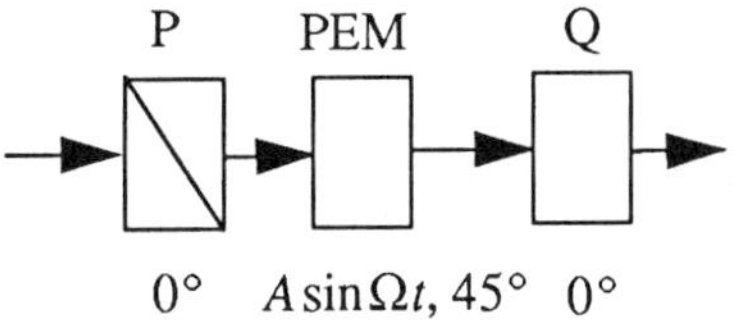	$\frac{1}{2}\begin{bmatrix} 1 \\ c_{A\sin\Omega t} \\ s_{A\sin\Omega t} \\ 0 \end{bmatrix}$
$(P/PEM_1/PEM_2)_{PSG}$	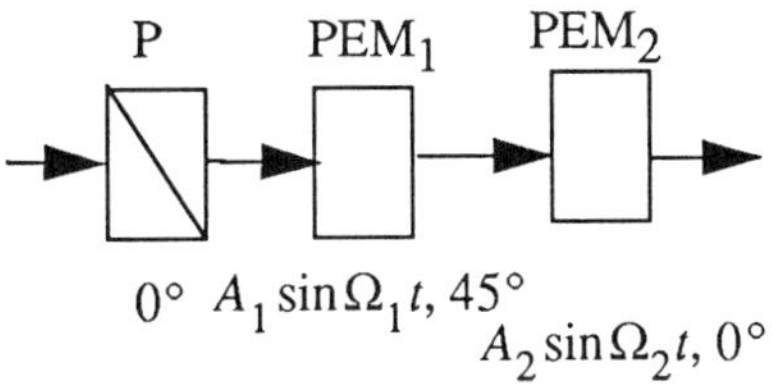	$\frac{1}{2}\begin{bmatrix} 1 \\ c_{A_1\sin\Omega_1 t} \\ s_{A_1\sin\Omega_1 t}s_{A_2\sin\Omega_2 t} \\ s_{A_1\sin\Omega_1 t}c_{A_2\sin\Omega_2 t} \end{bmatrix}$

TABLE 8.3 PSA Designs

NAME	Design	$\mathbf{S}_{PSA}$
$(-)_{PSA}$		$\begin{bmatrix} 1 \\ 0 \\ 0 \\ 0 \end{bmatrix}$

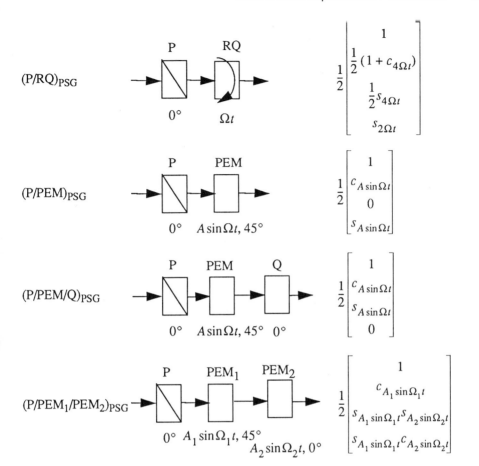

$(P_\theta)_{PSA}$

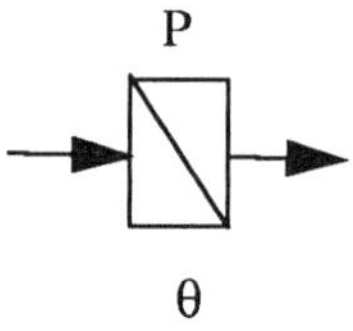

$$\frac{1}{2}\begin{bmatrix}1\\c_{2\theta}\\s_{2\theta}\\0\end{bmatrix}$$

$(Q/P)_{PSA}$

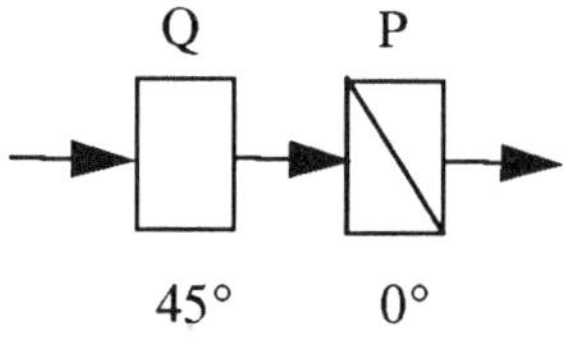

$$\frac{1}{2}\begin{bmatrix}1\\0\\0\\1\end{bmatrix}$$

$(CP)_{PSA}$

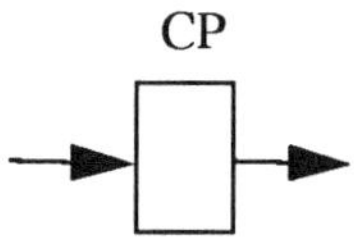

$$\begin{bmatrix}1\\0\\0\\\pm1\end{bmatrix}$$

where the plus sign is for left and the minus sign is for right circular polarizers.

$(RQ/P)_{PSA}$

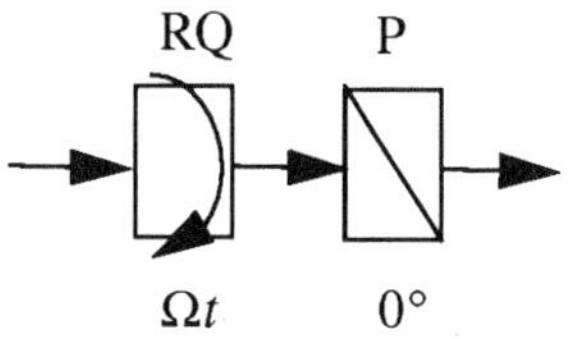

$$\frac{1}{2}\begin{bmatrix}1\\\frac{1}{2}(1+c_{4\Omega t})\\\frac{1}{2}s_{4\Omega t}\\-s_{2\Omega t}\end{bmatrix}$$

$(RP)_{PSA}$

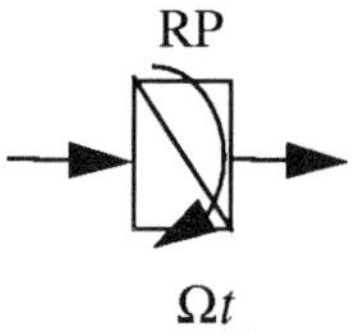

$$\frac{1}{2}\begin{bmatrix}1\\c_{2\Omega t}\\s_{2\Omega t}\\0\end{bmatrix}$$

$(Q/RP)_{PSA}$

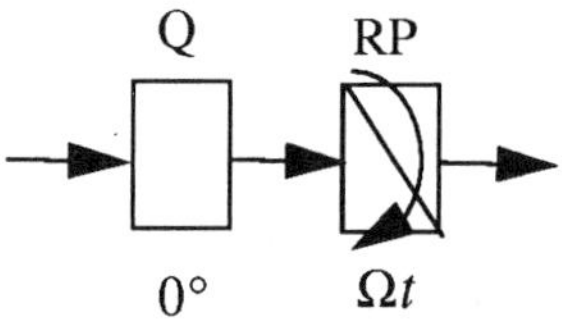

$$\frac{1}{2}\begin{bmatrix}1\\c_{2\Omega t}\\0\\s_{2\Omega t}\end{bmatrix}$$

$(PEM/P)_{PSA}$

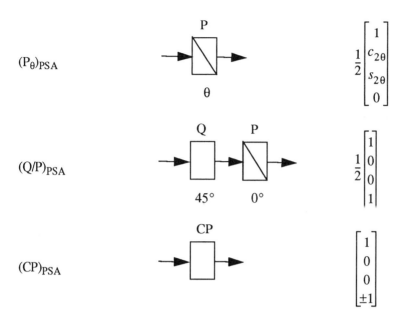

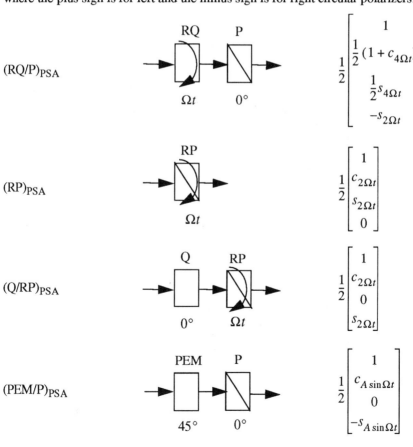

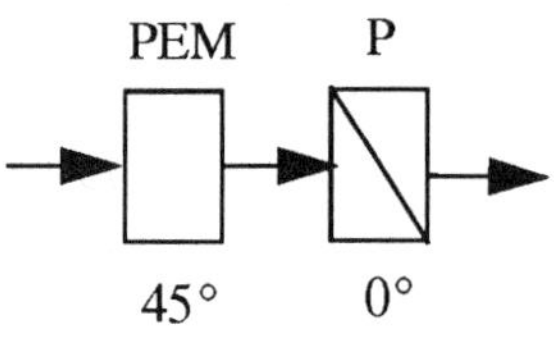

$$\frac{1}{2}\begin{bmatrix}1\\c_{A\sin\Omega t}\\0\\-s_{A\sin\Omega t}\end{bmatrix}$$

$(PEM_1/PEM_2/P)_{PSA}$	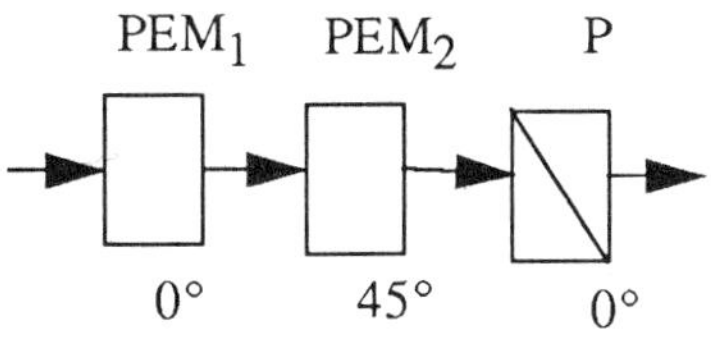	$\frac{1}{2}\begin{bmatrix} 1 \\ {}^{C}A_1 \sin\Omega_1 t \\ {}^{S}A_1 \sin\Omega_1 t\,{}^{S}A_2 \sin\Omega_2 t \\ -{}^{S}A_1 \sin\Omega_1 t\,{}^{C}A_2 \sin\Omega_2 t \end{bmatrix}$

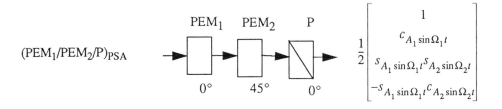

8.2 Fixed Element Systems

In the following subsections, several designs of optical arrangements are described for the purpose of measuring optical anisotropies. In these designs, the optical elements in the PSG and PSA are either fixed in orientation and possess constant optical properties or are adjusted manually.

8.2.1 The Crossed Polarizer System

This simple arrangement was previously discussed in section 2.6 and consists of two crossed polarizers that bracket the sample. In that section the sample was taken to be both birefringent and dichroic. These anisotropies were also taken to be coaxial and oriented at an angle θ. Using the notation describing the PSGs and PSAs in table 8.2 and table 8.2, this polarimeter is referred to as " $(P_\alpha)_{PSG}/\left(P_{\alpha+\frac{\pi}{2}}\right)_{PSA}$ " and the intensity is

$$I = \frac{I_0}{4}\sin^2[2(\theta-\alpha)]\,(\cosh\delta'' - \cos\delta')\,. \tag{8.4}$$

The angle α defines the orientation of the pair of orthogonal polarizers relative to a laboratory frame.

Crossed polarizer systems have been used extensively for birefringence measurements. However, equation (8.4) reveals a number of shortcomings of this simple method. First, the measurement will be equally affected by both birefringence and dichroism and should only be used if dichroism can be properly taken into account or neglected. Second, the measurement is not capable of determining the sign of either the birefringence or the orientation angle. Third, if both the birefringence and orientation angle are unknown, this single measurement is not sufficient to determine them simultaneously.

The last limitation can be simply overcome by performing two separate measurements at different values of the angle α. This procedure was first used by Osaki and coworkers [26]. They performed separate measurements of intensity for $\alpha = 0°$ and $\alpha = 45°$ relative to the flow direction. These intensities are denoted I^0 and I^{45}, respectively. Taking the dichroism to be negligible, ratios of these two intensities to the incident intensity would be

$$i_0 = \frac{2I^0}{I_0} = \sin^2 2\theta \sin^2\left(\frac{\delta'}{2}\right),\ i_{45} = \frac{2I^{45}}{I_0} = \cos^2 2\theta \sin^2\left(\frac{\delta'}{2}\right). \tag{8.5}$$

The sample retardance and orientation angle are then

$$\delta' = \frac{2\pi\Delta n' d}{\lambda} = \sin^{-1}\left(\sqrt{i_0 + i_{45}}\right), \tag{8.6}$$

$$\tan(2\theta) = \sqrt{i_0/i_{45}}. \tag{8.7}$$

Equations (8.5) bring out an additional problem that is inherent in any birefringence experiment, and that is that the measured intensity is a sinusoidal function of the retardance. This means that there can be ambiguity regarding the absolute magnitude of the retardance, since it will not have a unique value for a given measurement of the intensity. The intensity will oscillate and pass through a maximum whenever $\delta' = n\pi$, where $n = 1, 3, \ldots$ and will be zero when $n = 0, 2, \ldots$. This problem represents a serious limitation, and care must always be taken when analyzing data when the retardance passes through orders of π. In principle, one can always precisely determine the order of the retardance by simply allowing the system to relax. Whereas the birefringence and the retardance should monotonically decay to zero, the intensity will oscillate if multiple orders in δ' have been achieved. This procedure, however, cannot be applied to solid materials or to materials such as liquid crystals that are optically anisotropic at rest.

Recently Burghardt and coworkers [27] have developed a crossed polarizer apparatus that can analyze systems with multiple orders in retardation. The idea is to use a white-light source that generates light with wavelengths over the entire visible spectrum. A schematic of the instrument is shown in Figure 8.3. After passing through the polarizers and the sample, the light intensity is measured using a monochromator that is a combination of a prism and a one-dimensional array detector. The purpose of the prism is to separate the beam according to wavelength so that individual elements on the array detector are identified with specific wavelengths.

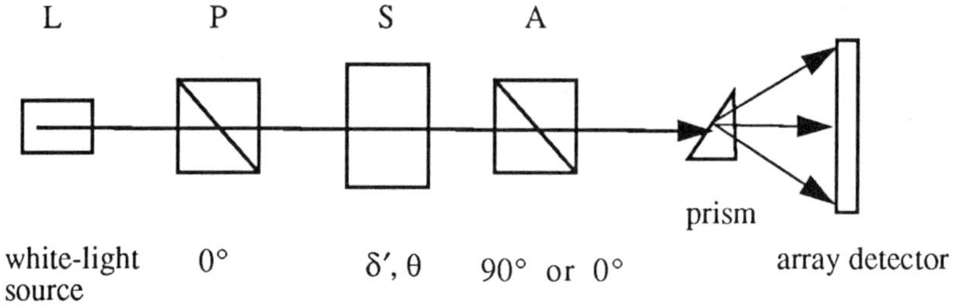

Figure 8.3 White-light apparatus for multiple orders in retardation.

Operation of the experiment requires making two separate measurements, one with the second polarizer crossed with respect to the first, and another with the second polarizer oriented parallel to the first. These two intensity measurements are identified as I_1 and I_2, respectively. With the orientation angle α of the first polarizer set to $0°$, these intensities are

$$I_1 = \frac{I_0}{2}\sin^2(2\theta)\sin^2\left(\frac{\delta'}{2}\right), \tag{8.8}$$

$$I_2 = \frac{I_0}{2}\left[1 - \sin^2(2\theta)\sin^2\left(\frac{\delta'}{2}\right)\right]. \tag{8.9}$$

These results can be combined to yield

$$\sqrt{\frac{I_1}{I_1 + I_2}} = \sin 2\theta \sin\left(\frac{\pi \Delta n' d}{\lambda}\right), \tag{8.10}$$

which can be used to measure the birefringence at any order in retardation and the orientation angle, simultaneously. This is possible since the sine function in equation (8.10) is measured as a function of wavelength. A fit of the data to a sine function will produce a frequency, from which the birefringence can be determined, and an amplitude, from which the orientation angle is found. It is necessary, however, to either assume that the birefringence is not a function of wavelength, or to measure its dispersion with wavelength. This technique becomes increasingly more accurate as the retardation increases and becomes multiply ordered. As the retardation becomes small, this method looses its accuracy. Again, neither the sign of the birefringence nor the orientation angle can be determined using this method. An additional limitation is the timescale for data acquisition, which can be long since some amount of time is required to measure the full spectrum of the light.

The method of Osaki described above requires that two distinct measurements be taken in succession with the cross polarizers being rotated in unison by $45°$ between each run. Such a procedure, besides being time consuming, demands that the successive experiments be dynamically identical, especially if transient phenomena are being considered. To overcome this difficulty, Chow and Fuller [28] developed the method of two-color flow birefringence, which makes it possible to simultaneously determine δ' and θ in a single measurement. This is accomplished by using the optical train described in Figure 8.4. The strategy behind this design is to produce two superimposed beams that are polarized in directions $45°$ apart. An argon ion laser with its prism removed is used as the light source. Such a laser will produce two intense beams of light with wavelengths $\lambda_1 = 488$ nm (blue) and $\lambda_2 = 514.5$ nm (green). These beams are effectively split into their respective wavelengths using laser line color filters, which have the property of transmitting one wavelength and reflecting the majority of the second wavelength. The blue filter CS_{blue} then reflects the green light and transmits the blue light. The blue and green beams are then directed along different optical paths where they are separately polarized by polarizers P_0 and P_{45}, respectively. The orientations of P_0 and P_{45} are such that the blue light is polarized at an angle α with respect to the laboratory axis (normally the flow direction if a flow cell is used), whereas the green light is polarized at a direction $45°$ from the blue. A second blue filter serves to recombine the blue and the green beams and direct them through the sample. On the other side of the sample, a blue filter reflects the green light to a green filter, polarizer P_{90}, and detector D_2. It is important to note that the blue color filters are oriented at low angles of incidence ($< 5°$) to minimize any polarization changes upon reflection and transmission by these devices. The blue light is passed by a blue color filter and transmitted through P_{-45} and measured by D_1. The signals measured by D_1 and D_2 are

$$i_1 = \frac{I_1}{I_0^1} = \cos^2 2(\theta - \alpha) \sin^2 \frac{\delta'_1}{2}, \tag{8.11}$$

$$i_2 = \frac{I_2}{I_0^2} = \sin^2 2(\theta - \alpha) \sin^2 \frac{\delta'_2}{2}, \tag{8.12}$$

where I_0^j, $j = 1, 2$ are the incident light intensities and $\delta'_j = (2\pi\Delta n' d)/\lambda_j$ are the retardances of the blue and green light beams. Equations (8.11) and (8.12) must be solved simultaneously to find the unknown birefringence and orientation angle. Using the small parameter, $\eta = (\lambda_2 - \lambda_1)/\lambda_2$, the following perturbation solution is obtained:

$$\Delta n' = \frac{\lambda_2 \delta'}{2\pi d}, \tag{8.13}$$

$$\tan^2 2(\theta - \alpha) = \frac{i_2}{i_1} \frac{\sin^2\{(\delta')/[2(1-\eta)]\}}{\sin^2((\delta')/2)}, \tag{8.14}$$

$$\delta' = 2\sin^{-1}(\sqrt{i_1 + i_2})\left(1 - \eta \frac{i_1}{i_1 + i_2}\right). \tag{8.15}$$

This technique has the advantage of very fast response times but cannot directly measure the birefringence when the retardation is of multiple order.

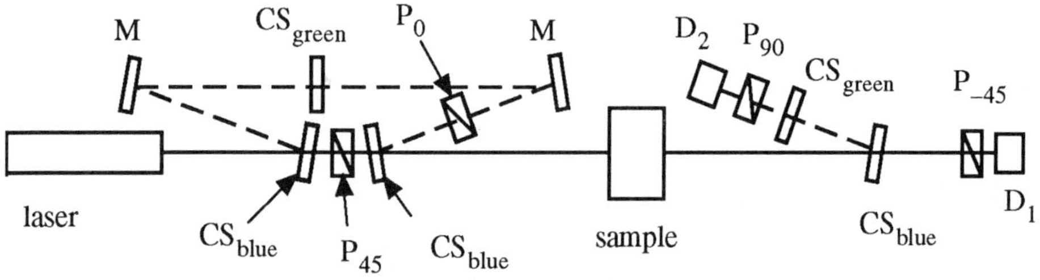

Figure 8.4 Two-color flow birefringence apparatus. CS_{green} : color splitter passing the green line; CS_{blue} : color splitter passing the blue line; P_θ : polarizer oriented at an angle θ ; D_i : detector i ; M: mirror.

8.2.2 Crossed Polarizers/Quarter-Wave Plate System

There are a number of applications where the orientation angle is known *a priori*. This is the case whenever uniaxial fields such as electric and magnetic fields are applied, or when purely extensional flows are used. In that case the experiments are normally performed by aligning the sample so that its principal orientation direction is at $45°$ with respect to the polarizers. If the sample is weakly birefringent, so that $|\delta'| \ll 1$, and if dichroism is negligible, then (8.4) predicts the intensity to be

$$\frac{I}{I_0} \approx \frac{\delta'^2}{16} . \tag{8.16}$$

Having the signal proportional to the square of the retardance is an inconvenience that can be removed by simply adding a quarter-wave plate immediately following the sample. The optical axis of this plate is then rotated by a small angle, ϕ, with respect to the first polarizer. This arrangement is shown in Figure 8.5. The light measured by the detector is then

$$\mathbf{S} = \mathbf{M}_{11}(90°) \cdot \mathbf{M}_{12}(\phi) \cdot \mathbf{M}_9(\delta', 45°) \cdot \mathbf{M}_{11}(0°) \cdot \mathbf{S}_0, \tag{8.17}$$

and the intensity is

$$\begin{aligned}\frac{I}{I_0} &= \frac{1}{4}(1 - \cos^2 2\phi \cos\delta' + \sin 2\phi \sin\delta') \\ &\approx \phi^2 + \frac{\delta'^2}{8} + \frac{\phi\delta'}{2} \approx \frac{\phi\delta'}{2}\end{aligned} \tag{8.18}$$

where the final approximation assumes that $\delta' \ll \phi \ll 1$. In equation (8.17), the Mueller matrices are found in Appendix I.

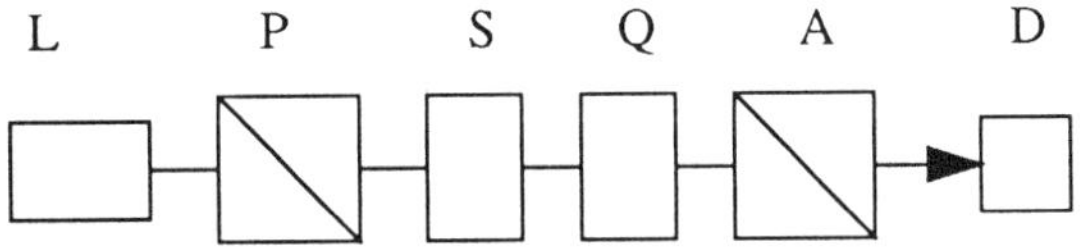

Figure 8.5 Crossed polarizer/quarter-wave plate system.

8.2.3 Null Methods

We shall consider here only methods used for the measurement of birefringence. In general, and especially for systems subject to flow, there will be two unknowns: the birefringence, $\Delta n'$, and the average angle of orientation of the sample relative to a laboratory frame, θ. Using a null technique, the measurement is normally carried out using a two-part procedure [29]. In the first step, the angle θ is determined using the simple crossed polarizer configuration shown in Figure 2.9. If the first polarizer is oriented at a direction α with respect to the laboratory frame, the intensity is given by

$$I = \frac{I_0}{2} \sin^2 [2(\theta - \alpha)] \sin^2\left(\frac{\delta'}{2}\right). \tag{8.19}$$

Both polarizers are rotated in unison while retaining their crossed orientation. At the point where $\alpha = \theta$, the intensity will tend to zero (or a sharp minimum, in practice). This completes the measurement of θ, which is often referred to as the extinction angle. To measure the retardation or birefringence, both polarizers are rotated so that the first polarizer is at an angle of $45°$ relative to the sample orientation. At this orientation, the intensity is now at a maximum value. A compensator is then placed in the experiment immediately following the sample. This element is simply a retardation plate of either fixed or variable retardation. For large retardations in the sample, a variable compensator, such as a Babinet compensator is useful. Normally, however, a quarter-wave plate (often referred to as a Senarmont compensator) is used. This element is placed so that its optical axis is either parallel or orthogonal to the first polarizer. At this orientation of the quarter-wave plate, the intensity emitted through two crossed polarizers will be zero if the sample is isotropic. For a birefringent sample, which causes the light to be elliptically polarized, insertion of the quarter-wave plate at this orientation returns the light to a state of linear polarization, but at a different orientation relative to the first polarizer. The second polarizer can then be rotated to finally extinguish the light. The rotation angle required to do this is can be shown to be precisely one-half the retardation angle. This can be demonstrated by calculating the intensity measured after the second polarizer. Using Mueller calculus, this intensity is found to be

$$\frac{I}{I_0} = \frac{1}{4}[1 - \cos(2\phi - \delta')] , \tag{8.20}$$

where ϕ is the angle by which the second polarizer has been rotated from orthogonality with respect to the first polarizer. Rotating the second polarizer so that $\phi = \delta'/2$ extinguishes the light. This procedure can be very accurate and is capable of determining the sign of the birefringence. In addition, one can show that the presence of a coaxial dichroism will not affect the values of δ' and θ evaluated by this method.

8.3 Polarization Modulation Methods

The measurement of the polarization properties of light can be automated and improved by introducing a modulation of the polarization. Here a regular, time-dependent variation is introduced onto the optical properties of certain devices within either (or both) the PSG or PSA sections of the instrument. The modulation can be one of two types: rotation of an optical element with fixed optical properties, or the modulation of the optical properties (retardation, for example) of an element with a fixed orientation. These are referred to as rotary modulators or field effect modulators, respectively. The latter name reflects the use of external fields (stress, electric or magnetic) to impart the modulation in these devices. In any case, a periodic oscillation is introduced into the signals that are measured that can effectively isolate specific optical properties in the sample.

The selection of a particular modulation scheme starts by determining which of the sample's Mueller matrix components need to be measured. This decision can be guided by examining the form of the Mueller matrix given in equation (I.18), which contains all optical anisotropies, but in the limit of small magnitudes. This matrix is reproduced below with annotations that identify matrix elements dominated by particular optical effects.

linear dichroism

circular dichroism

$$\mathbf{M}_{19}(\delta', \delta'', \delta'_{circ}, \delta''_{circ}, \theta', \theta'') = \begin{bmatrix} 1 & -\delta'' c_{2\theta''} & -\delta'' s_{2\theta''} & \delta''_{circ} \\ -\delta'' c_{2\theta''} & 1 & \delta'_{circ} & \delta' s_{2\theta'} \\ -\delta'' s_{2\theta''} & -\delta'_{circ} & 1 & -\delta' c_{2\theta'} \\ \delta''_{circ} & -\delta' s_{2\theta'} & \delta' c_{2\theta'} & 1 \end{bmatrix} \tag{8.21}$$

circular birefringence

linear birefringence

For example, linear birefringence is principally contained in the matrix elements m_{42}^{S} and m_{43}^{S}. This matrix, however, is only a guide, and when the magnitudes of the optical anisotropies become large, most of the matrix elements will be superposition of multiple effects.

Once the matrix elements of interest have been identified, equation (8.2) is used to specify which matrix elements of the PSG and the PSA need to be highlighted. In the subsections to follow, different modulation schemes are presented for the purpose of meeting specific requirements.

8.3.1 Rotary Polarization Modulators

A rotary polarization modulator simply consists of an optical element that rotates uniformly at a frequency Ω about the transmission axis of light. In practice, retardation plates and polarizers are used. In either case, the Mueller matrix of such a device is found by simply replacing the angle θ by Ωt in the equations listed in Appendix I. Typical PSGs based on rotary modulators and the associated Stokes vectors, $\mathbf{S}_{PSG}$, that are produced are listed in table 8.2.

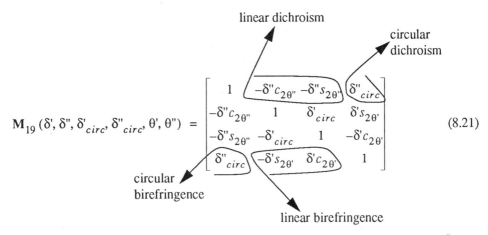

These designs reveal how modulation serves to isolate specific elements of the Stokes vectors that are generated. In the $(P/RP)_{PSG}$ design, for example, analysis of the signal at a frequency of 4Ω will highlight either the m_{21}^{PSG} or m_{31}^{PSG} components of the vector $\mathbf{S}_{PSG}$ defined in equation (8.2), depending on whether the "out-of-phase" Fourier component (the $\cos 4\Omega t$ term) or the "in-phase" Fourier component (the $\sin 4\Omega t$ term) is extracted. Evidently, all of the rotary modulator designs in table 8.2 can highlight the m_{21}^{PSG} and m_{31}^{PSG} components. The $(P/RQ)_{PSG}$ design, however, has the important advantage of being able to also isolate the m_{41}^{PSG} component by analyzing the signal for the in phase harmonic at the frequency 2Ω.

Modulation can also be incorporated into the PSA section. table 8.2 contains descriptions of several possible designs.

8.3.2 Field Effect Polarization Modulators

The application of an external field onto many materials will induce optical anisotropy. If the applied field oscillates, a time-dependent modulation of the polarization of the light transmitted by the device will result. Modulators of this sort include photoelastic modulators (PEM) [30,31], Faraday cells [32], Kerr cells [32], and Pockel cells.

With a PEM, imposition of an oscillatory stress is used to induce an oscillating birefringence. One of the earliest and most successful designs was described independently by Jasperson *et al.* [31] and Kemp [30]. This device is pictured in Figure 8.6 and consists of a rectangular block of fused quartz (other amorphous materials are used, depending on the desired wavelength), which is attached to a piezoelectric driver. The application of an ac voltage on the driver induces a sinusoidal stress birefringence within the block. The block is mechanically mounted in such a way that a standing stress wave is induced by the driver that is tuned to the natural frequency of vibration of the block. The retardation produced by the modulator is then given by

$$\delta'_{PEM} = A \sin\Omega t , \tag{8.22}$$

where $A = (2\pi\Delta n'_{max} d)/\lambda$ and $\Delta n'_{max}$ is the amplitude of the induced birefringence.

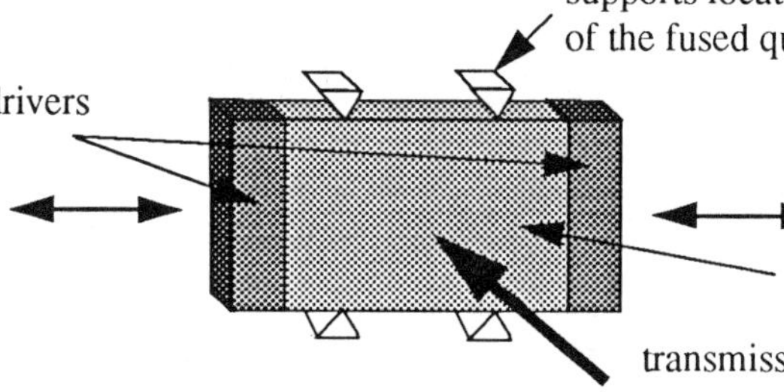

Figure 8.6 The photoelastic modulator.

Typically quite large retardations can be achieved ($\delta'_{max} > \pi$) and at frequencies in the range from 30 to 100 kHz. These frequencies are sufficiently high to permit most transient rheological responses to be followed. Other, simpler photoelastic modulators have been used in the past and one good example is the acoustically driven glass beam developed by Janeschitz-Kriegl and coworkers [33]. This device consists of a thin bar of strain-free glass onto which strips of glass are glued along the sides. The beam is supported at its nodal points and natural frequencies of oscillation in the range of 1 kHz are driven using a small acoustic speaker. The magnitude of the retardation produced with this device, however, is quite small and as will become apparent later on, it is usually desirable to achieve higher retardations.

The Faraday effect refers to the induction of circular birefringence following the application of a magnetic field upon an otherwise isotropic material. Such materials have gyrotropic symmetry and can be represented by the following constitutive relation for the dielectric tensor [3]:

$$\underline{\varepsilon} = \begin{bmatrix} \varepsilon & i\kappa\varepsilon & 0 \\ -i\kappa\varepsilon & \varepsilon & 0 \\ 0 & 0 & \varepsilon_z \end{bmatrix}. \tag{8.23}$$

The coefficient κ is a measure of the Faraday effect and is proportional to the applied magnetic field. For plane waves propagating along the z axis, the components of the electric field can be shown to be

$$\begin{bmatrix} E_x \\ E_y \end{bmatrix} = e^{i\alpha z}\mathbf{R}(\beta z) \cdot \begin{bmatrix} E_{0x} \\ E_{0y} \end{bmatrix}, \tag{8.24}$$

where $\alpha = \left(k\sqrt{1+\sqrt{1-\kappa^2}}\right)/\sqrt{2}$ and $\beta = \left(k\sqrt{1-\sqrt{1-\kappa^2}}\right)/\sqrt{2}$.

Operationally, an ideal Faraday cell driven by an AC magnetic field can be represented as an oscillating optical rotator and would be described by the Jones and Mueller matrices given by equations (I.16). In practice, however, the glass rods which are normally used have nonzero residual birefringence. Faraday cells are most often placed either immediately preceding or following a polarizer so that an effective departure in the orientation angle of the polarizer of $\Delta\theta(t) = \phi_0 + \phi_1 \sin(\Omega t)$ is produced. Here ϕ_0 is the rotation due to the residual birefringence, ϕ_1 is the amplitude of the Faraday effect, and Ω is the frequency of the magnetic field. The magnitudes of the orientations that can be produced are typically quite small. Descriptions of the use of such modulators can be found in the books by Azzam and Bashara [5] and O'Konski [32].

The Kerr effect is the result of applying an electric field to produce birefringence. This phenomena is commonly observed for both colloidal and polymeric liquids and is used in the characterization of the structure of these materials. Alternatively, by using an AC electric field, a modulation of the polarization of light can be affected. Such devices have rarely been used as modulators but do have the potential of allowing higher frequencies than the more common photoelastic devices.

The Pockel's effect [3] refers to an electro-optical process wherein the application of large electric fields onto crystals lacking a center of symmetry can lead to nonlinear polarization effects and optical rotation. Pockel cells can be used in place of photoelastic modulators and can achieve very high modulation frequencies but often have the undesirable property of a nonzero birefringence in the absence of an applied field.

The designs discussed in this monograph are restricted to the use of PEMs, and the Mueller and Jones matrix for this device are simply obtained by replacing the retardation in (I.9) by $A\sin\Omega t$. Various optical arrangements for PSGs and PSAs based on PEM's are found in tables 8.2 and 8.2.

As in the case of rotary modulators, field effect modulators can highlight specific components of the vectors $\mathbf{S}_{PSG}$ and $\mathbf{S}_{PSA}$. Because the retardation is sinusoidal, the

Stokes vector will contain components of the form $\sin(A\sin\Omega t)$ and $\cos(A\sin\Omega t)$. Consequently, the temporal response is rather complicated. To complete the analysis of the Fourier content of the signal the following expansions are required:

$$\cos(A\sin\Omega t) = J_0(A) + 2\sum_{m=1}^{\infty} J_{2m}(A)\cos(2m\Omega t), \tag{8.25}$$

$$\sin(A\sin\Omega t) = 2\sum_{m=0}^{\infty} J_{2m+1}(A)\sin[(2m+1)\Omega t], \tag{8.26}$$

where $J_m(A)$ are Bessel functions and depend on the amplitude of the modulated retardation. These coefficients are obtained by simple calibration procedures, but it must be noted that they depend on wavelength. This latter point is particularly important for spectroscopic applications. When a monochromatic light source is used, it is normally an advantage to adjust the magnitude of the modulation so that $J_0(A) = 0$. Ideally, this occurs at the first zero of this Bessel function so that $A \approx 2.4$.

8.4 Polarimeter Designs Based on Polarization Modulation

In the following subsections, designs are offered that have been formulated to measure specific combinations of Mueller matrix components.

8.4.1 Linear Dichroism Measurements

A sample containing linear dichroism, $\Delta n''$, at an angle θ'' can be analyzed using the designs pictured in Figure 8.7 [34,35].

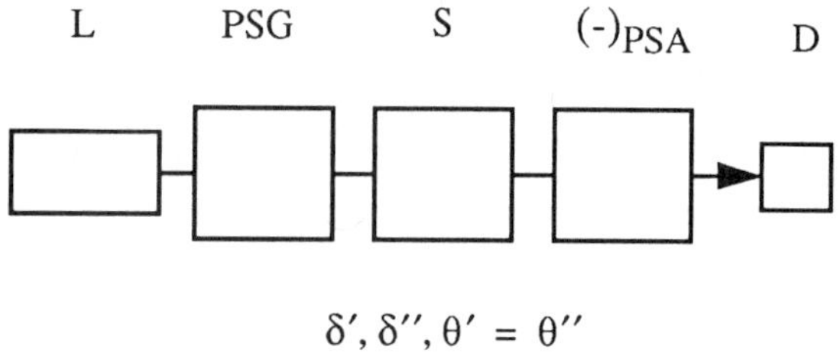

Figure 8.7 Polarimeter for linear dichroism measurements. Any of the PSGs listed can be used along with $(-)_{PSA}$.

As noted in the figure, these designs can accommodate samples with coaxial birefringence and dichroism. It can be shown that the presence of coaxial birefringence will not appear in the measured signals. All of the designs produce signals of the form,

$$\frac{I}{(I_0/p)} = \underline{C_{\delta''}(B + D\cos 2\Omega t)} - (E + F\cos 2\Omega t + G\cos 4\Omega t)\, S_{\delta''}c_{2\theta''} - (H\sin\Omega t + K\sin 2\Omega t + L\sin 4\Omega t)\, S_{\delta''}s_{2\theta''} \tag{8.27}$$

where the constants are supplied in table 8.4.

Of these designs, only the $(P/PEM)_{PSG}$ fails to produce a signal that is proportional to the term $S_{\delta''}s_{2\theta''}$. Since it only returns the term $S_{\delta''}c_{2\theta''}$, there is insufficient information to simultaneously determine the extinction, δ'', and the orientation angle, θ''. For that reason, it can be used only in cases where θ'' is known. In this case, the optimal signal is obtained by setting the orientation of the sample to $\theta'' = 45°$ relative to the initial polarizer of the PSG. It is evident that the coefficient B is simplified by setting the modulation amplitude so that $J_0(A) = 0$ when a PEM is used.

TABLE 8.4 Constants for the Intensity Signal for the Dichroism Polarimeter

PSG	*p*	*B*	*D*	*E*	*F*	*G*	*H*	*K*	*L*
(P/RH)	2	1	0	0	0	1	0	0	1
(P/RP)	4	1	1	1/2	1	1/2	1	1	1/2
(P/Q/RP)	4	1	0	0	1	0	1	1	0
(P/RQ)	2	1	0	1/2	0	1/2	0	0	1/2
(P/PEM)	2	M	0	0	N	0	0	0	0
(P/PEM/Q)	2	M	0	0	N	0	P	0	0

where $M = 1 + J_0(A)\, T_{\delta''}c_{2\theta''}$, $N = 2J_2(A)$, and $P = 2J_1(A)$.

Once the intensity has been measured, it is left to analyze the Fourier content of the signal to extract the coefficients multiplying the desired terms, $S_{\delta''}c_{2\theta''}$ and $S_{\delta''}s_{2\theta''}$. It is also necessary to isolate the underlined term in equation (8.27) for the purposes of normalization so that the coefficient $(I_0 C_{\delta''})/p$ can be eliminated.

Two methods are used to perform the Fourier analysis. If the signal is digitally recorded over at least one period of the lowest frequency, a fast Fourier transform (FFT) can be applied [36]. This will produce the time independent, or "DC" component, $(BC_{\delta''} - ES_{\delta''}c_{2\theta''})$, and the coefficients multiplying the harmonics of the signal (the terms multiplying $\cos 2\Omega t$, $\cos 4\Omega t$, $\sin\Omega t$, $\sin 2\Omega t$, and $\sin 4\Omega t$).

The second approach is to send the signal to analog, phase sensitive detectors, or lock-in amplifiers, that will analyze the harmonic content. The "DC" component, in turn, can be extracted using a low pass filter.

For example, if the $(P/RH)_{PSG}$ is used, extraction of the terms

$$
\begin{aligned}
I_{DC} &= I_0 C_{\delta''}/2 \\
I_{c_{4\Omega t}} &= -I_{DC} T_{\delta''} c_{2\theta''}
\end{aligned}
\tag{8.28}
$$

can be accomplished by measuring the "DC" signal along with the coefficients multiplying $\cos 4\Omega t$ and $\sin 4\Omega t$. Normalization by the "dc" term leads to the following ratios,

$$
\begin{aligned}
R_{c_{4\Omega t}} &= -T_{\delta''} c_{2\theta''}, \\
R_{s_{4\Omega t}} &= -T_{\delta''} s_{2\theta''},
\end{aligned}
\tag{8.29}
$$

where $T_{\delta''} = \tanh \delta''$. These ratios can be used to solve for the extinction and orientation angle as,

$$
\begin{aligned}
\delta'' &= -\mathrm{sgn}\,(R_{c_{4\Omega t}}) \tanh^{-1}\left(\sqrt{R^2_{c_{4\Omega t}} + R^2_{s_{4\Omega t}}}\right), \\
\theta'' &= \frac{1}{2}\tan^{-1}(R_{s_{4\Omega t}}/R_{c_{4\Omega t}}).
\end{aligned}
\tag{8.30}
$$

An identical procedure can be used for the $(P/Q/RP)_{PSG}$ with the exception that the harmonic content at the frequency 2Ω must be used.

If the $(P/PEM/Q)_{PSG}$ is used, and if $J_0(A) = 0$, a Fourier analysis of the signal yields the coefficients,

$$
\begin{aligned}
I_{DC} &= ((I_0 C_{\delta''})/4), \\
I_{c_{4\Omega t}} &= -2I_{DC} J_2(A)\, T_{\delta''} c_{2\theta''}, \\
I_{s_{4\Omega t}} &= -2I_{DC} J_1(A)\, T_{\delta''} s_{2\theta''},
\end{aligned}
\tag{8.31}
$$

which can be used to solve for the extinction and orientation angle.

The $(P/RP)_{PSG}$ and $(P/RQ)_{PSG}$ designs differ from the other designs due to the presence of the coefficient E in table 8.4. Since this term is time independent, the "DC" component will unavoidably contain some influence of the sample dichroism. This can adversely affect the ratios of the harmonics with the "DC" component since the numerator will no longer be a constant. In the majority of cases, however, $|\delta''| \ll 1$, and this complication can be neglected. Assuming this to be true, for either system,

$$\begin{aligned} I_{\mathrm{DC}} &= ((I_0 C_{\delta''})/p), \\ I_{c_{4\Omega t}} &= -(1/2)\, I_{\mathrm{DC}} T_{\delta''} c_{2\theta''}, \\ I_{s_{4\Omega t}} &= -(1/2)\, I_{\mathrm{DC}} T_{\delta''} s_{2\theta''}, \end{aligned} \tag{8.32}$$

and the extinction and orientation angle can be obtained simultaneously.

8.4.2 Linear Birefringence Measurements

A sample characterized by a retardation, δ', an extinction, δ'', and coaxial orientation angle, $\theta = \theta' = \theta''$, can be analyzed by the optical arrangement shown below [34,37,38].

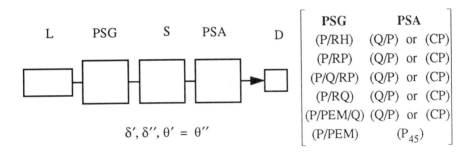

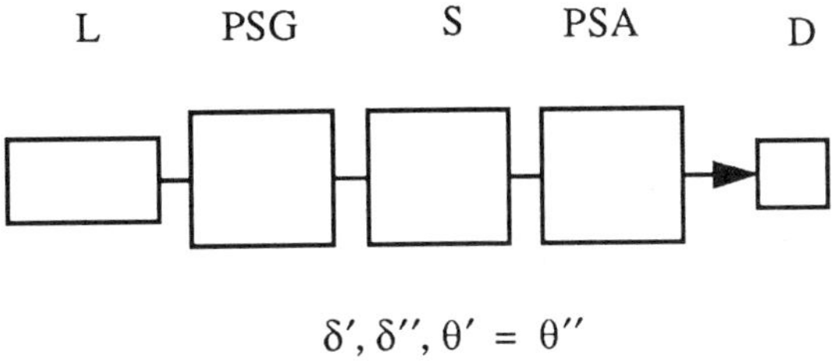

PSG	PSA
(P/RH)	(Q/P) or (CP)
(P/RP)	(Q/P) or (CP)
(P/Q/RP)	(Q/P) or (CP)
(P/RQ)	(Q/P) or (CP)
(P/PEM/Q)	(Q/P) or (CP)
(P/PEM)	(P_{45})

Figure 8.8 Polarimeter for linear birefringence measurements.

For each of these optical trains, except the $(P/PEM)_{PSG}$ system, the intensity measured at the detector is

$$\begin{aligned} \frac{I}{(I_0/(2p))} &= C_{\delta''}(B + D\cos 2\Omega t) \\ &\quad - (E + F\cos 2\Omega t + G\cos 4\Omega t)\,(s_{\delta'} s_{2\theta} + S_{\delta''} c_{2\theta}) \\ &\quad - (H\sin\Omega t + K\sin 2\Omega t + L\sin 4\Omega t)\,(-s_{\delta'} c_{2\theta} + S_{\delta''} s_{2\theta}), \end{aligned} \tag{8.33}$$

where the constants $p, B, D, \ldots, L$ are again supplied in table 8.4 and a right circular polarizer has been used as the PSA.

In the absence of dichroism, this equation is similar in form to equation (8.27) and the extraction of the retardation and orientation angle follows the same strategy as described for the measurement of dichroism in the previous subsection. However, when dichroism cannot be neglected, the measured signal will depend on both optical anisotropies. In this case, two separate measurements will be necessary: one using the optical trains of the previous section (which simply has the PSA section of the instrument replaced by $(-)_{PSA}$) to determine the dichroism, and a second measurement to obtain the combined effect of $\Delta n'$ and $\Delta n''$. Once these two measurements are available, the difference between the two signals uniquely determines the birefringence [37,40].

The $(P/PEM)_{PSG}$ produces the following intensity,

$$\frac{I}{(I_0/4)} = C_{\delta''} - 2J_2(A)\,c_{2\Omega t}S_{\delta''}c_{2\theta} - S_{\delta''}s_{2\theta} + 2J_2(A)\,c_{2\Omega t}s_{2\theta}c_{2\theta}(C_{\delta''} - c_{\delta'}) + 2J_1(A)\,s_{\Omega t}c_{2\theta}s_{\delta'}. \tag{8.34}$$

Unlike the previous designs, this arrangement is not convenient when dichroism is not negligible. This is because the contribution of dichroism is not easily subtracted out as it is with equation (8.33). When dichroism is set to zero we have

$$\begin{aligned}\frac{I}{(I_0/4)} &= 1 + 2J_2(A)\,c_{2\Omega t}s_{2\theta}c_{2\theta}(1 - c_{\delta'}) + 2J_1(A)\,s_{\Omega t}c_{2\theta}s_{\delta'} \\ &= 1 + J_2(A)\,R_{2\Omega t}c_{2\Omega t} + 2J_1(A)\,R_{\Omega t}s_{\Omega t}.\end{aligned} \tag{8.35}$$

The following ratios can be measured by analyzing the Fourier content of the signal and with the knowledge of the calibration constants, $J_1(A)$ and $J_2(A)$:

$$\begin{aligned}R_{2\Omega t} &= s_{4\theta}(1 - c_{\delta'}), \\ R_{\Omega t} &= c_{2\theta}s_{\delta'}.\end{aligned} \tag{8.36}$$

From these expressions, the orientation angle can be found using,

$$\cos\left[4\theta - \tan^{-1}(R_{2\Omega t}/R^2_{\Omega t})\right] = \frac{2R^2_{\Omega t} + R^2_{2\Omega t}}{2\sqrt{R^4_{\Omega t} + R^2_{2\Omega t}}}. \tag{8.37}$$

Once θ is known, the expression in equation (8.36) for $R_{2\Omega t}$ can be rearranged to give the following expression for the retardation:

$$\sin^2\left(\frac{\delta}{2}\right) = \frac{R_{2\Omega t}}{2\sin 4\theta}. \tag{8.38}$$

In general, this configuration is not preferred for birefringence measurements except if the orientation angle is known *a priori*. In that case the sample orientation should be set to $\theta = 0$ to produce the simplest result.

In addition to offering a form that allows for an easier subtraction of the dichroism, the resulting signals in equation (8.27) give retardation terms in the form of $\sin\delta' \sin 2\theta$ and $\sin\delta' \cos 2\theta$. As seen previously in the discussion of the stress optical rule, these particular combinations of retardation and the orientation angle are very similar to the expressions for the shear stress and the first normal stress difference. Indeed, in the limit of small retardations, the results produced by equation (8.27) are precisely proportional to these stresses. This feature of the optical designs leading up to equation (8.27) can make the results of experiments easier to interpret when the stress optical rule applies.

8.4.3 Linear Birefringence and Linear Dichroism: Coaxial and Noncoaxial Materials

The fact that equation (8.27) contains terms in dichroism identical to those in equation (8.33) suggests the scheme depicted in Figure 8.9 for the purpose of simultaneously measuring birefringence and dichroism [37]. Here any of the PSGs in Figure 8.7 can be used with the exception of the $(P/PEM)_{PSG}$ design. Following the sample, the light is sent to a beam splitter that reflects approximately one half of the light to detector D2 and transmits the remaining light to detector D1. The beam splitter for this purpose can be a partially silvered mirror that is inserted into the light path at a sufficiently small angle of incidence so that it does not affect the polarization of the light. For detector D2, the $(\text{-})_{PSA}$ is used so that this detector primarily responds to dichroism. Detector D1 receives the light following either a $(Q/P)_{PSA}$ or a $(CP)_{PSA}$ and produces a signal that is a mixture of birefringence and dichroism information. Indeed, when the material has a coaxial birefringence and dichroism, these detectors produce intensities given by equations (8.27) and (8.33) for D2 and D1, respectively.

When the material is noncoaxial, the optical train in Figure 8.9 can still be used and provides sufficient information to solve for all four unknowns, $\Delta n'$, $\Delta n''$, θ', and θ''. To demonstrate this capability, the analysis shall be worked out for the $(P/RH)_{PSG}$ design. The Mueller matrix for a noncoaxial birefringent and dichroic material was developed in section 2.4.7 and is given by equation (I.18) with $\delta'_{circ} = \delta''_{circ} = 0$. The intensities produced by detectors D1 and D2 are

$$\frac{I_1}{I_0^1/4} = (m_{11}^S - m_{41}^S) + (m_{12}^S - m_{42}^S)\, c_{4\Omega t} + (m_{13}^S - m_{43}^S)\, s_{4\Omega t}, \tag{8.39}$$

and

$$\frac{I_2}{(I_0^2/4)} = m_{11}^S + m_{12}^S c_{4\Omega t} + m_{13}^S s_{4\Omega t}, \tag{8.40}$$

where I_0^1 and I_0^2 are the incident light intensities for each detector. The Mueller matrix coefficients, m_{ij}^S, for the sample are given in equation (I.18) of Appendix I with $\delta'_{circ} = \delta''_{circ} = 0$. In principle, measurement of the "in phase" and "out-of-phase" components of these intensities at the frequency 4Ω will provide sufficient information to uniquely determine δ', δ'', θ', and θ'' and in the case where $\theta'' = \theta'$, the analysis is particularly simple. Other limiting cases are very useful in practice [37] and are presented below.

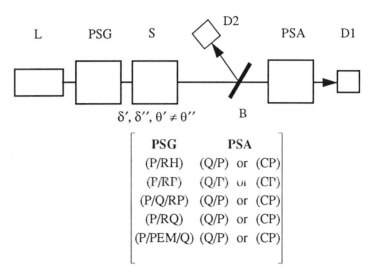

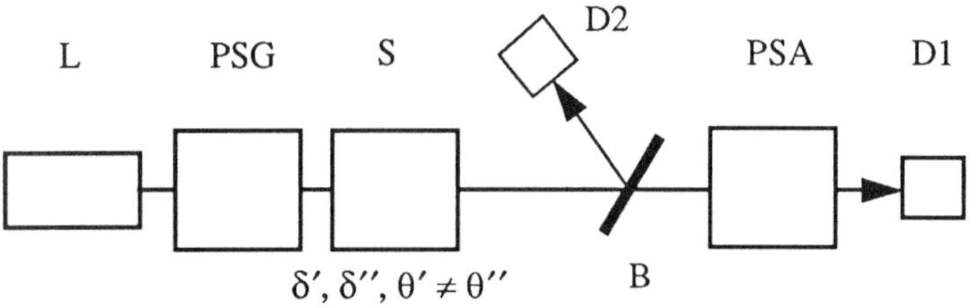

PSG	PSA
(P/RH)	(Q/P) or (CP)
(P/RP)	(Q/P) or (CP)
(P/Q/RP)	(Q/P) or (CP)
(P/RQ)	(Q/P) or (CP)
(P/PEM/Q)	(Q/P) or (CP)

Figure 8.9 Optical train for the measurement of noncoaxial birefringence and dichroism.

8.4.3.1 Case 1: $|\delta'| \ll 1, |\delta''| \ll 1$. If both the retardation and extinction are small, the Mueller matrix takes on the simple form given in equation (I.18). In this limit, the signal measured by detector D2 is

$$\frac{I_2}{I_0^2/2} = 1 - \delta'' c_{2\theta''} c_{4\Omega t} - \delta'' s_{2\theta''} s_{4\Omega t}, \tag{8.41}$$

and only responds to the presence of dichroism. Furthermore, manipulation of the two detector intensities can yield the following expression that depends only on birefringence:

$$\frac{I_1}{I_0^1/4} - \frac{I_2}{I_0^2/2} = 1 + \delta' s_{2\theta'} c_{4\Omega t} - \delta' c_{2\theta'} s_{4\Omega t}. \tag{8.42}$$

Fourier decomposition of these two expressions will provide coefficients that can be used to determine the retardation and extinction of the sample, along with the associated orientation angles.

8.4.3.2 Case 2: $|\delta''| \ll 1, |\delta''| \ll |\delta'|$. There are numerous cases where the dichroism is small in magnitude but is combined with a large birefringence. Examples include colloidal particles dispersed within a birefringent polymer liquid and subject to flow. Orientation of the particles would cause a scattering dichroism and the angle associated with the particles and the polymer will, in general, be different. The result will be a sample possessing noncoaxial birefringence and dichroism and such systems have been studied in detail using this method [37,40]. In this approximation, the detector signals are

$$\frac{I_1}{I_0^1/4} = 1 + s_{\delta'} s_{2\theta'} c_{4\Omega t} - s_{\delta'} c_{2\theta'} s_{4\Omega t}, \tag{8.43}$$

and

$$\frac{I_2}{I_0^2/2} = 1 - \delta'' c_{2\theta''} c_{\delta'/2} \left(\frac{s_{\delta'/2}}{\delta'/2}\right) c_{4\Omega t} - \delta'' s_{2\theta''} c_{\delta'/2} \left(\frac{s_{\delta'/2}}{\delta'/2}\right) s_{4\Omega t}. \tag{8.44}$$

The procedure to extract both the birefringence and dichroism would entail first using the intensity information from D1 to determine the retardation and the angle θ'. These values can then be used in the Fourier decomposition of the intensity from I_2 to obtain δ'' and θ''.

8.4.4 Circular Dichroism Measurements

The measurement of circular dichroism, in the absence of other optical anisotropies, is particularly simple since it only requires extraction of the m_{14}^S Mueller matrix component of the sample. This is achieved using the designs shown in Figure 8.10.

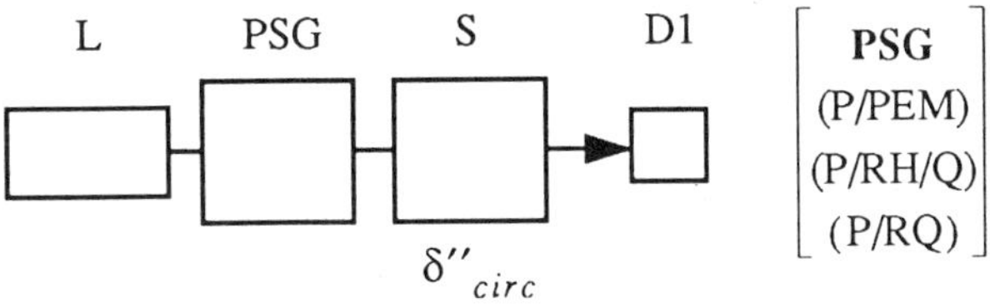

Figure 8.10 Polarimeter design for the measurement of circular dichroism.

The intensity measured at the detector is

$$\frac{I}{(I_0/2)} = C_{\delta''_{circ}} + (A' s_{\Omega t} + B' s_{2\Omega t} + C' s_{4\Omega t})\, S_{\delta''_{circ}}, \tag{8.45}$$

where the constants, A', B', and C' are given in the table below.

TABLE 8.5 Constants for the Circular Dichroism Polarimeter

PSG	A'	B'	C'
(P/RH/Q)	0	0	-1
(P/RQ)	0	1	0
(P/PEM)	$2J_1(A)$	0	0

8.4.5 Full Mueller Matrix Polarimeters

The previous designs targeted a limited set of Mueller matrix components. Inspection of the list of PSGs in table 8.2 indicates that the $(P/RQ)_{PSG}$ and the $(P/PEM_1/PEM_2)_{PSG}$ configurations produce vectors $\mathbf{S}_{PSG}$ with contributions in each of the four components. It is evident that combining these particular PSGs with PSAs that also fill each of the components in $\mathbf{S}_{PSA}$ can extract all of the components of a sample's Mueller matrix [25,41].

The design of a full Mueller matrix polarimeter based on a combination of a $(P/RQ)_{PSG}$ and a $(RQ/P)_{PSA}$ is shown in Figure 8.11. Here two rotary quarter wave plate modulators operating at different angular velocities, Ω_1 and Ω_2, are used. The intensity measured at the detector is of the form,

$$I = \frac{I_0}{4}\left[a_0 + \sum_{i=1}^{12}(a_i\cos\Theta_i + b_i\sin\Theta_i)\right], \tag{8.46}$$

where the angles, Θ_i, are the following combinations of the angular displacements of the two rotating quarter-wave plates, $\Omega_1 t$ and $\Omega_2 t$:

$$\Theta_1 = 2\Omega_1 t \qquad \Theta_7 = 2\Omega_2 t + 4\Omega_1 t$$
$$\Theta_2 = 4\Omega_1 t \qquad \Theta_8 = 4\Omega_1 t - 4\Omega_2 t$$
$$\Theta_3 = 2\Omega_2 t - 4\Omega_1 t \qquad \Theta_9 = 2\Omega_1 t - 4\Omega_2 t$$
$$\Theta_4 = 2\Omega_2 t - 2\Omega_1 t \qquad \Theta_{10} = 4\Omega_2 t$$
$$\Theta_5 = 2\Omega_2 t \qquad \Theta_{11} = 2\Omega_1 t + 4\Omega_2 t$$
$$\Theta_6 = 2\Omega_2 t + 2\Omega_1 t \qquad \Theta_{12} = 4\Omega_1 t + 4\Omega_2 t$$

The constants, a_i and b_i, are the Fourier coefficients of the intensity and are the following combinations of the Mueller matrix components of the sample:

$$u_0 = m_{11} + \frac{1}{2}m_{12} + \frac{1}{2}m_{21} + \frac{1}{4}m_{22},$$

$$a_1 = 0;\ b_1 = m_{14} + \frac{1}{2}m_{24},$$

$$a_2 = \frac{1}{2}m_{12} + \frac{1}{4}m_{22};\ b_2 = \frac{1}{2}m_{13} + \frac{1}{4}m_{23}$$

$$a_3 = -\frac{1}{4}m_{13};\ b_3 = -\frac{1}{4}m_{42},$$

$$a_4 = \frac{1}{2}m_{44};\ b_4 = 0,$$

$$a_5 = 0;\ b_5 = -m_{41} - \frac{1}{2}m_{42},$$

$$a_6 = -\frac{1}{2}m_{44};\ b_6 = 0,$$

$$a_7 = \frac{1}{4}m_{13};\ b_7 = -\frac{1}{4}m_{42},$$

$$a_8 = \frac{1}{8}m_{22} + \frac{1}{8}m_{33};\ b_8 = \frac{1}{8}m_{23} - \frac{1}{8}m_{32},$$

$$a_9 = -\frac{1}{4}m_{34};\ b_9 = \frac{1}{4}m_{24},$$

$$a_{10} = \frac{1}{2}m_{21} + \frac{1}{4}m_{22};\ b_{10} = \frac{1}{2}m_{31} + \frac{1}{4}m_{32},$$

$$a_{11} = \frac{1}{4}m_{34};\ b_{11} = \frac{1}{4}m_{24},$$

$$a_{12} = \frac{1}{8}m_{22} - \frac{1}{8}m_{33};\ b_{12} = \frac{1}{8}m_{23} + \frac{1}{8}m_{32}.$$

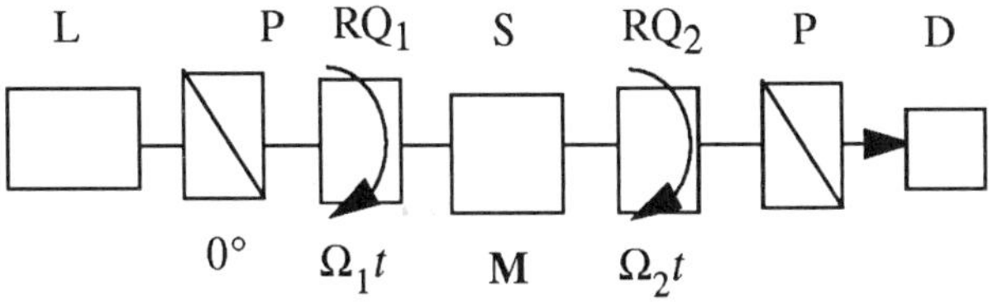

Figure 8.11 Full Mueller matrix polarimeter using rotating quarter-wave plates.

In this scheme, analog Fourier decomposition of the signal using lock-in amplifiers becomes impractical and a digital fast Fourier transformation is preferred. Such an analysis would lead to 22 linear combinations of the 16 Mueller matrix components of the sample. This is an overspecified system, and the extra information can be used as either an internal check of consistency or discarded. It is important to make proper choices of the angular velocities, Ω_1 and Ω_2. Clearly, these frequencies cannot be simple multiples of one another and a common choice is the fractional relationship, $\Omega_1 = 4\Omega_2/5$.

In practice, the quarter-wave plates will possess some imperfection in retardation. Furthermore, the phase angle of the plates may not be zero relative to the mechanical rotation device. Both sources of error can be taken into account. For example, an imperfect quarter-wave plate with retardation $\delta' = \pi/2 + \beta$ and a phase offset of ϕ would produce the following Stokes vector for the $(\mathrm{P/RQ})_{PSG}$,

$$\mathbf{S}_{PSG} = \frac{I_0}{2}\begin{bmatrix} 1 \\ \frac{1}{2}(1+\beta) + \frac{1}{2}c_{4\Omega t}(1-\beta) - 2\phi s_{4\Omega t} \\ \frac{1}{2}s_{4\Omega t}(1-\beta) + 2\phi c_{4\Omega t} \\ s_{2\Omega t} + 2\phi c_{2\Omega t} \end{bmatrix}, \tag{8.47}$$

where it has been assumed that $|\beta| \ll 1$ and $|\phi| \ll 1$. Similarly, an imperfect quarter-wave plate used in a $(\mathrm{RQ/P})_{PSA}$ analyzes the following Stokes vector,

$$\mathbf{S}_{PSA} = \frac{I_0}{2}\begin{bmatrix} 1 \\ \frac{1}{2}(1+\beta) + \frac{1}{2}c_{4\Omega t}(1-\beta) - 2\phi s_{4\Omega t} \\ \frac{1}{2}s_{4\Omega t}(1-\beta) + 2\phi c_{4\Omega t} \\ -(s_{2\Omega t} + 2\phi c_{2\Omega t}) \end{bmatrix}. \tag{8.48}$$

Using these vectors, the intensity signal can be readily calculated for a sample with an arbitrary Mueller matrix and the result is of the same form as equation (8.46). The coefficients a_i and b_i will again be linear combinations of the sample's Mueller matrix components, but will be slightly more complex due to the appearance of the constants β and ϕ. These imperfection constants are easily obtained by removing the sample and analyzing the Fourier content of the signal produced by the two rotating quarter-wave plates.

A full Mueller matrix polarimeter can also be fashioned from the $(\mathrm{P/PEM_1/PEM_2})_{PSG}$ and the $(\mathrm{PEM_3/PEM_4/P})_{PSA}$ since these designs, like the rotating quarter-wave plate, produce Stokes vectors with all of the elements filled with linearly independent Fourier modes. This polarimeter is pictured below and is analyzed in detail in reference 42.

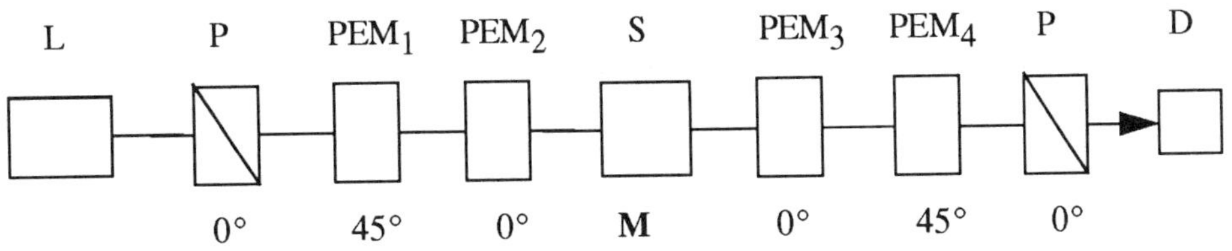

Figure 8.12 Full Mueller matrix polarimeter using photoelastic modulators.

8.5 Design of Scattering Experiments

8.5.1 Wide-Angle Scattering Experiments

In Chapter 4 the equations describing the scattering of light were developed. Regardless of the approximations that were used in the description of the scattering system, the observables in such a measurement were shown to be projections of the structure onto the scattering vector, $\mathbf{q} = \mathbf{k}_s - \mathbf{k}_i$. Because of the role of this vector in any scattering experiment, it is critical that its orientation relative to the principal axes of deformation and orientation induced using an external field be systematically controlled.

In Figure 8.13, a typical schematic diagram of a light scattering experiment is shown [8,43]. A laser is normally used as the light source and is essential for dynamic light scattering experiments where coherent light is required. A lens is then used to focus the beam at a point within the sample. Light scattered at an angle θ is observed through two pinholes. Alternatively, a lens can be used in place of the first pinhole. Such a lens would be positioned so that it images the scattering volume, V, onto the second pinhole, B. Using either scheme, the scattering volume will be proportional to

$$V \approx d^2 D (\sin\theta)^{-1}, \tag{8.49}$$

where d and D are respectively the diameters of the incident and scattered beams. The geometric correction factor, $(\sin\theta)^{-1}$, must be applied to the intensities measured in total intensity light scattering experiments.

The other geometric consideration is the relative orientation of the scattering vector, $\mathbf{q}$, and an applied field, such as a flow or electric field. For example, as seen in Chapter 6, the use of homodyne dynamic light scattering to measure velocity gradients requires that $\mathbf{q}$ be oriented parallel to the component of the velocity of interest. This is simply accomplished by having that velocity component directed parallel to the vector bisecting $\mathbf{k}_i$ and $\mathbf{k}_s$ [44].

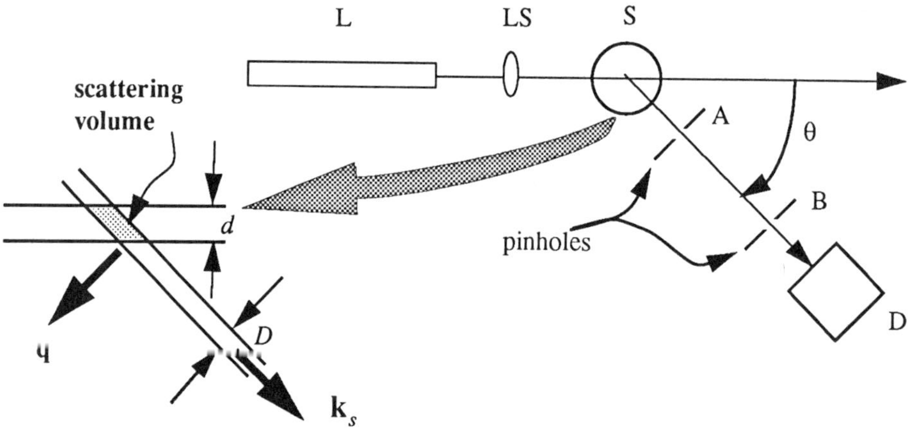

Figure 8.13 Wide-angle light scattering experiment and the scattering volume.

When using total intensity light scattering to measure the radius of gyration or the structure factor in the presence of an applied field, it is critical that the magnitude of the scattering vector be varied in a manner preserving the relative orientation of $\mathbf{q}$ and the field. In a uniaxial field, such as an electric or elongational flow field, the principal directions of deformation and orientation of the sample will be known. In this case, all that is needed is a method of simultaneously varying the orientations of $\mathbf{k}_i$ and $\mathbf{k}_s$ to keep $\mathbf{q}$ aligned with the field (or perpendicular to it if the structure orthogonal to the field is desired).

In the case of a simple shearing flow, however, the mean angle of deformation and orientation of the microstructure may not be known. The experimental protocol requires that the scattering vector be kept within a chosen plane of the flow (the plane containing the flow and the velocity gradient directions, for example) and then the magnitude and orientation of $\mathbf{q}$ can be varied systematically. By maintaining the magnitude of $\mathbf{q}$ fixed and varying its orientation while maintaining this vector in the chosen plane, the mean angle of deformation and orientation of the structure can be obtained. This is accomplished by locating the orientation of the maxima and minima in the intensity of scattered light as a function of the orientation of $\mathbf{q}$. As shown in Figure 8.14, the mean angle of orientation of a macromolecule corresponds to a minima in the intensity. Such a procedure is documented in reference 46.

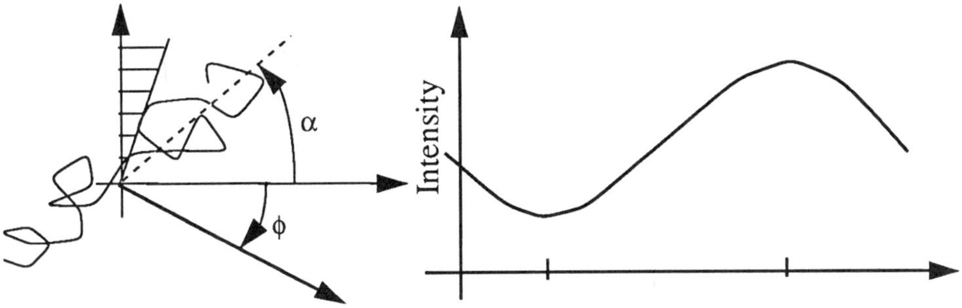

Figure 8.14 Determination of the principal axes of deformation and orientation in a total intensity light scattering experiment.

8.5.2 Small-Angle Light Scattering (SALS)

By definition, in a SALS experiment, information is gathered at small values of the scattering angle, θ. This results in two important simplifications in the experimental procedure. First, the light collected is confined to a small cone of scattering angles about the incident beam and can be gathered conveniently over the full range of desired angles, θ and φ, as shown in Figure 8.15. Early SALS experiments [47] simply photographed the pattern of scattered light falling on the screen A. It is required to eliminate the incident beam with either a beam stop or by inserting an aperture into the screen so that this light simply passes through the screen.

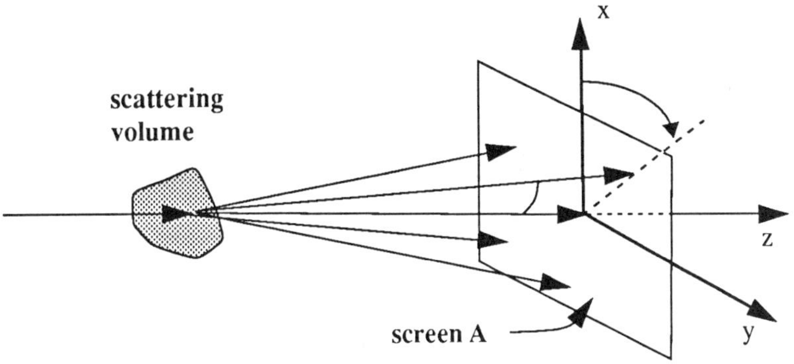

Figure 8.15 Small-angle light scattering geometry.

To obtain quantitative measurements of the scattered light, two-dimensional (2D) detectors can be used. Two optical trains for this purpose are shown in Figure 8.16. The train in Figure 8.16a simply captures the scattered light intensity falling onto a screen below the sample [48]. An aperture in the screen serves to eliminate the incident light. Because the 2D detector camera views the scattering pattern at an angle relative to the incident beam, a simple linear transformation of the data must be performed to remove any distortion caused by this geometric effect. This configuration has the distinct advantages over the scheme outlined in Figure 8.16b of simplicity and providing access to the incident, unscattered beam. This latter advantage allows for simultaneous polarimetry measurements of birefringence and dichroism by using the unscattered beam. Simultaneous turbidity measurements can also be made. In addition, the range of the scattering angle θ that is measured at the detector can be varied by simply adjusting the distance L from the sample to the screen. For the arrangement in Figure 8.16b, it is necessary to vary the focal length of the lenses to adjust the range of θ that is observed. It should be noted, however, that the light falling onto a point on the screen is the integrated effect of light scattered along the entire length of the sample, d. This effect, however, can be minimized by ensuring that the ratio of the length scales, L/d, is sufficiently large (see figure 8.16).

The arrangement described in Figure 8.16b measures the scattered light by sending it directly to the 2D detector [49,50]. This is accomplished by collecting the scattered light using lens L1. The light is then imaged onto a pinhole which ensures that the camera only receives light emitted from a localized spot with the sample. Following the pinhole, a second lens collimates the scattered light and sends it to the detector.

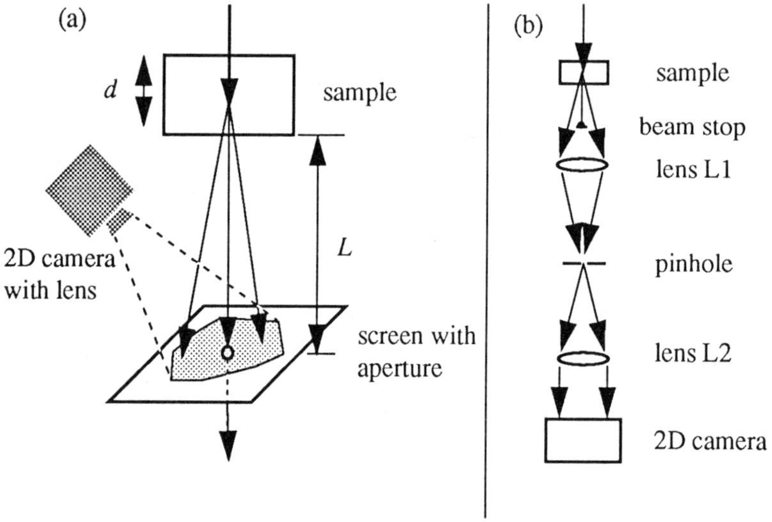

Figure 8.16 Small-angle light scattering arrangements.

The second simplification offered by SALS is that the scattering vector, $\mathbf{q}$, is primarily located in the plane of the screen, orthogonal to $\mathbf{k}_i$. This is due to the inherently small values of the scattering angle, θ, so that

$$\frac{\mathbf{q} \cdot \mathbf{k}_i}{|\mathbf{q}||\mathbf{k}_i|} \approx \frac{\theta}{2} \ll 1 . \tag{8.50}$$

This geometric simplification is an important advantage for experiments involving an external field. This is because it is possible to ensure that the scattering vector consistently remains in one plane relative to the orientation of the field as the scattering angle, θ, is varied over its range of interest.

8.6 Raman Scattering

The schematic diagram in Figure 8.17 shows the basic optical arrangement used for measuring Raman scattered light when a monochromator is used to analyze the spectrum that is emitted. Because Raman scattering is a second-order effect, a laser source with sufficient power (> 0.5W) is normally used. The choice of the wavelength of the laser is important, and care should be taken to ensure that it is removed from absorption and fluorescence regions of the sample. The latter can often be avoided by using sources in the near-infrared. By avoiding regions of absorption, Stokes and anti-Stokes Raman scattered light is produced instead of stimulated Raman scattering.

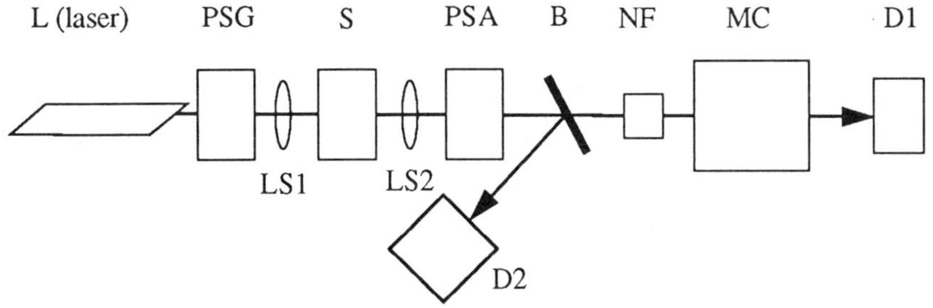

Figure 8.17 Optical train for a monochromator-based Raman scattering experiment.

Following the light source, a PSG is used to define the polarization of the light sent to the sample. In early experiments, this simply consisted of a polarizer set to specific orientations relative to the principal axes of the sample [17,24]. More recently, modulation techniques have been applied to this method [22] and are the focus of the discussion here.

The light scattered and transmitted by the sample is collected by lens LS2 and passes through a PSA . A beam splitter sends light to either detector D2 , for the purpose of birefringence measurements, or to the monochromator, MC . Before reaching the monochromator, a notch filter, NF , is used to reject light at and close to the frequency of the incident light. The light passed by this filter, therefore, contains the Raman scattered radiation. The monochromator selects a specific range of wavelengths for measurements and the intensity of this light is measured by the detector, D1 . Because the Raman scattered light is weak in its intensity, a photomultiplier tube is usually used for D1 .

A variety of choices are available for the PSG section of this experiment. As before, the selection is based on the Mueller matrix components of the sample that are sought. A convenient arrangement that has been used for samples subject to uniaxial deformation is described in detail in reference 22 and collects the Raman scattered light in the forward direction. The PSG section of the instrument consists of a polarizer oriented at zero degrees, and a photoelastic modulator at 45° . Following the sample, the PSA section consists of a polarizer oriented at 45° . The signal measured at the photomultiplier tube was shown to have the form:

$$\frac{I}{\eta^2 I_0} = \frac{1}{4}[R_{dc} + 2J_1(A)R_{\omega t}\sin(\omega t) + 2J_2(A)R_{2\omega t}\cos(2\omega t)], \tag{8.51}$$

where I_0 is the incident light intensity. The explicit form for the Fourier coefficients, $R_{\omega t}$ and $R_{2\omega t}$, will depend on the form of the Raman tensor describing the sample. For the simple choice that $\alpha'_{ij} = \eta(\delta_{ij} + \varepsilon u_i u_j)$, these coefficients are

$$R_{dc} = 1 + \varepsilon(\langle u_x^2\rangle + \langle u_z^2\rangle) + \frac{\varepsilon^2}{2}(\langle u_x^4\rangle + \langle u_z^4\rangle + 2\langle u_x^2 u_z^2\rangle) \tag{8.52}$$

$$R_{2\omega t} = \varepsilon(\langle u_x^2\rangle - \langle u_z^2\rangle) + \frac{\varepsilon^2}{2}(\langle u_x^4\rangle - \langle u_z^4\rangle), \tag{8.53}$$

$$R_{\omega t} = s[P(\beta, \delta')]\left(S\left(\frac{\beta\delta'}{2}\right)\{1 + \varepsilon(\langle u_x^2\rangle + \langle u_z^2\rangle) + \varepsilon^2\langle u_x^2 u_z^2\rangle\} + \varepsilon^2\frac{s\left(\frac{\beta\delta'}{2}\right)}{P(\beta, \delta')}\langle u_x^2 u_z^2\rangle\right) \tag{8.54}$$

where $\delta' = 2\pi\Delta n' d/\lambda_0$ is the retardation based on the incident wavelength, $\Delta n'$ is the birefringence, $P(\beta, \delta') = \delta'\left(1 - \frac{\beta}{2}\right)$, $S(x) = \frac{\sin x}{x}$, $s(x) = \sin x$, and $\beta = (1 - \lambda_0/\lambda_1)$. The wavelength of the Raman scattered light is λ_1. In making the calculation, the deformation is taken to be along the x direction, and the light is propagating along the y axis.

9 Selection and Alignment of Optical Components

The successful design of an optical experiment requires the correct choice and installation of its component parts. This chapter discusses the principles of operation, fabrication, and alignment of the most common optical elements.

9.1 Polarizing Optical Elements

9.1.1 Polarizers

For light in the visible range, the most commonly used, high-quality polarizers use birefringent crystals in combination with the phenomena of total internal reflection to produce linearly polarized light. The Glan-Thompson polarizer is one such design [51] and is pictured in Figure 9.1. This polarizer consists of two triangular pieces of a transparent crystal arranged in the square configuration shown below with an air gap existing along their mutual interface. The crystalline material (calcite is often used, for example) is characterized by two indices of refraction, n_e and n_o. For the polarizer in Figure 9.1, therefore, light polarized along the y axis will see the refractive index, n_e, whereas horizontally polarized light will respond to n_o.

Such a crystal is "double refracting" since the two values of the refractive index will cause the two polarizations to propagate along different paths. Separation of the two polarizations occurs at the interface where the horizontally polarized light is totally internally reflected. The vertically polarized light, on the other hand, is transmitted across the interface where the second half of the polarizing cube returns this polarized beam to its original path. Typical extinction ratios for this type of polarizer are greater than 10^3 and as high as 10^6. The same concept can be used to fabricate a polarizing beam splitter, which has the attractive feature of producing two separate beams of orthogonal polarizations.

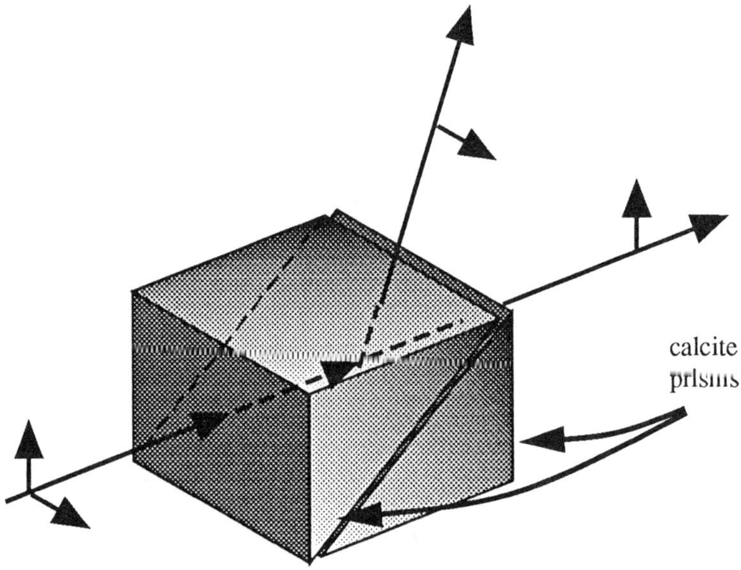

Figure 9.1 Glan-Thompson polarizer.

Also relying on light reflection to polarize light are Brewster angle windows based on the effect discussed in section 1.6. This type of window is also used in most laser cavities to produce linearly polarized light with extinction ratios in the neighborhood of 10^3.

Since the polarizers discussed above involve light reflection combined with the real part of the refractive index tensor, they can be used effectively over a broad spectral range about a central wavelength. Calcite Glan-Thompson polarizers, for example, operate successfully over the entire visible spectrum. When fabricated of crystalline quartz, these polarizers can be used to polarize ultraviolet light as well as visible light.

A second class of polarizers uses dichroism to produce linearly polarized light. Techniques to produce dichroic polarizing sheets were pioneered by Land [52]. These are made by dissolving a strongly dichroic, small molecule into an amorphous, transparent polymer. The dichroic molecules are then oriented by a uniaxial stretching of the polymer matrix. Since this is accomplished below the polymers glass transition temperature, this orientation is "frozen in." Polarizers having extinctions on the order of 10^3 can be produced in this manner. Furthermore, these are relatively inexpensive polarizers to produce and can be fabricated into large sheets which is necessary for photographic applications.

Polarizers for infrared applications include Brewster reflection-effect devices and wire grid polarizers. Figure 9.2 is a schematic of a Brewster angle reflection polarizer. The light is first reflected at the Brewster angle by a mirror. Only horizontally polarized light is reflected to a second mirror (normally made of copper to efficiently conduct away heat from the infrared source). This mirror sends the light to a second Brewster angled mirror, the purpose of which is to further polarize the light and to return it to its original path. These polarizers are characterized by very large extinction coefficients and wide spectral ranges.

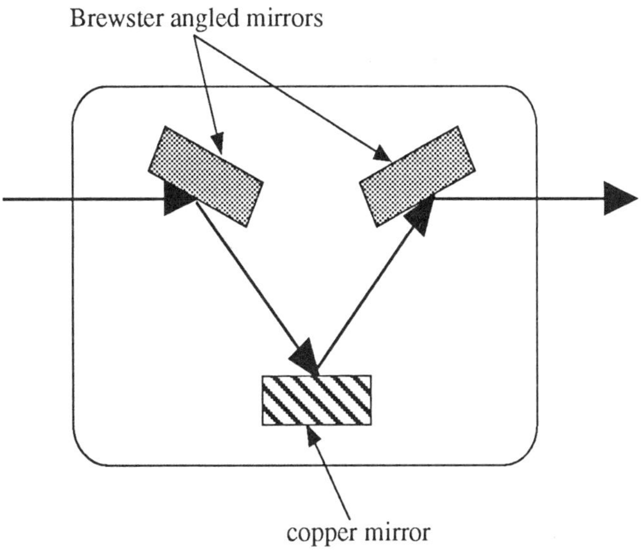

Figure 9.2 Brewster angle polarizer for infra-red light.

The infra-red polarizer described above has the disadvantages of high cost and large physical size. Another device that is commonly used is the wire grid polarizer [51] shown in Figure 9.3. This polarizer is fashioned from an amorphous disk that is transparent in the spectral range of interest. On the front surface of the disk, parallel lines of a conducting material are placed. One method of accomplishing this is to first etch parallel grooves on the disk surface and then deposit the conducting material at a glancing angle. This procedure will tend to accumulate the conductor on the "windward" side of the etched grooves. It is important to note that the spacing of the conducting lines must be comparable to the wavelength of the light at the center of the spectral range of interest.

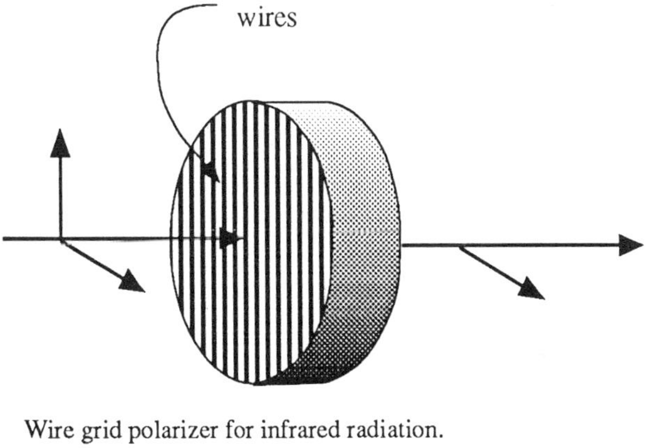

Figure 9.3 Wire grid polarizer for infrared radiation.

As is shown in Figure 9.3, a wire grid polarizer will attenuate light polarized parallel to the conducting wires. This is a result of attenuation of the electric field oscillating parallel to the conductors by generating a current within the lines. The orthogonal polarization, on the other hand, will be transmitted largely unaffected. Extinctions on the order of 10^2 are possible with wire grid polarizers but it must be emphasized that these devices only operate effectively over a limited range of wavelengths, in the vicinity of the size of the conductor spacing.

9.1.2 Retardation Plates

Retardation plates are frequently used in polarimetry instruments. Quarter-wave and half wave plates are most often used and may be fabricated in either wavelength-dependent or achromatic forms.

Quarter-wave plates are used to generate circularly polarized light and to convert elliptically polarized light to linearly polarized light (see section 8.2.3). For an instrument employing a monochromatic light source (a laser, for example), quarter-wave plates can be fabricated accurately using materials of known birefringence manufactured to a thickness,

$$d = \frac{\pi}{2}\left(\frac{\lambda}{2\pi\Delta n'}\right). \tag{9.1}$$

Crystalline materials with precisely known birefringences are normally used for this purpose, with mica being commonly used. It is difficult, however, to machine most birefringent crystals so that a retardation of precisely $\pi/2$ is produced. This is because the thicknesses of the crystal calculated using equaiton (9.1) may be too thin to be practical. Instead, multiple ordered quarter-wave plates can be fabricated with retardation, $\delta' = 2\pi j + \pi/2$, where $j = 1, 2, 3, \ldots$ If a *single*-order quarter-wave retardation is required, this can be achieved by combining two multiple-order quarter-wave plates of retardation $\delta' = 2\pi j + \pi/2$ and $\delta' = 2\pi j + \pi + \pi/2$, at a relative orientation of $90°$.

Half wave plates are used to rotate the principal axis of the polarization ellipse (see section 2.4.2). Half wave plates for applications involving single wavelengths can be fabricated in precisely the same manner as quarter wave plates, but with thicknesses that are twice those specified by equation (9.1).

Achromatic retardation plates that operate properly over a wide range of wavelengths are required in many applications. One example of an achromatic retarder is the Fresnel rhomb, pictured in Figure 9.4a. This element is fashioned from a transparent, amorphous material that is in the form of a parallelogram. The principle by which a quarter-wave retardation is created is by the two total internal reflections of the light at the points (a) and (b) within the device. By properly selecting the angle α, each reflection can cause a relative retardation of the phase of p and s polarized light to be $\pi/4$. The total retardation for the two reflections will then be $\pi/2$, which is required for a quarter wave plate. The design equation for this requirement is that

$$\left(\frac{R_{pp}}{R_{ss}}\right)^2 = \tan^2\psi e^{i2\Delta} = e^{i\frac{\pi}{2}}, \tag{9.2}$$

where the fact that $\tan\psi = 1$ for a total internal reflection at a nonabsorbing interface has been used. Using the reflection coefficients given in section 1.6, this equation reduces to

$$\Delta = \frac{\pi}{4} = 2\tan^{-1}\left[\frac{\cos\alpha\sqrt{\sin^2\alpha - n^{-2}}}{\sin^2\alpha}\right], \qquad (9.3)$$

which can be used to solve for the angle α for a choice of the refractive index of the material, n.

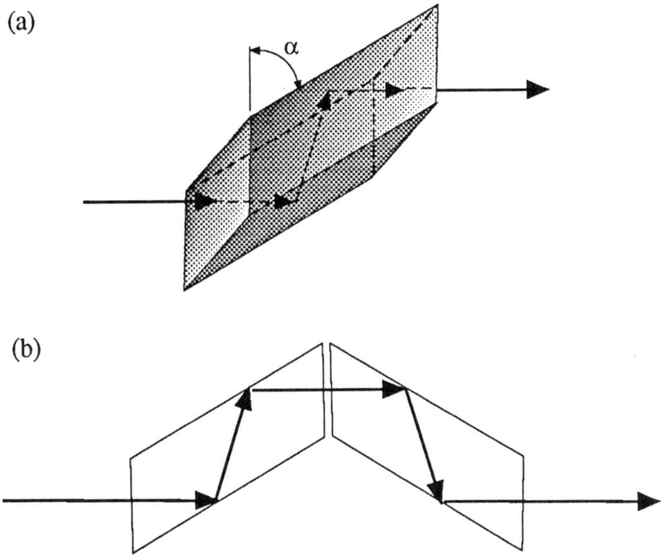

Figure 9.4 (a) Fresnel rhomb for quarter-wave retardation. (b) Double Fresnel rhomb arrangement for a half-wave retardation.

Achromatic half-wave plates can be easily made from the combination of two quarter-wave Fresnel rhombs shown in Figure 9.3b. This geometry has the nice feature of removing the translation of the light beam that is associated with a single rhomb.

Fresnel rhombs have the disadvantage of being large in construction and difficult to align. In the case of quarter-wave Fresnel rhombs, a substantial displacement of the light beam is induced. For that reason, a second class of achromatic retarders based on laminates of wavelength-dependent retardation plates are often more convenient. Figure 9.5 shows the scheme by which a laminate retarder is constructed. The objective here is to stack retardation plates of retardation δ'_i; $i = 1, 2, \ldots$ and orientation angles θ_i so that the following requirements are met:

(1) $\delta' = \delta'^*$; at $\nu = \nu^*$,

(2) $\left(\frac{\partial\delta'}{\partial\nu}\right)_{\nu = \nu^*} = 0,$ (9.4)

(3) The Jones matrix of the laminate, $\mathbf{J}_L$,

has the form of a true retardation plate.

Here δ' is the retardation of the laminate, ν^* is the central frequency, and δ'^* is the desired retardation (normally $\pi/2$ or π). In other words, the end result should be a retardation plate with the correct retardation in the center of the spectral region of interest, and its properties should be slowly changing with frequency or wavelength. The latter requirement is achieved by forcing the derivative of the retardation with frequency to be zero at $\nu = \nu^*$.

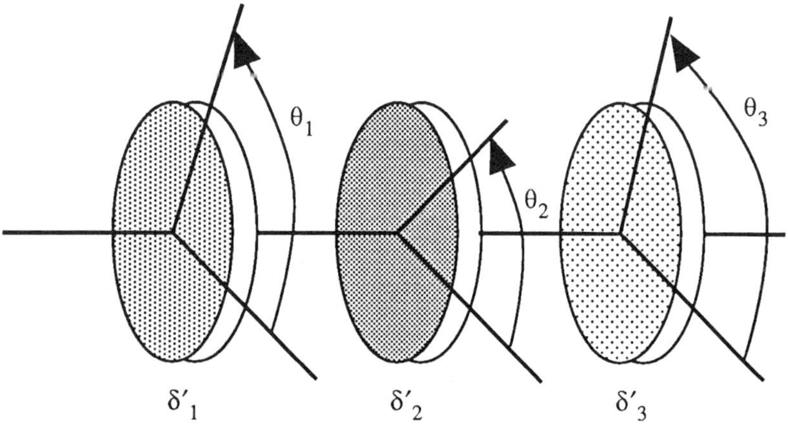

Figure 9.5 Composite laminate of three retardation plates to form a single, achromatic retarder.

The procedure to satisfy these requirements was first described by Pancharatnam [53] and later developed by Title and coworkers [54]. The presentation here closely follows the discussion given in reference 54. The requirement that the laminate behave like a true retarder means that it must have the form given by equation (I.9) in Appendix I. In other words, its Jones matrix should be

$$\mathbf{J}(\delta', \theta) = \begin{bmatrix} (c_{\delta'/2} + ic_{2\theta}s_{\delta'/2}) & is_{2\theta}s_{\delta'/2} \\ is_{2\theta}s_{\delta'/2} & (c_{\delta'/2} - ic_{2\theta}s_{\delta'/2}) \end{bmatrix}, \tag{9.5}$$

which can be written as

$$\mathbf{J}(\delta', \theta) = c_{\delta'/2}\mathbf{I} + is_{\delta'/2}\mathbf{E} \cdot \mathbf{R}(2\theta), \tag{9.6}$$

where $\mathbf{E} = \begin{bmatrix} 1 & 0 \\ 0 & -1 \end{bmatrix}$. Evidently, the real part of the Jones matrix resulting from the multiplication of the individual Jones matrices representing the elements of the laminate must be isotropic.

We consider here the case of a laminate consisting of three retardation plates with retardations δ'_1, δ'_2, and δ'_3, and orientation angles θ_1, θ_2, and θ_3. Multiplying the three Jones matrices for these elements leads to

$$\begin{aligned}
\mathbf{J}_T &= \mathbf{J}(\delta'_3, \theta_3) \cdot \mathbf{J}(\delta'_2, \theta_2) \cdot \mathbf{J}(\delta'_1, \theta_1) \\
&= \cos(\delta'_3/2)\cos(\delta'_2/2)\cos(\delta'_1/2)\,\mathbf{I} \\
&\quad -\{\cos(\delta'_3/2)\sin(\delta'_2/2)\sin(\delta'_1/2)\,\mathbf{R}[2(\theta_1-\theta_2)] \\
&\quad + \cos(\delta'_2/2)\sin(\delta'_2/2)\sin(\delta'_1/2)\,\mathbf{R}[2(\theta_1-\theta_3)] \\
&\quad + \cos(\delta'_1/2)\sin(\delta'_2/2)\sin(\delta'_3/2)\,\mathbf{R}[2(\theta_2-\theta_1)]
\end{aligned} \tag{9.7}$$

$$\begin{aligned}
&+ i\mathbf{E} \cdot \{-\sin(\delta'_3/2)\sin(\delta'_2/2)\sin(\delta'_1/2)\,\mathbf{R}[2(\theta_3-\theta_2+\theta_1)] \\
&\quad + \sin(\delta'_3/2)\cos(\delta'_1/2)\cos(\delta'_1/2)\,\mathbf{R}[2\theta_3] \\
&\quad + \sin(\delta'_2/2)\cos(\delta'_3/2)\cos(\delta'_1/2)\,\mathbf{R}[2\theta_2] \\
&\quad + \sin(\delta'_1/2)\cos(\delta'_3/2)\cos(\delta'_2/2)\,\mathbf{R}[2\theta_1]\}.
\end{aligned}$$

The requirement that the real part of $\mathbf{J}_T$ be isotropic means that its off-diagonal components must be zero. This leads to

$$\sin[2(\theta_2-\theta_3)] = \frac{\tan(\delta'_1/2)}{\tan(\delta'_3/2)}\sin 2\theta_2 + \frac{\tan(\delta'_1/2)}{\tan(\delta'_2/2)}\sin 2\theta_3, \tag{9.8}$$

where θ_1 has been arbitrarily set to zero, since its value simply prescribes the orientation of the laminate as a whole. Using the angle addition formula for the sine implies that

$$\cos 2\theta_3 = \frac{\tan(\delta'_1/2)}{\tan(\delta'_3/2)} \qquad \cos 2\theta_2 = -\frac{\tan(\delta'_1/2)}{\tan(\delta'_2/2)}. \tag{9.9}$$

The angles θ_2 and θ_3 are fixed and independent of wavelength so that the ratios on the right-hand side of (9.9) must also be independent of wavelength. Since the retardations are functions of wavelength, this requirement can be satisfied by either of the following constraints: $\delta'_1 = \pm\delta'_2$ or $\delta'_1 = \pm\delta'_3$. The first constraint, however, leads to $\theta_2 = \theta_1 = 0$, which is a trivial result since this has the second retarder parallel to the first element. Using the second constraint, we have $\theta_3 = \theta_1 = 0$ and, from equation (9.8),

$$\begin{aligned}
\mathbf{J}_T(\delta', \theta) &= (\cos\delta'_1\cos(\delta'_2/2) - \sin\delta'_1\sin(\delta'_2/2)\cos 2\theta_2)\,\mathbf{I} \\
&\quad + i\mathbf{E}\cdot\left[-\sin^2(\delta'_1/2)\sin(\delta'_2/2)\,\mathbf{R}(-2\theta_2)\right. \\
&\quad \left. + \cos^2(\delta'_1/2)\sin\delta'_2\mathbf{R}(-2\theta_2) + \sin\delta'_1\cos(\delta'_2/2)\,\mathbf{I}\right].
\end{aligned} \tag{9.10}$$

The next step is to determine the values of δ'_1, δ'_2, and θ_2 that minimize the dependence of the overall retardation, δ', and overall orientation angle, θ, with respect to wavelength or frequency. In addition, the retardation at the central frequency must be equal to δ'^*. Comparing equation (9.10) to (9.6), these overall properties are

$$\begin{aligned} \cos(\delta'/2) &= \cos\delta'_1\cos(\delta'_2/2) - \sin\delta'_1\sin\delta'_2\cos 2\theta_2, \\ \tan 2\theta &= \mathrm{Imag}(J_{T21})/\mathrm{Imag}(J_{T12}). \end{aligned} \tag{9.11}$$

Differentiating these two expressions with respect to frequency (assuming that the birefringence is independent of wavelength), and setting the differentials to zero, Title found that a half-wave plate that is quite achromatic can be fashioned from a three-element laminate if $\delta'_1 = \delta'_2 = \delta'_3 = \pi/2$ and $\theta_2 = \pi/3$. Increasing the number of plates will obviously minimize the frequency dependence of the laminate, and Title has examined ten element combinations that are achromatic to within $1°$ over the visible range of wavelengths.

9.1.0 Circular Polarizers

It was seen in Chapter 8 that circular polarizers are required components in several polarimeter designs. By definition, a circular polarizer will extinguish circularly polarized light of one sense (either left- or right-circularly polarized light), but allow the other circular polarization to pass.

A circular polarizer can be constructed using a combination of a polarizer and a quarter wave retarder at a relative orientation of $45°$. This arrangement was discussed in section 2.4.2 where circularly polarized light was generated by placing the quarter wave plate *after* the polarizer. Left- or right-circular polarizers can be created with the same combination of elements, but with the quarter-wave plate positioned *before* the polarizer. The sense of circular polarization that is extinguished with such an arrangement is determined by setting the relative orientation of these two optical elements to either plus or minus $45°$. The Mueller matrix of such a combination is

$$M_{Q_{45}P} = \frac{1}{2}\begin{bmatrix} 1 & 0 & 0 & -1 \\ 1 & 0 & 0 & -1 \\ 0 & 0 & 0 & 0 \\ 0 & 0 & 0 & 00 \end{bmatrix}, \tag{9.12}$$

which is not strictly identical to the form for a true circular polarizer, shown below:

$$\mathbf{M}_{18,\,right} = \frac{1}{2}\begin{bmatrix} 1 & 0 & 0 & -1 \\ 0 & 0 & 0 & 0 \\ 0 & 0 & 0 & 0 \\ -1 & 0 & 0 & 1 \end{bmatrix}. \tag{9.13}$$

The difference in these two matrices is normally not important since only the elements of the top row affect the final intensity that is measured.

Recently, true circular polarizers have been developed and are available from commercial sources. These are fashioned by sandwiching a chiral nematic liquid crystal between parallel glass windows. These devices are characterized by a high transmission and extinction ratios greater than 1000. They operate over a fairly narrow range of wavelengths about a central wavelength.

9.1.4 Variable Retarders

It is often useful to install a retarder in an optical train that can be systematically varied in its retardation. Reasons for using such an element include nulling of parasitic window birefringence, and to determine the order of retardation. The earliest design of a variable retarder is the Babinet compensator, shown in Figure 9.6. Two wedges of birefringent material are sandwiched together. Variation of the retardation is accomplished by translating the bottom wedge relative to the top wedge, thereby changing the total retardation thickness, *d*.

Variable retardation can also be achieved using electric birefringence induced in liquid layers. Commercial devices are available that use this method.

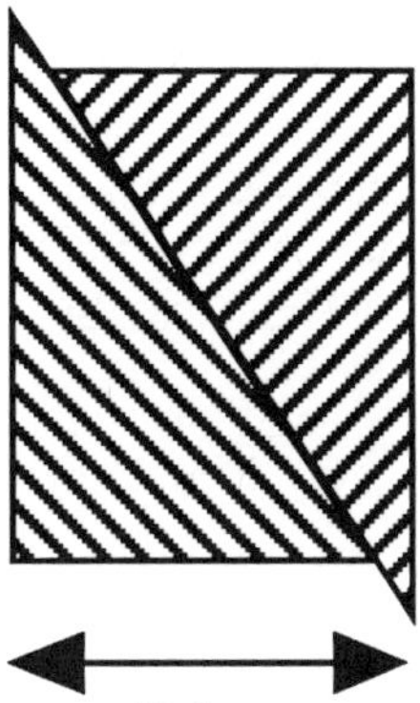

Figure 9.6 Babinet compensator.

9.2 Alignment of Polarizing Elements

The successful operation of any optical instrument will depend on the alignment of its component parts. Correct placement of a polarizing element requires attention to its orientation defined by the polar angle θ and the azimuthal angle ϕ in Figure 9.7. Combinations of translation stages, rotation stages and gimbal mounts may be used to adjust the position, the angle ϕ, and the angle θ, respectively.

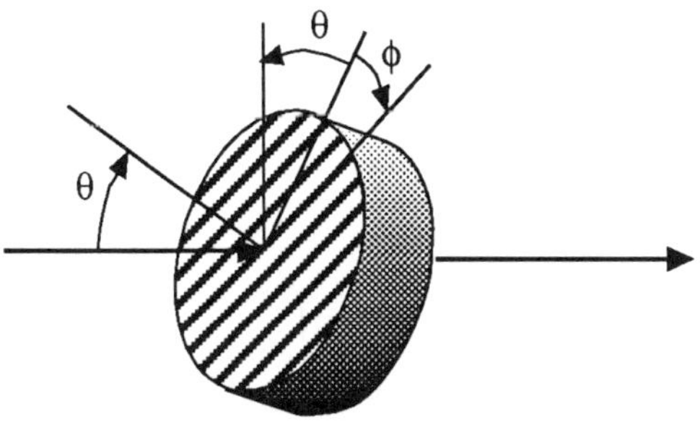

Figure 9.7 Orientation of an optical element. The lines on the front face are parallel to the principal optical axes of the element.

The first step in the alignment of an optical train is usually to orient the incident light beam so that its axis of propagation traverses the desired path. This path is most often defined by an optical railing or other surface onto which the optical elements are mounted. Not only must the light successfully strike a given target, but normally it must be oriented *parallel* to the surface. Both requirements can be met using the simple device pictured in Figure 9.8. This consists of a pinhole positioned at the target location and with a flat glass window attached in front. It is important to ensure that the normal vector that defines the surface of the window is *parallel* to the mounting surface and the desired path of the light. Once this element has been properly located, the light beam can be aligned using a simple two-part, iterative procedure. First the source is adjusted by translating it so that a maximum amount of light is transmitted through the pinhole. After this step, the light that is *back reflected* from the glass window is examined to determine whether it is returned directly back to the source. Misalignment of this property of the light is corrected by adjusting the orientation of the source so that the back-reflected light impinges on the source. At this point, the angle θ in Figure 9.7 is set to zero. Since this second step causes the light to no longer target the pinhole, the first step will have to be repeated. Using this trial and error procedure, the direction of the incident beam can be established.

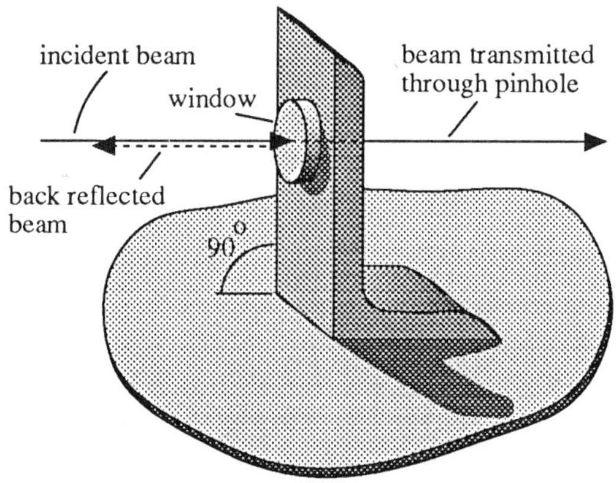

Figure 9.8 Alignment tool for positioning of a light beam.

Forcing the back-reflected light onto the incident propagation axis is a common method of aligning many optical elements. This can be accomplished by fastening the element onto a gimbal mount. Care must be taken, however, when sending the back reflected light back onto the incident propagation axis if a laser source is used. This is due to instabilities in the operation of a laser if light is injected inside of the laser cavity.

It is essential to define the reference orientation axis of the polarization vector in an instrument. A convenient way to determine this reference is to use a Brewster angle window, shown in Figure 9.9. The goal of this procedure is to establish the orientation of the incident polarization relative to the plane AB . It is assumed that the beam has been aligned so that it is parallel to this plane using a method such as the one outlined above. The physical arrangement shown in Figure 9.9 will ensure that only light polarized perpendicular to the plane of incidence will be reflected to point C. The orientation of the incident light polarization is then adjusted by rotating the polarizer P until the reflected light intensity is extinguished. At this point the polarization is parallel to the plane of incidence.

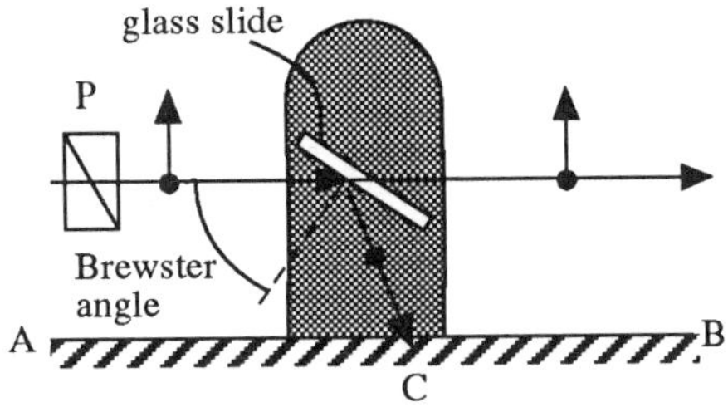

Figure 9.9 Alignment of the polarization vector using a Brewster angle window.

9.3 Calibration of the Sign of Dichroism and Birefringence

Once an optical polarimeter has been aligned, its ability to correctly determine the sign of birefringence and dichroism should be checked. For birefringence, the sign can be calibrated using a stretched polymer film. Using such a sample, oriented with its stretching direction parallel to the angle defined as the "zero" orientation angle, the instrument should produce a birefringence that is the same sign as the stress optical coefficient for that polymer.

A polarizer can be used to calibrate a polarimeter in its ability to determine the sign of dichroism since, by definition, a polarizer has a *negative*, infinite dichroism. Using such an element in place of the sample, and orienting it parallel to the laboratory axis from which orientation angles are to be measured, the instrument should produce a negative dichroism.

9.4 Calibration of the Flow Direction Axis in a Couette Shear Flow Cell

Figure 9.10 shows the geometry of a simple shear flow where the flow direction is slightly misaligned with respect to the axis from which orientation angles are measured. The angle of misalignment is given by ϕ. When a shear is applied, the constituent particles or molecules making up the sample will orient at an unknown angle, α, that must be determined.

A method that can be used to determine α and ϕ is to perform two measurements with shear rates of equal magnitude, but that are opposite in sign [28]. In Figure 9.10a, where the flow is directed toward the right, an angle β_1 is measured for the average orientation in the sample. This angle is simply $\beta_1 = \alpha + \phi$. Applying the flow in the opposite direction (towards the left) produces the result shown in Figure 9.10b, and an angle $\beta_2 = -\alpha + \phi$ is measured. From these two results, the angles α and ϕ are easily determined using

$$\alpha = \frac{\beta_1 - \beta_2}{2}; \; \phi = \frac{\beta_1 + \beta_2}{2}. \tag{9.14}$$

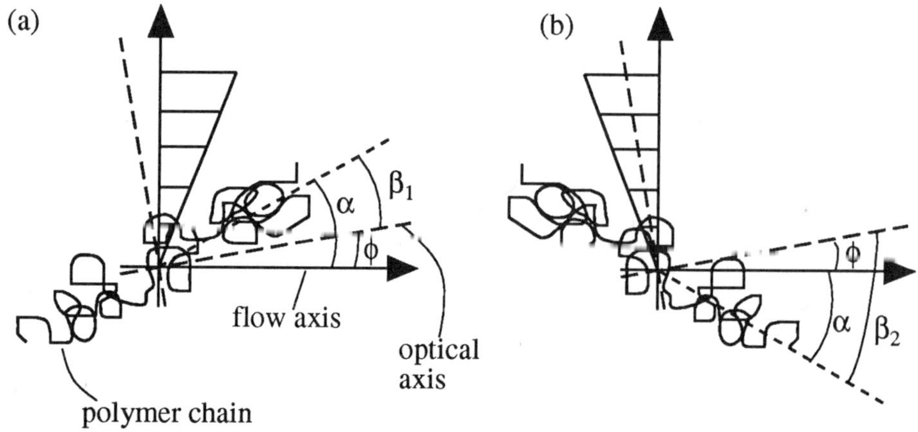

Figure 9.10 Determination of flow-induced orientation angles in the presence of misalignment between the optical axis and the flow axis.

10 Applications and Case Studies

This chapter reviews past research that has applied methods in optical rheometry to solve problems in the structure and dynamics of complex liquids. This review is organized according to the types of complex liquids that have been studied. In addition, several case studies are provided where the reader is introduced to specific applications that have been chosen to highlight various optical techniques and the interpretation of data. These studies chronicle the execution of the experiment, including the motivation of the choice of a particular technique, the design of the instrumentation, and the experimental procedures.

10.1 Polymeric Liquids

10.1.1 Verification of the Stress-Optical Rule

Early work on the use of optical methods on the dynamics of polymeric liquids focused on establishing the validity of the stress-optical rule. A comprehensive account of this research can be found in the books by Janeschitz-Kriegl [29] and Wales [84]. The majority of studies have considered simple shear flow, and the rule has been found to hold up to stresses up to 10^4 Pa. Such tests have not only considered the linearity of the shear stress with $\Delta n' \sin 2\theta$, but also the coaxiality of the stress and refractive index tensors by ensuring that the extinction angle, θ, obeys the relation: $\tan 2\theta = 2\tau_{xy}/N_1$. In a polymer melt, a violation of the rule would be expected if chain orientation became sufficiently high to cause a departure in gaussian chain statistics. Such large stresses, however, are rarely encountered in practice.

Extensive work investigating the stress-optical rule has also been performed on polymer solutions [101]. Here the rule can be successfully applied if the solvent contributions to the birefringence are properly subtracted. Care must be taken, however, to avoid form birefringence effects if there is a large refractive index contrast between the polymer and the solvent.

Investigations of the stress-optical rule in extensional flow are fewer in number due to the difficulty in establishing this flow. Using an extrusion device, van Aken and Janeschitz-Kriegl [102] produced extensional flows in melts by forcing the material through a lubricated die with a surface given by $|z|R^2 = \text{constant}$, where z is the axial compression axis and R is the radial outflow coordinate. By combining pressure drop measurements with birefringence, the stress-optical rule was examined and found to hold up to extensional stresses of 8×10^4 Pa. For polymer solutions, characterized by much lower entanglement densities, however, the rule was found by Talbott and Goddard [103] to break down at much lower stresses. Those researchers used a spinning apparatus to produce a uniaxial, extensional flow and simultaneously measured the tension on the filament, as well as its birefringence. This is a difficult geometry for birefringence measurements, however, since the dimensions of the filament continuously change along the filament and become very small at high strain rates.

The majority of studies of the stress-optical rule have considered steady state flow conditions. Comparisons of stress and birefringence measurements in transient shear flow were made by Gortemaker *et al.* [104] on a polystyrene melt where the stress and refractive index tensors were found to be coaxial as functions of time during the inception and cessation of flow. Several studies have also been made on oscillatory flow. The most extensive work along these lines is by Schrag and coworkers [105] on dilute polymer solutions. This work is distinguished by the large frequency range that could be accessed by the instruments that were used (10^5 Hz). At low frequencies in the "terminal region" where the loss modulus is linear in the applied frequency, the stress-optical rule was found to hold. At high frequencies, however, the relationship failed, even when the solvent contributions to the stress and the birefringence were properly removed. Explanations for this departure include the possibility a frequency dependence of the segmental constituents that contribute to chain orientation, and a frequency dependence of couplings between segmental and local solvent orientations.

Dynamic optical measurements on melts began with the work of Vinogradov and coworkers [106]. In this work, the frequencies were limited to the terminal region and the stress and refractive index tensors were found to be coaxial. At higher frequencies, however, the stress-optical relationship is expected to break down as the dynamics become sufficiently fast as to excite glassy modes of relaxation. This behavior is shown in Figure 10.1, where the storage and loss moduli of polybutadiene are plot as functions of reduced frequency [107]. In this figure, the solid curves were measured mechanically, and the symbols are data measured using birefringence. The data were measured at a series of temperatures and shifted using the WLF time-temperature superposition principle. The optical and mechanical measurements superimpose over the frequency range that includes the terminal and plateau modulus regions. At frequencies above the plateau, however, a deviation is observed, reflecting the onset of glassy modes of orientational relaxation. Similar departures from the stress-optical rule have were reported earlier by Osaki and coworkers [108], where the apparent stress optical coefficient for polystyrene was found to change sign from a negative value in the terminal and plateau regions, to a positive value above the glass transition frequencies. Such a change in sign suggests that different chemical constituents are involved in the orientation processes in these two frequency ranges.

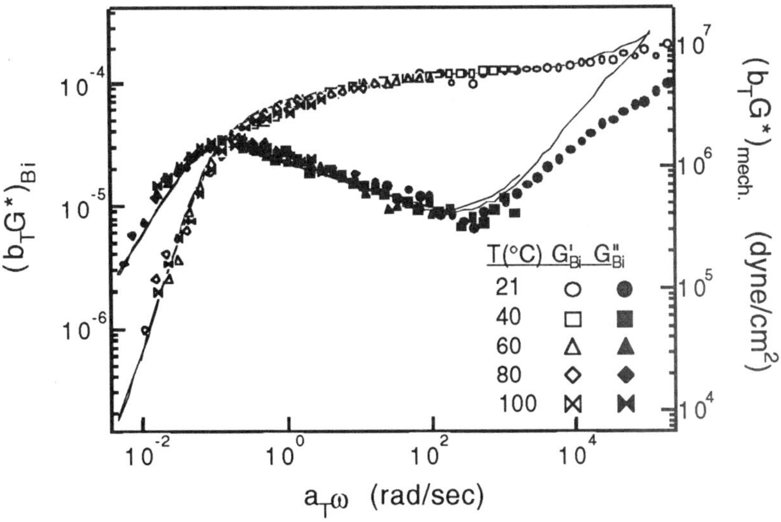

Figure 10.1 Storage and loss moduli of polybutadiene, MW 204,000. Solid lines are the mechanical moduli and the symbols are the optical measurements. (From reference 107, with permission).

As discussed in section 7.1.6.4, semidilute solutions of rodlike polymers can be expected to follow the stress-optical rule as long as the concentration is sufficiently below the onset of the isotropic to nematic transition. Certainly, once such a system becomes nematic and anisotropic, the stress-optical rule cannot be expected to apply. This problem was studied in detail using an instrument capable of combined stress and birefringence measurements by Mead and Larson [109] on solutions of poly(γ benzyl L-glutamate) in m-cresol. A pretransitional increase in the stress-optical coefficient was observed as the concentration approached the transition to a nematic state, in agreement of calculations based on the Doi model of polymer liquid crystals [63]. In addition to a dependence on concentration, the stress-optical coefficient was also seen to be dependent on shear rate, and on time for transient shear flows.

10.1.2 Rheometry of Polymeric Liquids

For systems where the stress-optical rule applies, birefringence measurements offer several advantages compared with mechanical methods. For example, transient measurements of the first normal stress difference can be readily obtained optically, whereas this can be problematic using direct mechanical techniques. Osaki and coworkers [26], using a procedure described in section 8.2.1 performed transient measurements of birefringence and the extinction angle on concentrated polystyrene solutions, from which the shear stress and first normal stress difference were calculated. Interestingly, N_1 was observed to overshoot its steady state value upon the inception of simple shear flow for sufficiently large velocity gradients. The result offers an important, qualitative test of rheological constitutive models, such as the Doi and Edwards reptation model, which fails to predict such an overshoot. Using the method of two-color flow birefringence (also discussed in section 8.2.1), Zebrowski and Fuller [110] also examined transient shear flow of concentrated polystyrene solutions and confirmed the existence of normal stress overshoots following the inception of strong flows. In addition, the extinction angle was observed to undershoot its steady state value during start-up, a phenomena that the Doi-Edwards model also fails to predict. Following the cessation of flow, the extinction angle decreases from the value it had during the flow, whereas molecular models normally predict that relaxation will occur at a constant value of this angle. This result, however, can be simply explained by considering polydispersity in the sample, which will cause the sample to be characterized by a distribution of extinction angles with higher molecular weight species achieving orientations closer to the flow direction than smaller chains. During relaxation following cessation of the flow, the smaller chains will relax first, causing the average orientation angle to decrease toward the values associated with the longer chains. More recently, Pearson and coworkers [111] have examined the problem of overshoot in the normal stress using measurements on polyisoprene dissolved in squalene (an oligomer of isoprene) and have used the results to test a modification of the Doi and Edwards reptation model that incorporates the possibility of stretching of the chain within the tube of entanglements.

Transient birefringence measurements were used by Larson *et al.* [112] to test the validity of the Lodge-Meissner relationship for entangled polymer solutions. This relationship states that the ratio of the first normal stress difference to the shear stress following a step strain is simply $N_1/\tau_{xy} = \gamma$, where γ is the strain. Those authors found the relationship was valid, except for ultrahigh molecular weight materials.

Optical measurements often have a greater sensitivity compared with mechanical measurements. Semidilute polymers, for example, may not be sufficiently viscous to permit reliable transient stress measurements or steady state normal stress measurements. Chow and coworkers [113] used two-color flow birefringence to study semidilute solutions of the semirigid biopolymer, collagen, and used the results to test the Doi and Edwards model discussed in section 7.1.6.4. That work concluded that the model could successfully account for the observed birefringence and orientation angles if modifications to the model proposed by Marrucci and Grizzuti [114] that account for polydispersity, were used.

10.1.3 Applications in non-Newtonian Fluid Mechanics

Knowledge of the distribution of stresses in non-Newtonian flows is important to the design of complex flow processes, and the underlying physical basis for such phenomena as extrudate instability. Spatially resolved measurements of flow birefringence in cases where the stress-optical rule applies can be used for such purposes and numerous applications of this method can be found in the literature. The birefringence measurements are either made in a pointwise fashion, or with a full field measurement using the method of isoclinics and isochromatics. The latter procedure is appropriate when the retardation of the material exceeds by many orders the value of π and is conducted by placing the sample between crossed polarizers. The diameter of the transmitted light is chosen to cover the field of the flow that is of interest and the resulting patterns are analyzed to extract the birefringence and extinction angle as a function of position. A detailed account of this method is given in reference 115.

Using the method isoclinics, Checker and coworkers [116] studied the flow of polyethylene and polypropylene melts through two dimensional contractions and used birefringence measurements to determine the relaxation behavior of these materials subject to complex flows.

In a series of papers, Lee and Fuller [117,118,119] used polarization modulation techniques to examine the problem of the spatial development of transient flow development in couette and eccentric cylinder flows. The first two papers concern the phenomena of "shear wave" propagation as first observed by Joseph and coworkers [120] using a mechanical technique. This phenomena refers to the tendency of stresses in viscoelastic fluids to propagate as waves instead of diffusively, as in the case of Newtonian liquids. By applying the stress optical rule, birefringence was used to make pointwise measurements of stress in the flow fields during the start-up and cessation of couette flow. In the third paper, spatially resolved measurements of birefringence were used to map the stress fields of concentrated polystyrene solutions subject to eccentric cylinder flow. These stress profiles, taken at different eccentricities of the cylinder spacing, were compared against numerical simulations of these flows using the Geisekus constitutive equation [64]. The comparison between the numerical simulations and the experimental measurements was found to be acceptable, except in regions of the flow characterized by strong extensional components. This suggests that the constitutive model, although offering good predictions of shear flow behavior, may not account for the extensional flow rheology of the solutions.

Two dimensional contraction flows were also studied using birefringence measurements by Galante and Frattini [121]. Those authors used polarization modulation and used the stress optical rule to measure stress distributions in poly(dimethyl siloxane) melts. Particular attention was paid to measurements of stress near the re-entrant corners and the measurements were compared against predictions using the second-order fluid model. A principal conclusion of this work was that the measured stresses did not increase near the corners as dramatically as predicted by numerical simulation of this flow [122]. The planar contraction flow was also examined by Kiriakidis *et al.* [123]. in a study comparing birefringence measurements (using the isoclinic method) to numerical simulations employing the K-BKZ integral constitutive equation. Those authors reported a qualitative agreement between the experimental and simulated results.

Baird and coworkers [124] compared numerical simulations of the stress field generated by flow over a rectangular slot with birefringence measurements on a polystyrene melt. The simulations modeled the melt using the White-Metzner constitutive equation and good agreement was found with the birefringence measurements (using the method of isoclinics) for the spatial variation of the shear and normal stresses.

10.1.4 Spectroscopic Investigations of Polymer Melts and Blends

Spectroscopic methods play an important role in optical rheometry since may important systems involve multiple components, each having separate dynamical responses. An important account of earlier work using spectroscopy for the orientation of polymeric materials can be found in the book by Ward [17]. This provides an excellent account of earlier applications of infrared dichroism, Raman scattering, and fluorescence polarization. More recently, the laboratory of L. Monncrie has made important advances in the development and application of spectroscopic techniques to study orientation in polymeric materials. A summary of the work of this group can be found in reference 125. Among the novel experiments initiated by Monnerie and coworkers is a determination of the dependence of chain relaxation on the molecular weight of its surrounding matrix [126]. Here it was found that the degree of orientation increased with both increasing matrix molecular weight and the rate of stretching. This group also applied infrared dichroism techniques to the study of compatible polymer blends, and for the system of poly(ethylene oxide) and poly(methyl methacrylate), found that the PEO failed to orient in blends where this component was of lower molecular weight than the PMMA [127].

The problem of orientational coupling has also attracted the attention of Monnerie and coworkers [128,129,130]. In this respect, the Monnerie group was the first to examine this effect in polymer melts and blends subject to transient deformations. This phenomena causes segments within a polymeric system to adopt an orientation that is proportional to the bulk orientation. For example, this causes the stress optical coefficient of polymer networks to drop with the addition of symmetrically shaped solvent molecules. This effect was first explained by Stein [85] by noting that the addition of such solvent molecules would screen orientational couplings between chains in a stretched network, thereby reducing the level of orientation. Direct evidence of orientational coupling (often referred to as nematic interaction) was offered by Deloche and Samulski using deuterium NMR on strained elastomers swollen with labeled solvents. In that study, the solvent molecules were observed to adopt an orientation proportional to the elastomers, in spite of the fact that they were not mechanically coupled to the networks [131].

The majority of work using spectroscopic techniques to study polymer orientation and dynamics has applied large strains to the samples in order to achieve measurable signals. Furthermore, most past studies did not make *in situ* measurements during flow, but normally quenched the samples during deformation or relaxation processes prior to making an optical measurement. Using polarization modulation techniques for both infrared dichroism and Raman scattering, Fuller and coworkers have offered among the first spectroscopic measurements in the limit of linear viscoelasticity and during transient dynamic conditions. The infrared dichroism instrument employed by this group follows the design shown in Figure 8.9 and therefore allowed for simultaneous dichroism and birefringence measurements [132]. The light source used in these infrared experiments generated light at the wavelength of absorption of the stretching vibration of a carbon-deuterium bond (2180 cm^{-1}). The dichroism measurements, therefore targeted the orientation dynamics of polymer species tagged with deuterium. The birefringence measurements, on the other hand, are not significantly influenced by the presence of deuterium labels, and reflect the orientation of the bulk.

Using the infrared polarimeter described above, Kornfield and coworkers [132] studied the orientation dynamics of bidisperse polymer blends subject to step strains in the limit of linear viscoelasticity. These experiments were designed to offer quantitative comparisons against the constraint release model of Doi and coworkers [133]. In addition to confirming the predictions of this model, the measurements revealed evidence of orientational coupling among the different chains the blend. These data motivated a molecular model of orientational coupling that properly accounted for this effect, while explaining why the stress-optical rule is unaffected by the presence of this coupling [133]. These results were further verified by studying the same bidisperse systems subject to dynamic simple shear flow [134].

The Doi-Edwards, reptation based model makes specific predictions for the relaxation dynamics of different portions of a polymer chain. Specifically, the relaxation of the chain ends is predicted to be substantially faster than the relaxation of the center. This is a result of the reptation dynamics, which have the ends first leaving the confines of the tube. Using polymer chains that were selectively deuterated either at the ends or at the middle, Ylitalo and coworkers [135] examined this problem and found that the Doi-Edwards model was able to successfully predict the observed behavior once the effects of orientational coupling was included. The same group further explored the phenomena of orientational coupling in papers that focused on its molecular weight [136] and temperature [137] dependences. The quantitative values of the coupling parameters found using infrared dichroism and NMR were compared directly by Ylitalo *et al.* [138].

Linear viscoelastic measurements using infrared dichroism on the compatible blend poly(ethylene oxide) and poly(methyl methacrylate) were reported by Zawada *et al.* [139]. Unlike Monnerie and coworkers [127], who reported seeing only orientation in the PMMA component, and none in the PEO, Zawada *et al.* observed alignment in the PEO. However, since the PEO was of lower molecular weight (as was the case for Monnerie and coworkers), its relaxation timescales were substantially faster than the PMMA. This may explain the lack of any measurable orientation by Monnerie and coworkers, who studied quenched samples, since their preparation may have allowed the PEO to relax prior to testing.

Using a novel experimental design, called "2-dimensional infrared spectroscopy," Noda and coworkers [140,141,142] have studied the dynamics of solid polymers and blends subjected to oscillatory elongation. This instrument is based on a dispersive infrared spectrometer employing a blackbody radiation light source fed to a monochromator for wavelength selection. Dichroism is measured by modulating the polarization of the light prior to introducing it to the sample. A photoelastic modulator is used for this purpose. The "2-dimensional" character of the experiment refers to the treatment of the data acquired at different incident frequencies, ω, of the light. If the dichroism signal measured as a function of time at an incident frequency, ω_i, is referred to as $\Delta n''(t;\omega_i)$, the following correlation function can be formed from dichroism signals measured at a frequency, ω_j:

$$C_{ij}(\tau) = \int_0^T dt\ \Delta n''(t;\omega_i)\,\Delta n''(t-\tau;\omega_j)\,, \tag{10.1}$$

where T is the period of the mechanical oscillation. If $i = j$, then $C_{ij}(\tau)$ is an autocorrelation function. If $i \neq j$, then $C_{ij}(\tau)$ measures the correlation between two different absorption frequencies in the spectrum. Furthermore, if $\tau = 0$, the correlation function measures the strength of couplings that are synchronous with one another. However, if $\tau/T = \pi/2$, then the correlation between asynchronous dynamics is calculated. The result is a very powerful diagnostic tool for the examination of complex dynamics.

10.1.5 Dynamics of Polymeric Liquids in Extensional Flow

Extensional flows applied to polymeric liquids have received considerable attention due to the large orientations and deformations that are possible with such strong flows. Furthermore, since these flows are inherently transient and inhomogeneous, the speed and spatial resolving power of optical methods are particularly useful. Among the first applications of optical methods to extensional flows was the work of Keller and coworkers investigating polymer solutions subjected to flows generated by opposed nozzles [144], four roll mills [145], and cross-slot devices [146]. These researchers used birefringence to measure the degree of chain orientation in stagnation point flows. Such flows are characterized by points where the velocity is zero, but where the velocity gradients are finite. Since the residence times in the vicinity of a stagnation point are very large, molecules can achieve large strains, and approach steady state conformations. This is manifested by highly localized lines or planes of birefringence near stagnation points or stagnation lines. For example, in an opposed nozzle flow cell, where the flow is created by aspirating liquid into two opposed orifices, a stagnation point will develop at the geometric center of the system. Flexible chains, which require a finite strain to develop an appreciable degree of deformation and orientation will only produce birefringence along a very thin filament spanning the two jets. Elsewhere in the flow, the chains will not experience sufficient strain to cause a measurable birefringence.

Dilute solutions of flexible chains in extensional flow are predicted to undergo a "coil-to-stretch" transition when the strength of the flow reaches a critical value relative to the inverse of the longest relaxation time, τ, of the chains. This criteria is met when [147]

$$g^{+} \geq 1/\tau , \tag{10.2}$$

where g^{+} is the largest positive eigenvalue of the velocity gradient tensor. For example, for a hyperbolic, extensional flow field generated using a four-roll mill, the velocity field is $\mathbf{v} = \dot{\varepsilon}\,(x, -y, 0)$ and $g^{+} = \dot{\varepsilon}$. At this point, simple molecular theories suggest that the chains will deform and become highly oriented [70,71]. Such transitions have been observed in four-roll mill flows [72] and in opposed nozzle flows [144]. By performing measurements using polymers with narrow molecular weight distributions, the dependence of the relaxation time on molecular weight was determined to be $\tau \approx aM^{1.5}$, regardless of the solvent quality [72,144]. This result is surprising since it is generally accepted that the relaxation time for polymers in good solvents should scale as $\tau \sim M^{1.8}$. Odell and coworkers have used this connection between the molecular weight and the relaxation time (and therefore the critical strain rate induce birefringence) to develop a technique for determining molecular weight distributions [144].

The flow-strength criteria stated in equation (10.2) has been examined experimentally by Fuller and Leal [72] and Dunlap and Leal [149] using four- and two-roll mills, respectively. These devices allow one to systematically vary the flow type (the relative amount of pure extension to pure rotation). The birefringence was measured for dilute and semidilute polymer solutions as a function of both the magnitude and type of the flow. Simple molecular models of flexible polymer chains suggest that such data, when plotted as a function of g^{+}, should collapse onto a single curve regardless of the amount of extension relative to rotation. Such a correlation was found to exist for the full range of flow types accessible in these flow cells.

Birefringence measurements in dilute solution respond to local, segmental orientation, but do not directly reveal the extent of overall chain deformation. Total intensity light scattering, however, performed as a function of the scattering angle can be used to determine the radius of gyration and has been employed in extensional flows. Hoagland and coworkers [148] measured the extension of polystyrene chains subject to opposed jets flow. This group reported only moderate deformation of the chains (ratios of the radius of gyration in flow to its equilibrium value on the order of 2), even though strain rates in excess of the values necessary to saturate the birefringence were used. This suggests that chains in extensional flows are able to fully orient without substantially elongating. These results are in agreement with the conclusions of Cathey and Fuller [83], who combined birefringence with measurements of the extensional viscosity of dilute solutions subject to opposed jets flow. They found that as the strain rate increased and the birefringence approached an asymptotic value, the extensional viscosity went through a maximum. Since the extensional viscosity is proportional to the largest length scale of the chains squared, this result implies that as the strain rate is increased, the mean residence time available to the chains is proportionately decreased, thereby limiting their ability to extend. Orientation of polymer segments, which is achieved with lower strains, does occur and causes the birefringence to saturate.

Extensional flow deformation of flexible chains can lead to substantial extensional viscosities which, like the birefringence, is highly localized near the stagnation point. Such an occurrence can lead to polymer-deformation-flow-interactions that can dramatically alter the flow. Using birefringence, Odell and coworkers [150] have observed distinct transitions in the appearance of the localized birefringence patterns in opposed jets flow. As the flow was increased single birefringent strands were found to be transformed to a "pipe" appearance with a dark, less birefringent region enveloped by a sheath of higher birefringence. These experiments were recently analyzed by Harlen, Hinch, and Rallison [151], who solved for the flow field existing in the presence of a strand of high extensional viscosity and demonstrated that it causes a decrease in the strain rate along the center stagnation line. This decease can be sufficient to diminish the birefringence in the center.

10.1.6 Field-induced Phase Transitions

There are numerous examples of polymer solutions and mixtures that undergo apparent changes in phase. Optical techniques are an obvious choice to measure the resulting structures and elucidate the mechanisms for these phenomena. Indeed, one of the first observations of this phenomena was an increased turbidity in the flow of a polystyrene solution in dioctyl phthalate (DOP) [152]. It was observed that when a semi-dilute solution was forced to flow through a contraction, the liquid became cloudy in the vicinity of the region of greatest extension. Later work by Rangel-Naifle *et al.* [89] explored this phenomena in greater detail and offered quantitative measurements of the critical shear rates necessary to induce the transition from a clear to a cloudy solution at various temperatures above the cloud point. These data were interpreted as evidence of an actual flow-induced phase transition and these authors proposed a simple mechanical model that predicted a relationship between the observed apparent increase in the cloud point as a function of the normal stresses in the polymer liquids.

An alternative explanation of the observed turbidity in PS/DOP solutions has recently been suggested simultaneously by Helfand and Fredrickson [92] and Onuki [93] and argues that the application of flow actually induces enhanced concentration fluctuations, as derived in section 7.1.7. This approach leads to an explicit prediction of the structure factor, once the constitutive equation for the liquid is selected. Complex, butterfly-shaped scattering patterns are predicted, with the "wings" of the butterfly oriented parallel to the principal strain axes in the flow. Since the structure factor is the Fourier transform of the autocorrelation function of concentration fluctuations, this suggests that the fluctuations grow along directions perpendicular to these axes.

The first measurements of the butterfly-shaped structure factors in flowing polymer liquids were made by Hashimoto and coworkers [153]. These measurements were made using a cone-and-plate flow cell, constructed of quartz. By sending the light parallel to the axis of rotation of the flow cell, the scattering vector in the limit of small scattering angles will principally lie in the plane defined by the flow and vorticity axes (normally referred to as the (1,3) plane). The scattering patterns were measured by simply photographing the scattered light falling onto a screen, and butterfly shapes were observed with lobes oriented along the flow direction. This observation was confirmed by scattering dichroism measurements made by Yanase *et al.* [154] on PS/DOP solutions in parallel plate flow. These researchers reported flow-induced dichroism oriented parallel to the vorticity axis, which confirmed that the fluctuations themselves are aligned in the same direction. At higher rates of shear, the orientation angle was observed to abruptly switch from the vorticity direction to the flow direction. This latter transition was found to precede a pronounced flow-instability accompanied by dramatic fluctuations in the turbidity and dichroism.

Quantitative light scattering measurements using a couette cell were first made by Wu *et al.* [155] on a PS/DOP solution. By performing measurements with the scattering vector residing within the plane of the flow, rotation of the butterfly patterns with the vorticity of the flow was observed. At low velocity gradients, the lobes of the patterns are oriented at $45°$ relative to the flow direction. As the flow strength was increased, the lobes rotated in the same sense as the vorticity, toward the flow direction.

Transient SALS and scattering dichroism measurements were reported by van Egmond [48] following the inception and cessation of simple shear flow. By sending the incident beam parallel to the vorticity axis, SALS measurements made at sufficiently low scattering angles will restrain the scattering vector to remain in the plane of the flow. The optical arrangement used in this work is shown in Figure 8.16(a), where the scattered light is collected on a screen placed below the sample. The transmitted beam was passed through an aperture in the screen and used for simultaneous dichroism measurements. Upon inception of the shear flow, the SALS patterns, which were originally circular, developed a butterfly pattern with the lobes initially oriented at $45°$ relative to the flow direction. As time proceeded, the lobes rotated in the same sense as the vorticity, until a steady state condition was reached. The dichroism was found to be positive, and initially was oriented at $135°$ to the flow axis, which is in accord with the SALS measurements, which are related to Fourier transformations of the structure factor. Again, as the flow continued, the orientation angle of the dichroism rotated with the vorticity of the flow and achieved a steady state value that was approximately orthogonal to the orientation of the lobes of the SALS patterns.

Van Egmond and Fuller [60] also considered the effect of extensional flows on the growth of concentration fluctuations in polymer solutions. In these experiments, the light was transmitted along the stagnation point of a four-roll mill. These measurements were accompanied by calculations using the Helfand-Fredrickson model [92], which was able to qualitatively predict the shape of the observed patterns. The calculation employed a simple second-order fluid equation of state to represent the non-Newtonian rheology of the polymeric liquid. The analysis follows the same procedure used to obtain equation (7.102). In dimensionless form, the following estimate for the structure factor was obtained:

$$S(\hat{\mathbf{k}}) \approx \frac{1}{1+\hat{k}^2 - D_{\Pi}\hat{\mathbf{l}}\hat{\mathbf{l}}:\frac{\partial}{\partial\psi}\frac{\underline{\tau}}{(\eta_p^0\dot{\varepsilon})}}, \tag{10.3}$$

where

$$\hat{\mathbf{l}}\hat{\mathbf{l}}:\frac{\partial}{\partial\psi}\frac{\underline{\tau}}{(\eta_p^0\dot{\varepsilon})} = 2(l_x^2 - l_y^2) + Wi'(4+8R_{\Psi}') - 8Wi^0(2+8R_{\Psi}^0)\,l_x^2 l_y^2 . \tag{10.4}$$

Here $\dot{\varepsilon}$ is the strain rate and D_Π is the ratio of the stress, $\eta_p^0 \dot{\varepsilon}$, to the osmotic pressure. The scattering vector, $\mathbf{k}$, has been made dimensionless using the correlation length defined in equation (7.99), and $\hat{\mathbf{l}} = \hat{\mathbf{k}}/|\hat{\mathbf{k}}|$. Several dimensionless groups are present in equation (10.4), which measure the relative magnitudes of shear and normal stresses in the fluid. R_Ψ is the ratio of the second to the first normal stress difference coefficient and is normally a negative number. Wi is the Weissenberg number, which is the ratio of the first normal stress to twice the shear stress. The superscript "0" refers to the average value of these parameters in solution, whereas the primed values signify gradients with respect to concentration. This calculation was carried out for a two-dimensional extensional flow with the principal axis of strain in the x direction. The compression axis is the y axis.

Two regimes were observed. At low values of the Weissenberg number, the SALS patterns developed a butterfly shape with lobes that were oriented parallel to the flow direction. This is in agreement with the model, which predicts that concentration fluctuations will grow orthogonal to the principle axis of strain, and the maximum intensity of scattered light will then be parallel to this axis. Such a development is predicted even for a constitutive model that only includes a Newtonian shear stress. At higher values of the Weissenberg number the last term in equation (10.4) becomes important and a four-fold symmetry of scattered light intensity is superimposed onto the butterfly pattern. In other words, the "wings" of the butterfly became decorated with two bright spots. The predicted form of the SALS pattern is shown in Figure 10.2 for one choice of the model parameters.

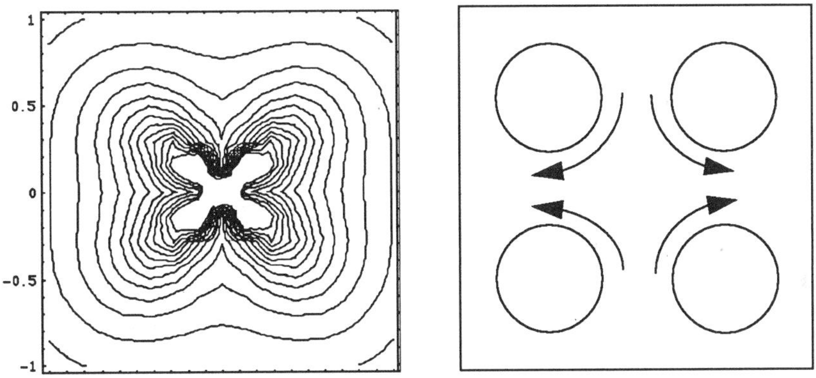

Figure 10.2 Small-angle light scattering pattern predicted using the Helfand-Fredrickson theory for an extensional flow generated using a four-roll mill (shown on the right). Note the four-fold symmetry at small scattering angles. $D_\Pi = 0.1$, $Wi^0 = Wi' = 3$, $R_\Psi^0 = R_\Psi' = -0.333$.

Distinct butterfly patterns have also been observed when inhomogeneous gels are deformed by uniaxial deformation. The length scales in such systems, however, are usually much smaller that those encountered for polymer solutions subject to flow, and small-angle neutron scattering has been used to measure the structure of concentration fluctuations in deformed gels [156].

Electric fields can also distort concentration fluctuations and, unlike hydrodynamic fields, can induce actual changes in the phase behavior of complex mixtures. Wirtz *et al.* [157] used both SALS and scattering dichroism to analyze the influence of electric fields on the structure of polystyrene/cyclohexane solutions close to the phase boundary. Not only did electric fields deform the structure factor of solutions close to the critical point, but they also induced a shift in the phase boundary. In this case elliptical SALS patterns were measured upon application of electric fields, and the major axes of the ellipses were found to be perpendicular to the field. This means that the concentration fluctuations were deformed parallel to the field. Furthermore, for an upper critical solution temperature system, application of electric fields for systems slightly in the two-phase region was seen to cause a "remixing" into a single phase. Upon removal of the field the system relaxed back into a two-phase state. Conversely, lower critical solution temperature systems, just within the single-phase region, were induced to phase separate upon application of an electric field of sufficient strength. In other words, whether the system is characterized by an upper or a lower critical solution temperature, the phase boundary was observed to drop in the presence of an electric field. These experimental observations were successfully modeled using a simple, modified Cahn-Hilliard model containing a specific coupling between the electric field and concentration fluctuations.

10.1.7 The Dynamics of Polymer Liquid Crystals

Liquid crystals represent an important class of complex materials, and display unusual rheological phenomena. Their striking optical appearance makes the use of optical methods a natural choice for examining their microstructural response during flow and there is a large body of literature devoted to this topic. A recent review by Navard summarizes many of these techniques applied to polymer liquid crystals, including polarimetry, small-angle and X-ray scattering, and optical microscopy [158].

As the name implies, liquid crystals are materials that can sustain a flow when subject to an applied stress. Their microstructure, on the other hand, is characterized by a long-range *orientational* order. The particular molecular organization possessed by various liquid crystals has them organized into different categories, according to the particular *positional* long-range order that different phases may lack. For example, the simplest case, the nematic phase, lacks any long-range order in the center of mass of the constituent molecules. Another classification concerns whether the transition from a liquid crystalline phase to an isotropic state occurs as a result of changing the concentration of nematogens in solution, or by changing the temperature of a pure liquid, or melt. The former class of materials are termed "lyotropic," whereas temperature-induced liquid crystals are referred to as thermotropics.

This brief account of past investigations is concerned with work on polymer liquid crystals (PLCs), where the majority of work has considered nematic materials. Among the unique rheological responses associated with PLCs are

1. The first normal stress difference is observed to change sign from positive, to negative, and back to positive as the shear rate is increased [159,160]
2. Transient shear stresses following either a flow reversal or a "step-up" in shear rate show pronounced oscillations, and the period of these oscillations scales with the shear rate [161].

3. Recoverable shear strain measurements following the removal of shear stress after the sample has been sheared at a rate $\dot{\gamma}_0$ collapse onto a single curve when the data are plotted against $\dot{\gamma}_0 t$ [162].

Theoretical models for the rheology of liquid crystals can be divided into continuum, macroscopic models, and microscopic, molecular models. The Leslie-Ericksen (LE) model is the most commonly used continuum model [163,164] and represents the material through a single vector, the director, $\mathbf{n}(\mathbf{x})$, which defines the average direction of molecular orientation as a function of position. In this model, the dynamics of the director are described through a torque balance that incorporates both hydrodynamic torques and a resistance to creation of distortions of the director field by elastic stresses. The hydrodynamic torques involve the viscosity of the liquid, and due to the intrinsic anisotropy of liquid crystals, six "Leslie viscosity coefficients," α_i, are defined, which correspond to specific velocity gradient directions applied to the principal axes of the material. Similarly, there are unique elastic restoring forces resisting different types distortion of the director and separate constants can be defined for bending, splay, and twisting deformations. These are the "Frank elastic coefficients."

The LE model can successfully reproduce many important macroscopic dynamics, including the tendency of liquid crystals to either undergo tumbling or shear aligning when subjected to a simple shear flow. The former type of motion has the director undergo periodic oscillations in its orientation with a period that is inversely proportional to the shear rate and a simple function of a subset of the viscosity coefficients. Shear aligning, on the other hand, is characterized by a director dynamic that evolves monotonically in time to a fix orientation at steady state. The tendency of a liquid crystal to either tumble or flow align is controlled by the sign of the ratio α_2/α_3: positive values lead to flow aligning and negative values cause tumbling.

The constitutive equation derived from the LE theory is restricted to low flows since it is a linear function of the velocity gradient tensor. It cannot predict shear thinning of the shear viscosity or the transition observed in the sign of the normal stress differences as a function of flow strength. Because it incorporates Frank elasticity, however, it does offer the potential of describing polydomain structure, where defects in a liquid crystal can induce strong, local distortions in the director field. Such a theory based on the LE model has recently been proposed by Larson and Doi [165], which is able to successfully reproduce the oscillatory, damped shear stresses following flow reversals observed experimentally. In this theory, the LE equations are assumed to apply on a local length scale that is much smaller than that associated with the domain structure. These equations are then averaged over the domain structure and coupled to dynamic equations representing the response of the domain size, the average domain orientation, and the stress tensor. The model also predict the proper transient recoverable strain response observed in PLCs.

The molecular theory of Doi [63,166] has been successfully applied to the description of many nonlinear rheological phenomena in PLCs. This theory assumes an untextured monodomain and describes the molecular scale orientation of rigid rod molecules subject to the combined influence of hydrodynamic and Brownian torques, along with a potential of interaction (a Maier-Saupe potential is used) to account for the tendency for nematic alignment of the molecules. This theory is able to predict shear thinning viscosity, as well as predictions of the Leslie viscosity coefficients used in the LE theory. The original calculations by Doi for this model employed a preaveraging approximation that was later removed by Marrucci and Maffettone [167], and Larson [168], who were able to predict the sign reversal in normal stresses and explain the phenomena as a transition from tumbling to shear aligning dynamics.

Optical rheometric experiments have been applied to PLCs to elucidate the microstructural origins of their complex flow behavior and to test the models mentioned above. Among the earliest sets of experiments is the work of Asada and coworkers [169], where the light transmitted either through crossed or parallel polarizers was measured when a PLC [a racemic mixture of poly(γ-benzyl glutamate)] was sheared between parallel plates. In these experiments, the direction of the flow cell was either parallel or at 45° to the principal axis of the initial polarizer. Data were reported both during the start-up of flow, and following the cessation of flow. Such a simple experimental arrangement, however, can be very difficult to interpret because multiple orders of retardation are normally present in liquid crystalline samples, so that oscillatory responses will occur in transient flows. These oscillations will mask oscillations that are known to accompany the tumbling behavior of PLC's subject to transient flows. Furthermore, since PLC's are anisotropic at rest, the precise level of retardation is difficult to ascertain from a simple crossed polarizer arrangement. Finally, the birefringence will generally be a complex combination of both intrinsic and form effects. The latter contribution arises due to the presence of scattering from the defect structure in these materials, as well as from orientational fluctuations in the director.

Using the spectroscopic method described in section 8.2.1 [27], Burghardt and coworkers were able to eliminate the ambiguity in the order of retardation associated with previous crossed polarizer experiments. Their studies on poly(benzyl glutamate) indicate that during steady state flow in the linear regime (at shear rates where the viscosity of the PBG is constant), the birefringence is approximately one-half of its monodomain value. This was in contrast to the predictions of the Larson-Doi polydomain model [165], which predicts values that are between 80% and 90% of the monodomain level. In a companion paper, Hongladarom and Burghardt reported birefringence measurements during transient flow experiments. During flow reversals, the birefringence was observed to oscillate with a period that scaled with the shear rate, in the same manner as did the shear stress. Surprisingly, the birefringence was found to *increase* during relaxation from its value attained during shear to levels that were essentially the same as the monodomain levels. This latter finding was in direct contrast to predictions of the Larson-Doi polydomain model.

Scattering dichroism measurements have also been made on PLC's [171,172]. This observable is thought to arise from scattering from disclinations in the material that cause substantial local gradients in the refractive index. Once subjected to flow, anisotropy in the spatial distribution of the defects will cause dichroism to occur. The dichroism, unlike the birefringence, is observed to be zero for a material at rest, and oscillates with a period that scales inversely with shear rate during flow reversal or during a "step-up" in shear rate. Measurements of scattering dichroism as a function of wavelength [172] during steady state shear flow revealed a maximum that shifted to lower wavelengths as the shear rate increased. This was interpreted as a decrease in the length scale of the polydomain structure by hydrodynamic forces.

Small-angle light scattering has also been extensively applied to PLCs subject to flow [173]. As in the case of scattering dichroism, SALS patterns arise principally from fluctuations in orientation, and these are strongest in the vicinity of disclinations, or defects in the director field. The experimental geometries used for SALS in liquid crystals normally use polarizers placed before and after the sample. The arrangements include "VV" scattering, where the polarizers are parallel to one another and aligned along the flow direction, and "HH" scattering, where both polarizers are perpendicular to the flow. In addition, depolarized scattering experiments can be performed where the two polarizers are crossed and oriented along various directions relative to the flow. For example, in "VH" scattering, the first polarizer would be placed parallel to the flow. Reporting on studies made on poly(benzyl L-glutamate) in m-cresol and (hydroxypropyl) cellulose in water and subject to parallel plate flow, Ernst *et al.* [173] found that at low shear rates the SALS patterns were generally elliptical, with the major axis of the ellipse perpendicular to the flow. This was interpreted as arising from a modest deformation of the spatial defect distribution. At higher shear rates, the patterns became increasingly complex, with VH experiments producing "clover leaf" shapes superimposed onto streaks of light oriented perpendicular to the flow. Ultimately, at sufficiently high rates, the SALS patterns disappear altogether, suggesting that the flow is able to reduce the defects in either size or number.

10.1.8 Applications to Thin Films

Processing of thin films can often involve hydrodynamic forces that can deform and orient the constituents of a film. Examples would include spin coating of polymer films and Langmuir-Blodgett deposition. Optical methods can be use for *in situ* and post-processing measurements of the structure and dynamics of these systems. In the fabrication of magnetic media from dispersions of magnetic particles, for example, Fuller and Nunnelley [174] used dichroism measurements to determine the degree of particle orientation in spun coated disks of magnetic media and obtained an excellent correlation between those measurements and direct visual observation using electron microscopy. The latter test, however, is destructive and the dichroism measurement offered the capability of making measurements directly on the disk itself. The instrument used in this study used a photoelastic modulation of light polarization and an optical train of the form shown in Figure 8.7, except that the light was measured following a reflection off of the disk at small angle of incidence. This was possible since the particle layer was deposited on a polished aluminum substrate that was highly reflective. The small angle of incidence minimized any alteration of polarization from the reflection.

10.2 Colloidal Dispersions

10.2.1 Dilute Systems

Applications of optical methods to study dilute colloidal dispersions subject to flow were pioneered by Mason and coworkers. These authors used simple turbidity measurements to follow the orientation dynamics of ellipsoidal particles during transient shear flow experiments [175,176]. In addition, the superposition of shear and electric fields were studied. The goal of this work was to verify the predictions of theories predicting the orientation distributions of prolate and oblate particles, such as that discussed in section 7.2.1.2. This simple technique clearly demonstrated the phenomena of particle rotations within Jeffery orbits, as well as the effects of Brownian motion and particle size distributions. The method employed a parallel plate flow cell with the light sent down the velocity gradient axis.

Although turbidity measurements are attractive because of their simplicity, they suffer from an important limitation: they cannot distinguish between changes in the integrated light intensity from changes in the *degree* of orientation and changes in the average *angle* of orientation relative to a laboratory frame. Both effects will occur during the transient shear flow of nonspherical particles. Dichroism measurements, however, when reported with simultaneous determinations of the average orientation angle of the imaginary refractive index tensor, can eliminate this ambiguity. The first reports of dichroism measurements on dilute colloidal particles subject to transient shear flow were made by Frattini and Fuller [35,177]. This work used photoelastic, polarization modulated dichroism to study the orientation of colloidal bentonite.

The influence of non-Newtonian rheology of a suspending fluid on the orientation of colloidal particles was studied using simultaneous birefringence and dichroism by Johnson *et al.* [37,178]. The optical train used here was described in section 8.4.3 and dichroism was used to track the orientation of prolate, iron oxide particles that were suspended in transparent, polymeric liquids. The birefringence was predominantly a result of orientation of the polymeric matrix fluid. The principal finding of this work was that sufficiently large ratios of normal stresses to shear stresses induced particle orientations along the vorticity axis.

Small-angle light scattering is also a very effective method for measuring particle orientation and has been used by Salem and Fuller [50,178] for larger particles than can be easily handled using scattering dichroism. In these studies, hardened red blood cells were used as model particles in both Newtonian and non Newtonian suspending fluids.

Ackerson and coworkers [179] have also used small-angle light scattering to examine the phenomena of shear-induced melting in "crystals" of dilute dispersions of spheres that are electrostatically stabilized. By effectively removing ions from the suspending fluids, the electrostatic repulsion between the spheres is made very long ranged and forces the spheres into a crystalline registry. Modest shearing forces, however, can disturb such a structure and light diffracted from these systems provide a clear indiction of the manner by which crystalline materials flow.

10.2.2 Structure in Concentrated Dispersions

There are relatively fewer rheo-optical studies on concentrated dispersions, primarily because of the large turbidity of these materials. By optically matching the refractive indices of the particles and the solvent, however, high volume fraction materials can be accessed. This technique was used by Wagner *et al.* [180] and by d'Haene *et al.* [181] to study dispersions approaching the maximum volume fraction. In both cases scattering dichroism was used to measure anisotropy in the pair distribution of spherical particles. The work reported by Wagner *et al.* demonstrated that flow-induced distortion was possible and used the Onuki and Doi theory [13] to calculate predictions of the dichroism from statistical mechanical descriptions of the structure of sheared, hard sphere dispersions. The paper by d'Haene *et al.* considered the microstructural origins of shear thickening viscosities in dense suspensions. By examining the relaxation timescale of dichroism following the cessation of flow both outside and within the region of shear thickening, it was concluded that the formation of "hydrodynamic clusters" were responsible for this effect.

10.3 Case Study 1: Flow-induced Phase Separation in Polymer Solutions

There are many examples of complex, multicomponent liquids that undergo real and apparent changes in phase when subjected to external fields. There are numerous observations of such phenomena for the case of flow fields, all showing a marked turbidity of the samples when a sufficiently strong flow field is applied. As discussed in section 10.1.6, the first studies of this phenomena analyzed this effect as the manifestation of a true, flow-induced phase transition, and correlated the onset of turbidity with the measured normal stresses in the fluids [89]. Such a correlation was suggested by models of the thermodynamics of the mixtures wherein the effects of the flow were included by adding a term to the free energy that was proportional to the deformation of the end-to-end distance of the polymer chains. This approach was later criticized by researchers using an alternative description that envisioned the appearance of turbidity as a consequence of a coupling of concentration fluctuations in the material to the stresses generated by the flow [92,93,94]. This latter approach has the advantage of leading to a prediction of the structure factor, which can be directly observed using small-angle light scattering. This procedure was outlined in section 10.1.6 and led to the SALS pattern shown in Figure 10.2 for the specific case of a two-dimensional, extensional flow field. This simple example modeled the rheology of a polymer liquid as a second-order fluid.

In this case study, the use of a combination of SALS and optical polarimetry (the measurement of birefringence and dichroism) to measure flow-induced structure in polymer solutions will be discussed. Furthermore, the connection of these structural observables to the mechanical properties of a polymer liquid showing this effect will be presented. Among the first decisions to be made in such an investigation is the choice and design of a flow cell to both generate the desired flow and accommodate the light used in the optical measurement. In this example, two flow cells were used: a four-roll mill to produce a two-dimensional flow, and a parallel plate flow cell for simple shear flow measurements. A schematic drawing of the four-roll mill is shown in Figure 10.3. Here four rollers located on the corners of a square are rotated to cause the fluid in the center region to execute an approximation to a two-dimensional flow with hyperbolic streamlines. A stagnation point, where the fluid velocity is precisely zero, exists at the geometric center between the rollers. The ratio of the center to center distance, d, shown in the figure, to the diameter of the rollers, D, is chosen to produce an approximation to a homogeneous hyperbolic flow. Recently, Higdon has analyzed this geometry for the case of a Newtonian liquid and has determined various choices for the ratio of d/D for given ratios of the outer cup diameter to the roller diameter that produce optimum velocity gradient profiles in the center region.[182]

In the side view in Figure 10.3, a glass rod is shown that is inserted between the rollers and is centered on the stagnation point. This rod serves to conduct the light downward into the sample, which only partially fills the device. The purpose for this configuration is to avoid any contact of the fluid with the upper, double bearings that support the rollers. Double bearings are used on the top to support the rollers and simple bushings (not shown) support them on the bottom. In this way, there is no danger of contamination of the sample by lubricants in the bearings. The bottom of the outer cup is fashioned from a single glass or fused quartz plate so that light that is scattered at small angles can be brought out of the cell for measurement. If only birefringence or dichroism is measured, a smaller window can be used. At the top of the cell, the rollers are connected to a drive mechanism that rotates them in the appropriate directions. If only an extensional flow is desired, this is most easily accomplished by using a single motor that drives one of the rollers. By connecting all four rollers together with intermeshing gears at the top of the cell, the rotation of a single roller will induce the desired motion of the assembly.

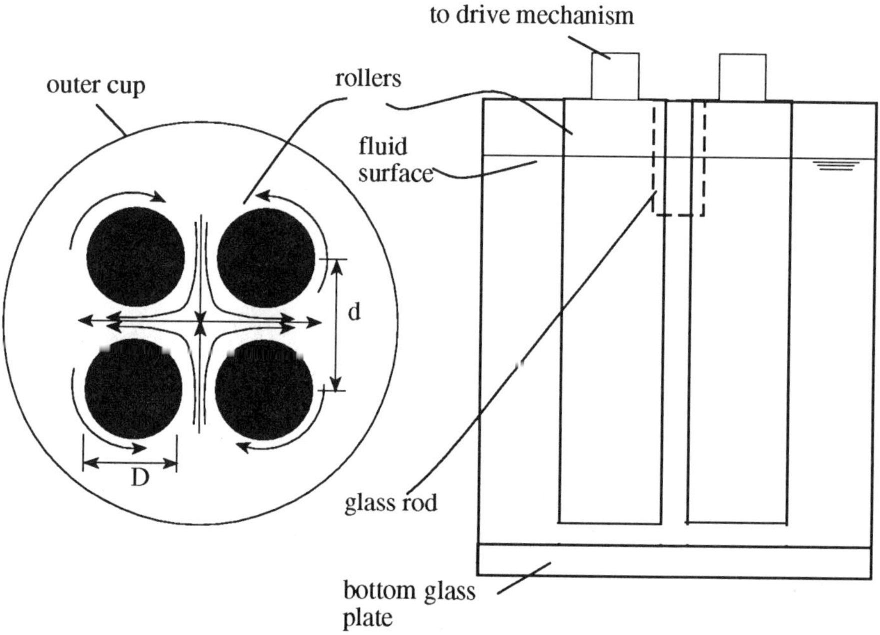

Figure 10.3 Schematic diagram for a four-roll mill flow device.

The light scattering instrument used in this study is shown in Figure 8.16a with the four-roll mill being inserted at the sample location. Because of the design of the mill, it must be oriented vertically, with the rollers parallel to the axis of gravity, and for that reason, the optical train is mounted in a vertical configuration. In these experiments, the light was generated using a helium-neon laser and, if desired, polarization modulation optics (a suitable PSG setup) could be added between the sample and the laser if simultaneous birefringence and dichroism measurements were needed. In that case, a PSA section and a photodetector would be added after the screen. Light transmitted through the aperture would be used for this purpose.

The sample was a solution of polystyrene (PS) dissolved in dioctyl phthalate (DOP). This system has a theta temperature of approximately 22°C [183] and has been the subject of most of the studies investigating flow-induced phase transitions in polymer solutions. The particular sample used here had a molecular weight for PS of 2 million, a polydispersity of $M_W/M_N = 1.06$, and a concentration of 6%. This results in a semidilute solution that is quite viscoelastic. The cloud point was determined to lie between 10°C and 10.5°C.

Figure 10.4 shows a measured SALS pattern for the PS/DOP system subjected to an extensional flow of a strain rate of 2 s^{-1} using a four-roll mill with its elongation axis at an angle of –45° relative to the horizontal axis. This pattern should be compared against the prediction shown in Figure 10.2 (note that the extension axis in Fig. 10.2 is along the horizontal). As predicted, the structure factor develops a butterfly-shaped pattern with the "wings" of the butterfly oriented parallel to the flow direction. Because the structure factor is the Fourier transform of the actual correlation function of concentration fluctuations, this result can be interpreted as evidence that the fluctuations principally exist perpendicular to the flow. This is a result of the coupling the concentration fluctuations to the stress tensor through the rheological material functions of the fluid. As discussed earlier, the basic butterfly shape can be predicted solely on the basis of the shear viscosity. The four-fold symmetry that exists both in the measured and predicted patterns arises from the normal stress coefficients, which measure the elasticity of the material.

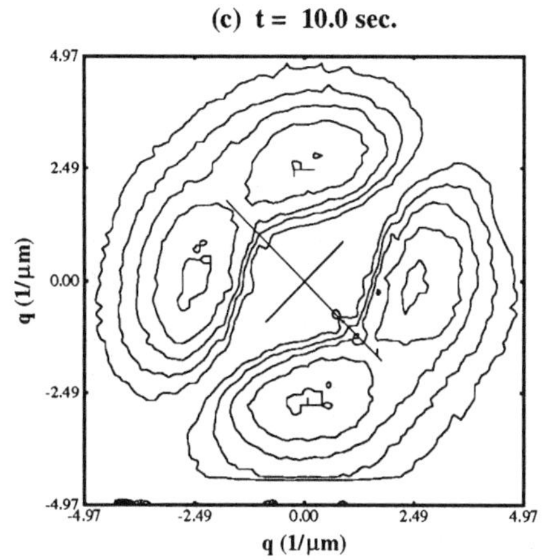

Figure 10.4 Small angle light scattering pattern for a 6% solution of polystyrene (molecular weight 2 million) in dioctyl phthalate. The flow was generated using a four-roll mill with the extension axis at $-45°$ relative to the horizontal and a strain rate of $2.0\ s^{-1}$.

For experiments using simple shear flow, several different flow devices are available. These would include couette, cone and plate, and parallel-plate geometries. In this study, a parallel plate geometry was use that was built into a controlled stress rheometer (a Rheometrics Dynamic Stress Rheometer was used). This instrument generates a shearing flow by driving the upper disk of a parallel plate cell with a programmable stress. The lower plate is stationary. Both plates consisted of disks of fused quartz that were 25 mm in diameter. The use of fused quartz is not essential, but is normally available with very uniform properties and is relatively free of birefringence. These qualities are desirable for the upper, rotating disk to minimize unwanted background fluctuations.

When a parallel-plate device is used, and if the light is transmitted at normal incidence to the plates, the angle of orientation that is measured is restricted by symmetry to being either parallel or orthogonal to the flow direction. This geometry is explained in Figure10.5.

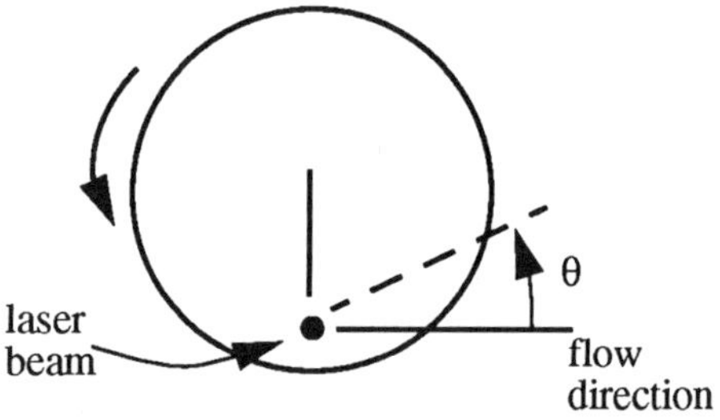

Figure 10.5 Top view of a parallel-plate flow cell where the light beam is directed normal to the disks. The orientation angle, measured relative to the flow direction, is either 0° or 90°.

The integration of the optics within the stress rheometer allows the simultaneous acquisition of optical and mechanical properties of the sample, which can be important for determining the relationship between a material's microstructure and its rheological response. The particular rheometer used here has an open construction that easily accommodates the insertion of an optical train for the purpose of birefringence and dichroism measurements. The optical train consisted of a helium-neon light source with a $(P/RH)_{PSG}$ module located below the rheometer. The light was sent through the flow cell at normal incidence to the parallel plates from below and was passed to a photodetector. For the case of dichroism measurements, the light intensity was measured directly after passing through the sample. In the case of birefringence measurements, a circular polarizer was inserted between the sample and the detector.

The intensity signal was recorded by an analog/digital converter operated by a personal computer. The data conversion process was timed using pulses generated from an optical encoder attached to the shaft containing the rotating half-wave plate. In this way, 128 intensity measurements were retrieved on each cycle of the half-wave plate. This wave form was then processed using a fast Fourier transformation which yields the coefficients multiplying the constants B, G, and L in equations (8.27) and (8.33). Using this instrumentation, the data shown in Figure 10.6 was obtained for a PS/DOP solution at room temperature. The molecular weight of the polymer was 4 million and the concentration of the polymer was 6%.

The mechanical and optical data reveal a remarkable transition in the properties of this sample once a critical shear rate in the neighborhood of 40 s^{-1} is surpassed. When this is achieved, the viscosity, which was shear thinning prior to this point, becomes sharply shear thickening. This is accompanied by a marked increase in the magnitude of the birefringence and a change of sign in the dichroism data. It is important to note that the dichroism data plotted in this figure is multiplied by the cosine of twice the orientation angle. As described in Figure 10.5, due to the symmetry of the flow in the plane probed by the polarization of the light, the orientation angle is either 0° or 90°. The cosine function is then either plus or minus unity for those two values, respectively. Since the dichroism is a result of scattering by concentration fluctuations, it is a form effect and is expected to be positive. Therefore, when the dichroism signal in Figure 10.6 is negative, it can be interpreted as sensing the anisotropy of structures that are oriented *perpendicular* to the flow. In the shear thickening region, the sign becomes positive, indicating that at large shear rates those structures become oriented in the flow direction.

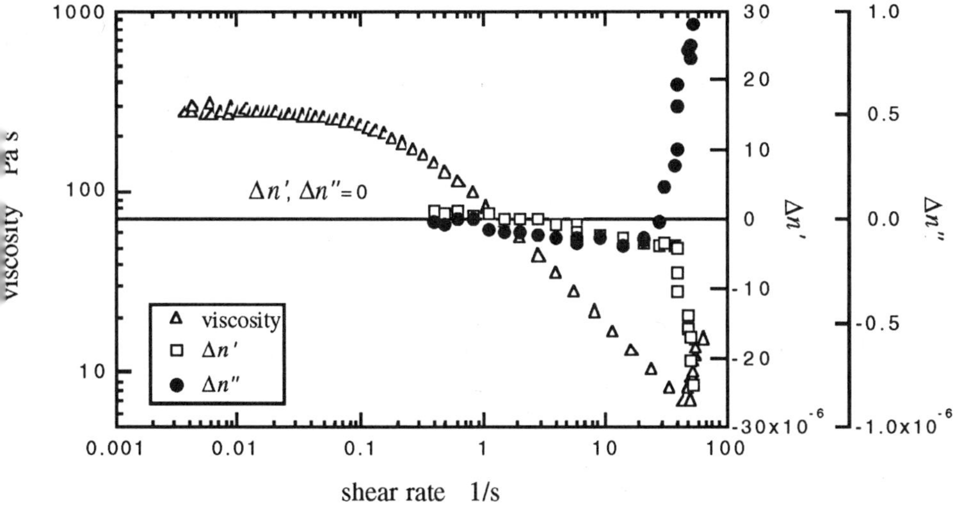

Figure 10.6 Mechanical and optical flow properties of a 6% solution of polystyrene (molecular weight 4 million) in dioctyl phthalate. *Triangles*: shear viscosity; *squares*: birefringence; *filled circles*: dichroism. Both the dichroism and birefringence are multiplied by the cosine of twice their respective orientation angles.

The measurement of a dichroism that is oriented perpendicular to the flow at lower shear rates agrees with the measurement of butterfly-shaped structure factors by small-angle light scattering. When those patterns have the wings of the butterfly oriented parallel to the flow, this also indicates orientation of scattering structures perpendicular to the flow. Evidently, at higher rates of shear, the wings ultimately become distorted themselves perpendicular to the flow, which means the structures become aligned with the flow.

The birefringence is also plotted as the product of the birefringence times the cosine of twice its orientation angle. The birefringence signal that is measured is negative, and this agrees with the fact that polystyrene has a negative stress optical coefficient. However, the birefringence is a combination of both intrinsic and form effects, and these two effects are difficult to separate. However, it is apparent that the intrinsic contribution is the dominant component, since the form effect shown be positive in the shear thickening region. The fact that the birefringence shows an increasingly negative value in this region suggests that the birefringence is predominantly a result of local, segmental orientation.

10.4 Case Study 2: Dynamics of Multicomponent Polymer Melts - Infrared Dichroism

Polydispersity is an important aspect of polymer rheology since the relaxation times of macromolecules are strongly dependent on molecular weight and its distribution. A useful model system that has been investigated to offer insights into the role of polydispersity is the bimodal distribution, where two narrow molecular weight chains of different sizes are blended together. If the two molecular weights are sufficiently well separated, mechanical measurements can offer a reasonable separation of the individual dynamics of the separate components. However, this separation is never perfect and becomes difficult to establish when the two molecular weights are too close together. Optical methods, when properly designed, can provide the capability of effecting such a separation, and spectroscopic techniques such as infrared dichroism and Raman scattering are particularly useful.

In this case study, which is extracted from reference 184, infrared dichroism is described as a means of separating the component dynamics in multicomponent polymer melts. What is necessary is the existence of distinct absorption peaks for at least one of the components. In the present problem, however, where two chains of identical chemistry but different molecular weights are mixed, there will not be any intrinsic differences in their absorption spectra. In this case it is necessary to label one of chains with a tag that will allow its presence in the blend to be revealed. For this purpose, deuteration of one of the chains is often used. This provides the labeled chain with an absorption of infrared light at the symmetric stretching vibration of the C-D bond, which occurs in the vicinity of 2180 cm^{-1}. Fortunately, the unlabeled polymer contains no absorption peak at this location. It is important, however, to determine that the presence of a label on one species will not alter the physical response of the sample at a level that will affect the phenomena under study. For example, the labeling should not induce phase separation or cause unwanted specific interactions.

The optical apparatus used in this study was designed according to the strategy described in section 8.4.3, which permits the simultaneous measurement of birefringence and dichroism. The source was a infrared diode laser that generates light at a wavelength in the range from 2180 to 2195 cm^{-1} with a power of 10 mW. This source was chosen instead of a blackbody radiation lamp because of its high power. This allows relatively thicker samples to be studied, which is an advantage in rheological experiments where well-defined flow fields must be applied. A blackbody source combined with a monochromator, on the other hand, has the advantage of allowing one to select the wavelength. The polarizing components were chosen to optimize the transmission of infrared radiation. Brewster angle polarizers of the design shown in Figure 9.2 were used and the photoelastic modulator was fabricated using a crystal of zinc selenide. The detectors were liquid nitrogen cooled, photovoltaic, indium-antimonide devices.

The polymer samples were polybutadienes (PB) with 1% of the repeating units modified with 4-phenyl-1,2,4-triazoline-3,4-dione (urazole) groups. The PBs were synthesized using anionic polymerization and two samples were prepared: PB109-PU1, which had a molecular weight of 109,000 and a polydispersity index of $M_W/M_N = 1.09$, and PB28-PU1, with $M_W = 28,000$ and $M_W/M_N = 1.05$. The lower molecular weight was also available in a deuterated form referred to as dPB29-PU1, with $M_W = 29,000$ and $M_W/M_N = 1.05$. Both the high and low molecular weight species are well above their entanglement molecular weight. Preparation of the blends consisted of dissolving the PB109-PU1, PB28-PU1, and the dPB29-PU1 in toluene. The weight fraction of the PB109-PU1 was determined by weighing out the appropriate amount of this component relative to the total amount of the sum total of the PB28-PU1 and dPB29-PU1. The relative amounts of PB28-PU1 and dPB29-PU1 in the blend was determined by adjusting the amount of dPB29-PU1 so that a measurable amount of dichroism could be detected without causing too much absorption of the light. In practice, this meant that the total amount of dPB29-PU1 in a blend was in the range of 5%-10%. The toluene was later removed through a combination of heat and vacuum.

The presence of the urazole groups leads to hydrogen bonding between the chains and the rheological properties are strongly affected by these interactions. These systems are models of "sticky" chains where the chain dynamics are coupled together and the effective friction factor between the chains is enhanced. Among the questions that was addressed by this study was the role of specific interactions on the phenomena of orientational coupling between polymer segments. This effect leads to a tendency of segments to achieve orientations that are proportional to the mean orientation of segments in their vicinity. For example, small, entangled solvent molecules dissolved within a cross-linked polymer will orient as the network is deformed, even though they are not mechanically coupled [131]. In a polydisperse polymer melt relaxing following a strain relaxation, low molecular weight chains will continue to remain oriented long after they have disengaged from their entanglements since their orientation will remain coupled to that of longer chains that have not fully relaxed [132]. The magnitude of orientational coupling is expressed by the coupling coefficient, ε, which is defined as

$$\langle \mathbf{uu} \rangle_{probe} = \varepsilon \langle \mathbf{uu} \rangle_{bulk}, \tag{10.5}$$

where $\langle \mathbf{uu} \rangle$ is the average of the second moment over the distribution function of unit vectors, $\mathbf{u}$, that define the orientation of single segments in the system.

This phenomena is most clearly seen by following the relaxation dynamics of low molecular weight chains comprising a binary molecular weight polymer blend. Infrared dichroism arising from labeled short chains will identify these dynamics. Simultaneous birefringence measurements will provide a direct measure of the average, bulk orientation of polymer segments in the mixture. However, it is necessary to provide absolute measurements of the averages $\langle \mathbf{uu} \rangle$ that appear in equation (10.5) to obtain a quantitative determination of the coupling parameter. From equations (7.24) for the birefringence and (5.24) for the dichroism, the measurement of these two optical anisotropies will provide only the second order moment to within constants characterizing the intrinsic anisotropies in the polarizability and absorption cross sections of the polymer. This problem could be addressed by determining these constants by calibration measurements performed on systems of known orientation (a fully oriented material, for example). In the present experiments, which involve fully entangled chains, a simpler, alternative approach can be used.

Immediately following a step strain deformation, all of the segments of a fully entangled melt can be assumed to have the same degree of orientation. In other words, both the short and the long chains will be characterized by identical functions $\langle \mathbf{uu} \rangle_{t = 0^+}$, where $t = 0^+$ refers to the time just following the step strain. For that reason, the initial birefringence, $\Delta n' (t = 0^+)$, and the dichroism, $\Delta n'' (t = 0^+)$, values correspond to the same degrees of segmental orientation and can be used to normalize the data measured at subsequent times. Clearly, at later times, these two functions will reflect different degrees of alignment since the dichroism corresponds to orientation of the labeled short chains, and the birefringence reflects the average orientation of the bimodal blend.

Figure 10.7 shows the measured birefringence and dichroism of a blend consisting of 70% PB109-PU1 and 30% PB28-PU1 at a temperature of $-20°C$. These data are plotted as functions of time following a step strain deformation and are in arbitrary units. The result is a significantly faster relaxing dichroism relative to the birefringence since this observable tracks the short chains. The birefringence relaxes in a manner reflecting the combined influence of both the high and low molecular weight species. However, at longer times, it is evident from these two curves that the dichroism ultimately relaxes at a rate comparable to the birefringence, and this is a consequence of orientational coupling of the short chain orientation to that of the bulk.

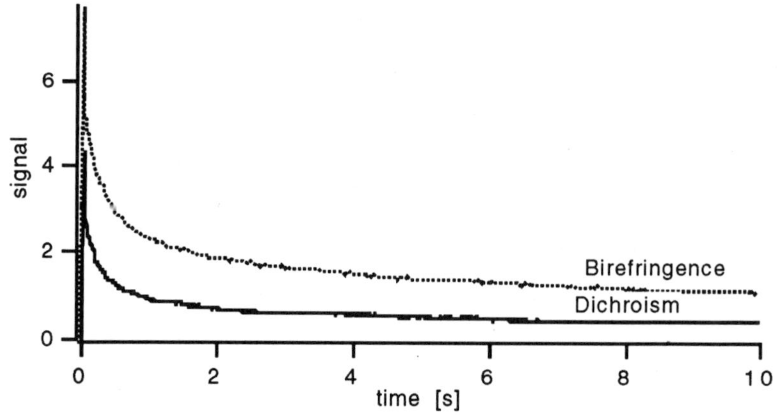

Figure 10.7 The relaxation of birefringence and dichroism following a step strain. The sample was a 70/30 blend of PB109-PU1/PB28-PU1 at a temperature of −20°C.

As motivated by the discussion above, the data in Figure 10.7 can be properly normalized by the initial values of the dichroism and birefringence. The result of such a normalization is that at time $t = 0^+$, these two functions will both start with values of unity, and will reflect identical degrees of segmental orientation. In Figure 10.8, the normalized dichroism and birefringence are plotted against each other creating a parametric plot in time. The point in the upper right-hand corner with a value of (1,1), is the initial point at time $t = 0^+$. As time proceeds in the experiment, the data move toward the origin in the plot, which represents the ultimate equilibrium state where both the birefringence and dichroism tend to zero. As is evident from the data in Figure 10.8, as the data tend toward the origin, the normalized dichroism and birefringence become proportional to one another and the slope of the curve is the coupling coefficient. A value of $\varepsilon = 0.72$ was determined for this system.

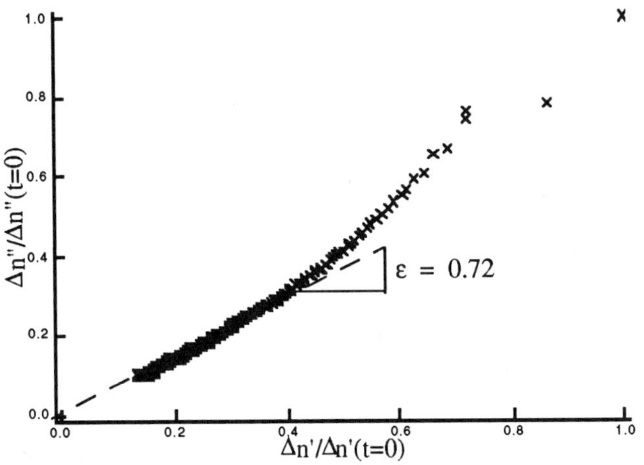

Figure 10.8 Determination of the coupling parameter, ε, for a 70/30 blend of PB109-PU1/PB28-PU1 at a temperature of $-20°C$. This value was determined to be 0.72 by calculating the limiting slope of the data near the origin. The dashed, straight line represents the least-squares fit of the data in the linear region near the origin.

10.5 Case Study 3: Orientation in Block Copolymers - Raman Scattering

Block copolymers consist of chemically distinct polymer chains that are tethered together to form a single macromolecule. If the individual blocks are immiscible when they are unattached, phase separation will also normally occur in the case of the copolymer, with morphologies that depend on the relative composition of the separate block species, and their manner of attachment (diblocks, triblocks, stars, etc.). This is a result of the physical connection of the blocks, which prevents them from separating over distances greater than the contour lengths of the respective blocks. The result is a microphase separation with adjacent domains that are richer in either of the chemical species.

As the blocks attempt to phase separate and reduce the enthalpic penalty that drives immiscibility, their connectivity also imposes conformation restrictions that will oppose large elongations of the chains because of entropic considerations. As a consequence of these two competing mechanisms, microphase domains tend to be on the order of the radius of gyration of the blocks. The transition from an amorphous state to a microphase separated state identifies the microphase separation temperature (MST). Above this temperature, Gaussian statistics are expected to apply to the chains of each block. Below this temperature, where strong segregation occurs, the interface between domains will sharpen. Furthermore, the constraint of incompressibility requires uniform space filling of the polymer segments and the available chain conformations become restricted. When this occurs, a departure from Gaussian statistics may occur, and this can be accompanied by chain stretching.

Recently, a theoretical description of microphase separation in block copolymer by Semonov has assumed significant chain stretching [185]. This work demonstrated that this assumption simplifies the description of the thermodynamics of block copolymers under conditions of strong segregation. This has been accompanied by experimental work aimed at investigating the degree of stretching in these systems. Neutron scattering measurements, conducted by Almadal *et al.* [186] were used to measure the block dimensions in a diblock system. These researchers found stretching, even for temperatures above the MST, where the system is homogeneous. Birefringence measurements have also been used by Lodge and Fredrickson [187] and Folkes and Keller [188], but these results appear to be contradictory, with the former group finding substantial intrinsic birefringence, and the latter group reporting a predominantly form birefringence result. Intrinsic birefringence would suggest segmental orientation. Orientation of microdomains, on the other hand, will produce form birefringence.

A spectroscopic method, capable of identifying the orientation of the individual blocks of a copolymer would remedy the ambiguity of birefringence measurements, where the effects of intrinsic and form effects cannot be easily separated. Such measurements of local bond level orientation would offer important insight into the question of chain stretching that would complement the neutron scattering results where overall block conformation is obtained. In this case study, which follows the paper by Archer and Fuller, [189] the use of Raman scattering as a vibrational spectroscopy capable of isolating the orientation of selected chemical species is described.

The material that was used in this study was a polystyrene-polybutadiene (PS-PBD) six-arm star block copolymer. This was supplied by Shell Development Company as a research-grade material having a molecular weight of 27,800 for each arm and a polydispersity index, $M_W/M_N = 1.3$. The styrene content was 30% by weight with the PBD occupying the inner portion of the star. Electron microscopy of thin films of this material cast from toluene reveal a cylindrical morphology for the microphase domain structure with PBD acting as the matrix. To identify the MST, both static birefringence and dynamic mechanical rheometry were used. This temperature was identified to be between 145 and 149°C.

To use Raman scattering as a method of isolating the orientation of the individual blocks of the sample, Raman scattering spectra were examine to ensure that each block yielded separate, identifiable peaks. For the PS-PBD system, these are provided by the C-C aromatic-aliphatic stretching vibration of the PS at a frequency of 1029 cm^{-1}, and at 1620 cm^{-1}, the PBD generates a peak corresponding to the C=C symmetric stretching vibration mode. Although other peaks could be used, these are well separated, and quite pronounced.

The objective of these experiments was to determine the level of local orientation of the individual blocks as the sample is cooled from the disordered, homogeneous state across the MST and into the strongly segregated region. However, when microphase separation normally occurs, a domain structure will appear where the PS cylinders will have a random, macroscopic orientation. Since the Raman experiment and birefringence measurements measure orientation that is averaged over the entire portion of the sample illuminated by the incident light, a polydomain sample would appear unoriented, even if there was a significant degree of local orientation. To overcome this difficulty, the single domain samples were first subjected to a constant compressional load at a temperature of 90°C and the the temperature was elavated to 120°C for a period of 48 hours. The effect of this procedure was to produce a monodomain material, where the axes of the PS cylinders were oriented macroscopically in a single direction.

The experimental geometry of the sample relative to the axis of the optical instrument is shown in Figure 10.9. In this configuration, the PS cylinders are oriented along the "y" axis and perpendicular to the incident light, which is along the "y" axis. Both the Raman scattering and birefringence experiments that were used to probe orientation are capable of determining the angle at which orientation occurs, and these were measured relative to the "x" axis. For this reason, the orientation angle of any form birefringence due to the orientation of the cylinders will appear at $90°C$. Local, segmental orientation, on the other hand, is expected to run from cylinder to cylinder and will produce an orientation angle of $0°C$. The form birefringence will be positive because it is proportional to the *square* of the difference between the refractive indices of the PS and PBD. Intrinsic birefringence can be either positive or negative, depending on the sign of the stress optical coefficients for the two blocks. For PBD this is a positive quantity. PS, however, has a negative stress optical coefficient above its glass transition temperature and a positive value below this point [190].

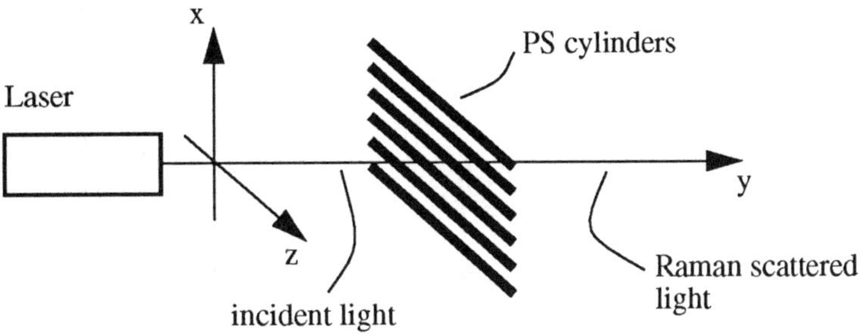

Figure 10.9 The geometry of the sample relative to the axis of the optical apparatus. The PS cylinders are oriented parallel to the "z" axis and perpendicular to the axis of the incident light beam, which is directed along the "y" axis. Angles of orientation are measured relative to the vertical, "x" axis.

The optical apparatus used in this work was described in section 8.6 and has the capability of providing both Raman scattering and birefringence measurements simultaneously. The Fourier expansion of the overall Raman scattering signal is given by equation (8.51), and the coefficients are given by equations (8.52) to (8.54). In these expression, a simple, uniaxial form for the Raman tensor was assumed. From these coefficients, the anisotropies in the second and fourth moments of the orientation distribution can be solved as

$$\langle u_x^2 - u_z^2 \rangle = \frac{15 - 15r_1 - 33r_2 + \varepsilon(10 - 10r_1 - 22r_2) + \varepsilon^2(1 - 4r_1 - 7r_2)}{\varepsilon(-29 - 4r_1 + 11r_2 - \varepsilon(11 + r_1 - 5r_2))}, \quad (10.6)$$

$$\langle u_x^4 - u_z^4 \rangle = \frac{30 - 30r_1 + \varepsilon(20 - 20r_1 - 24r_2) + 2\varepsilon^2(1 - 4r_1 - 7r_2) + 6r_2\varepsilon^3}{\varepsilon[-29 - 4r_1 + 11r_2 - \varepsilon(11 + r_1 - 5r_2)]}, \quad (10.7)$$

where $r_1 = R_\omega/\sin\delta$, $r_2 = R_{2\omega}$, and ε was the parameter that was introduced in the description of the uniaxial Raman tensor. The retardation, $\delta = 2\pi\Delta nd/\lambda$, is determined directly from the birefringence measurement.

In Figure 10.10 Raman spectra of the PS/PBD copolymer sample are displayed. These spectra were generated using the $R_{2\omega}$ signal described in equation (8.53). This is a convenient choice since this signal is only nonzero in the presence of orientation and is not affected by birefringence. As the temperature is brought below the MST, orientation in either the C-H or the C-C bonds does not become readily apparent until a temperature of $85°C$ is reached. Evidently, no local orientation is present as the material first crosses the MST, but only is achieved in the strongly segregated region.

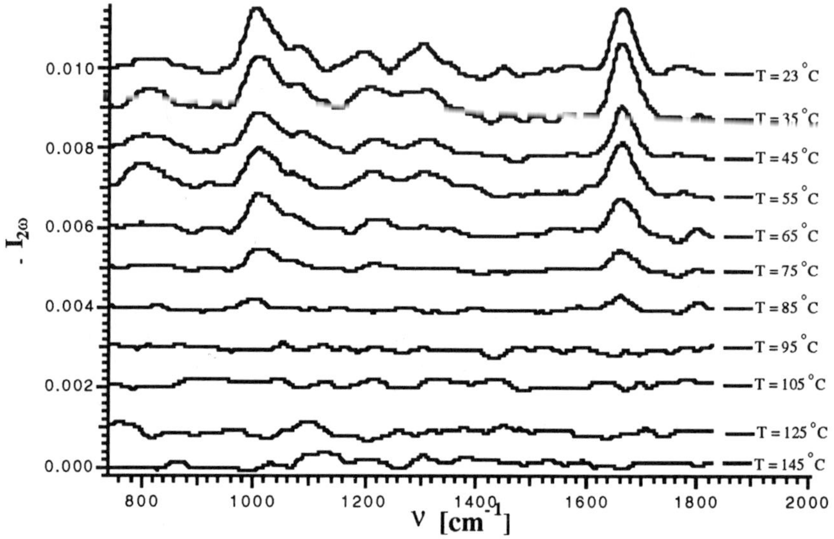

Figure 10.10 $I_{2\omega}$ Raman spectra of a PS/PBD star block copolymer as functions of temperature from 23 to $145°C$. The MST is approximately $140°C$.

The birefringence measured simultaneously with the Raman spectra are shown in Figure 10.11. In plotting this figure, the birefringence was multiplied by the cosine of twice the orientation angle of the birefringence. By symmetry, this angle is either $0°$ when the birefringence is oriented parallel to the "x axis in Figure 10.9 or $90°$ when the orientation is along the "z" axis. The cosine function will have values of +1 or -1, respectively, for these two conditions. As shown below, this product is first becomes negative when the sample is cooled from above the MST. This can be interpreted as the appearance of a cylindrical domain structure with the cylinders oriented parallel to the "z" axis since this would produce a *positive* form birefringence. Just below $100°C$ this product changes sign and becomes positive. This occurs at about the same temperatures where the Raman spectra indicate that segmental orientation occurs in both the PS and PBD blocks. This can be interpreted as the creation of a positive, intrinsic birefringence of segments oriented parallel to the "x" direction. The intrinsic birefringence is expected to be positive since this effect occurs in the vicinity of the glass transition of PS, where the stress optical coefficient becomes positive for this polymer. The intrinsic birefringence of PBD is always positive.

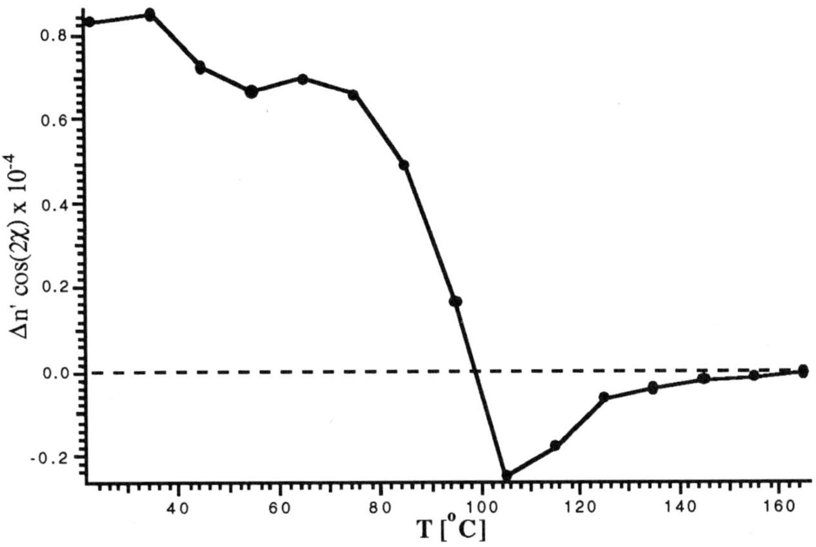

Figure 10.11 Birefringence multiplied by the cosine of twice the orientation angle of a PS/PBD star block copolymer as a function of temperature.

10.6 Case Study 4: Local Orientational Dynamics - Two Dimensional Raman Scattering

The two dimensional infrared technique of Noda and coworkers discussed in section 10.1.4 has proven to be a powerful tool for investigating the local dynamics of multicomponent polymer materials. This method of analyzing the orientational dynamics of different components in a complex liquid is not restricted to dichroism, however, and can be used for other spectroscopic methods, such as Raman scattering. Recently, in the laboratory of the author, two-dimensional Raman scattering has been developed as a tool to examine the local dynamics of polymeric liquids [143]. In this case, the synchronous and asynchronous correlation functions are constructed as described by equation (10.1), except that an appropriate Raman scattering orientation anisotropy signal is used in place of $\Delta n''$. A convenient choice would be the $R_{2\omega}$ signal given in equation (8.53) since this observable is not affected by birefringence.

As an example, we consider here the orientational dynamics of a simple polymer melt of poly (isobutylene). The sample used here is a room temperature melt of average molecular weight 300,000, which is well above the entanglement chain length for this polymer. It is first necessary to examine the Raman spectra for this material to determine the extent of sensitivity to the polarization of light by various vibrational modes. In Figure 10.12, both the polarized (with the analyzing polarizer parallel to the incident polarizer) and depolarized (with the analyzer crossed) spectra are shown for an undeformed sample. Examination of these data indicate two highly polarized peaks at approximately 700 and 2900 cm^{-1}. The peak at 700 cm^{-1} is assigned to the symmetric stretching vibration of carbon-carbon bonds on the chain backbone. The band at 2900 cm^{-1} is typical of symmetric and asymmetric carbon-hydrogen vibrations in polymeric materials. It must be emphasized that there are two primary C-H populations that can contribute to this band: C-H bonds along the backbone ($2900\ cm^{-1}$) and C-H bonds on the methyl groups ($2925\ cm^{-1}$).

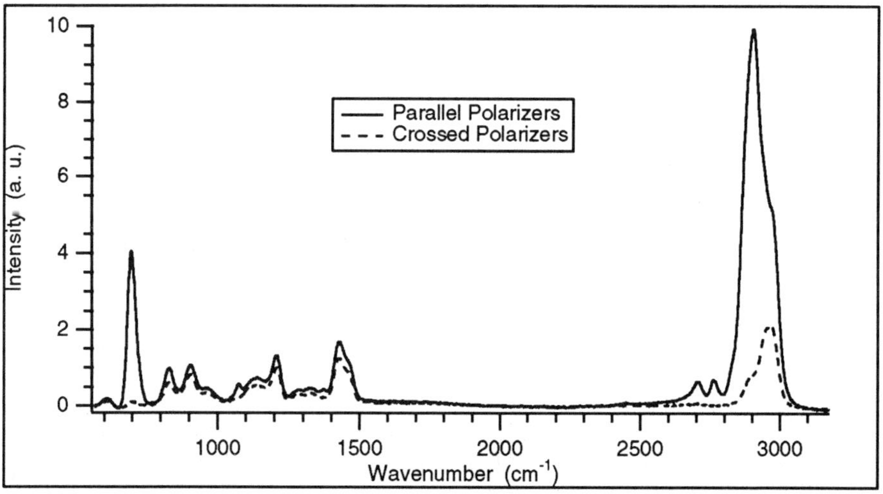

Figure 10.12 Polarized (parallel polarizers) and depolarized (crossed polarizers) spectra of undeformed polyisobutylene at room temperature.

The fact that these bands are highly polarized suggests that they should be sensitive to orientation of the groups that contribute to these vibrational modes and can be used to follow the local orientation. For this purpose, "dog bone" shaped samples of approximately 1 mm is thickness were tested by placing the specimen within the jaws of a simple stretching device that was capable of imparting a sinusoidal elongation. Strains of 1.5% were applied over a range of frequencies, and the Raman scattering signals were measured along with the birefringence of the sample.

In Figure 10.13 the dynamic birefringence and Raman anisotropy for the C-C bond are plotted as a function of frequency. Here the signals have been decomposed into the components that are respectively "in-phase" ($\Delta n'$ and Γ') and "out-of-phase" ($\Delta n''$ and Γ'') with the applied strain. Since this melt would obey the stress-optical rule, the birefringence signals are proportional to the mechanical storage and loss moduli. This relationship, however, would breakdown at frequencies above a mechanical glass transition, as mentioned in the discussion of Figure 10.6, where such a deviation was found for poly(butadiene). The upturn in the out-of phase birefringence data above a mechanical frequency of 10 Hz suggests that this may occur in the PIB system.

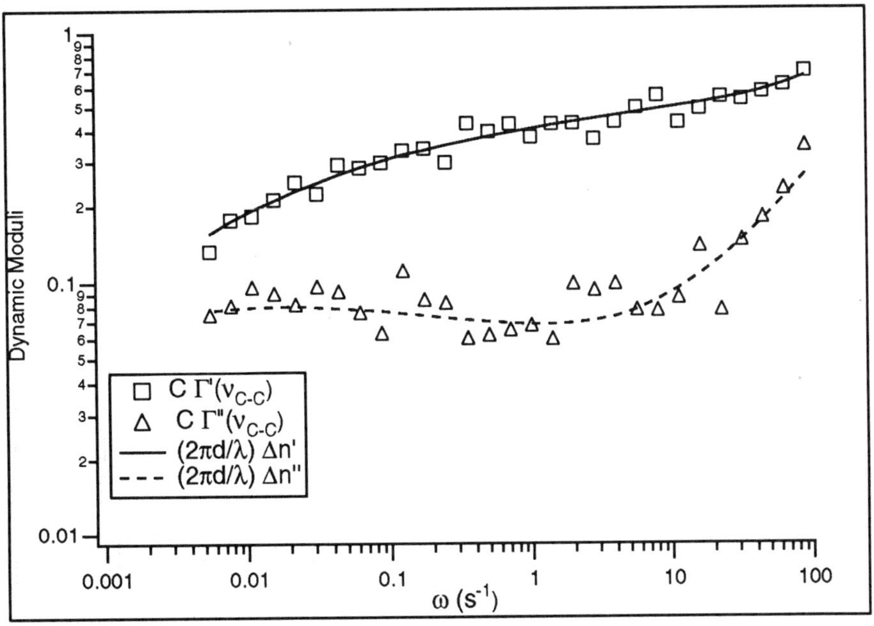

Figure 10.13 Dynamic in-phase (single prime) and out-of-phase (double prime) signals for the birefringence, $\Delta n'$, and the Raman anisotropy for the C-C bond, Γ, for a room temperature polyisobutylene melt subjected to a strain of 1.5%. The phase is relative to the applied strain.

In Figure 10.13 the Raman signals have been scaled by a single factor C that adjusts them to lie on top of the birefringence measurements. It is observed that the C-C bond dynamics follow the birefringence response, which is not surprising for a homopolymer melt.

The correlated orientation dynamics of different species along the chain can be explored by analyzing the Raman anisotropy data according to the two-dimensional strategy of Noda, and this is shown in Figures 10.14 and 10.15, where the synchronous and asynchronous two-dimensional spectra are plotted. The data in these figures were acquired using a 1.5% strain and a frequency of 82 Hz . The synchronous spectrum shows dominate features in the vicinity of 700 cm^{-1} and 2900 cm^{-1}, corresponding to the C-C and C-H bonds of the polymer. Peaks along the diagonal correspond to the synchronous autocorrelation function of orientation of vibrational modes. The off-diagonal peaks, on the other hand, provide a measure of the intensity of cross-correlated motions that are in phase with one another. the presence of strong peaks at the (700, 2900) and (2900, 700) positions indicate that the C-C and C-H bonds primarily orient in a mutually synchronous fashion.

As evidenced by Figure 10.15, some asynchronous motion does exist for this sample within the 2990 cm^{-1} band. The intensity of this behavior is an order of magnitude less than the synchronous motion, as noted by the scales of the maximum contour levels in both plots. This asynchronous data suggests that the two populations of C-H bonds (those on the backbone carbons, and those associated with the methyl groups) do not precisely orient in phase with one another. Since these data were collected at a frequency above the mechanical glass transition, this could be a manifestation of the appearance of glassy modes of relaxation.

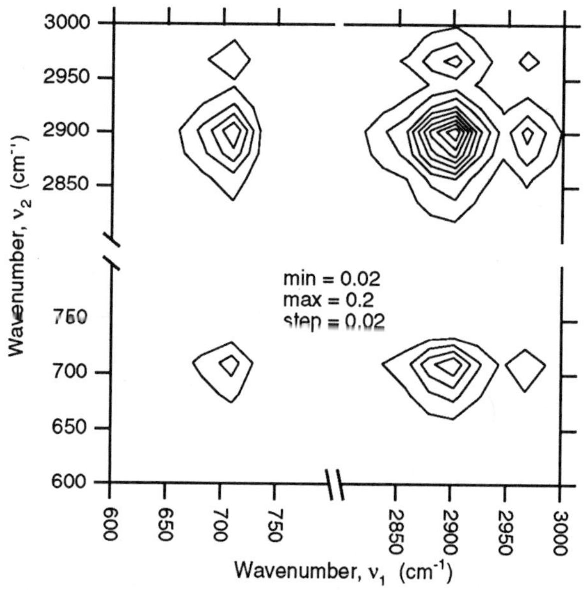

Figure 10.14 Synchronous, two-dimensional Raman spectrum for polyisobutylene subjected to a 1.5% strain at 82 Hz and at room temperature.

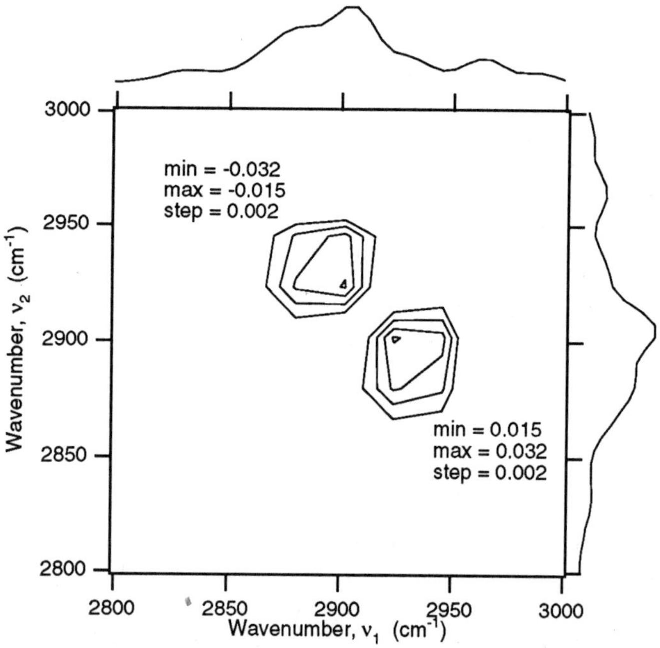

Figure 10.15 Asynchronous, 2-dimensional Raman spectrum for polyisobutylene subjected to a 1.5% strain at 82 Hz and at room temperature.

10.7 Case Study 5: Spatially Resolved Stress Measurements in Non-Newtonian Flows

A principal advantage of optical measurements is their ability to provide measurements of structure and orientation over small length scales. Scattering techniques can be used to follow spatial variations in three dimensions with a resolution that is limited by the size of the scattering volume. Using a well-focused laser and proper collection optics, length scales down to 5-10 microns can be accessed. The same resolution can be achieved using laser Doppler velocimetry for velocity measurements. Techniques such as birefringence and dichroism, however, are "line of sight" measurements, and the measured phenomena are the result of effects that are integrated along the path of the light residing within the sample.

In this case study, the combined techniques of flow birefringence and laser Doppler velocimetry are used to analyze the non-Newtonian flow of a polymeric liquid within a complex geometry. This represents the work of J. P. W. Baaijens of the University of Eindhoven; a full account can be obtained in reference 191. In such flows, the complex rheology of polymeric liquids can lead to profound differences in the velocity and stress fields compared with what would be anticipated from Newtonian fluids. To simulate non-Newtonian flows numerically requires the availability of a reliable constitutive equation that would connect the stress and velocity gradient tensors existing within the fluid. In a complex flow that is a superposition of shear and extension, it is important to test such simulations since constitutive models that can accurately represent simple shear flow rheology often fail to reproduce the response to flows with a significant extensional component. Furthermore, it is important to consider both the velocities and stresses, since the non-Newtonian character of the fluid will influence each field to different degrees.

The problem considered here was the flow of a polymeric liquid past a cylinder residing in a channel. The cylinder, however, was offset from the center of the channel so that the flow field is asymmetric relative to the axis of the channel. The geometry of the cylinder residing within the channel can be seen in Figure 10.17, which also contains the measured and simulated velocity fields, that will be discussed later. The optical configuration was designed to allow the simultaneous acquisition of velocity components using a laser Doppler velocimeter and flow birefringence using a polarization modulated device. The polymer liquid that was used obeyed the stress-optical rule, and this allowed the birefringence measurements to be converted directly to shear and normal stress components.

The optical train used to measure flow birefringence consisted of a $(\mathrm{P/RH})_{PSG}$ described in section 8.4.2. A schematic of the optical configuration is shown in Figure 10.16. This view shows the flow cell from the top. The laser doppler velocimeter utilizes crossed beams, as described in section 6.1, and produced a measuring volume of $50\mathrm{x}50\mathrm{x}200\ \mu\mathrm{m}^3$. To make a simultaneous measurement of birefringence, the light beam for that measurement is passed parallel to the axis of the cylinder and directly through the intersection point of the LDV beams. In this way, the two optical measurements sample the same point in the flow field, if that field can be approximated as being two-dimensional.

Because the birefringence measurement is an integrated effect along the path of the beam in the sample, it is important that the flow be two-dimensional with the neutral axis parallel to the axis of that beam. For that reason, the width of the flow channel (which defines the length of the beam in the sample) must be relatively large relative to its height. A rule of thumb that is often used is a factor of 10, and the value used in this work was approximately 30.

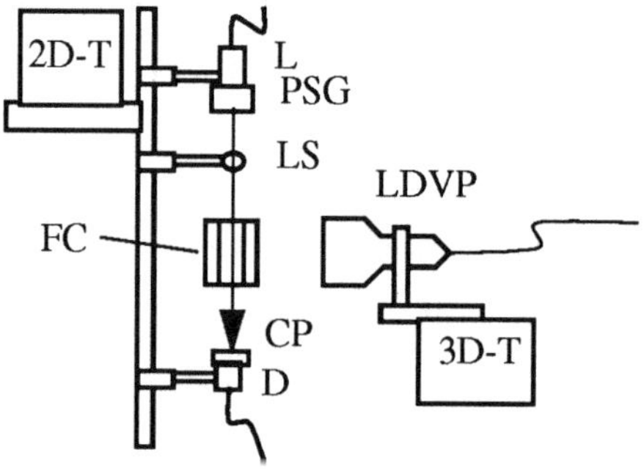

Figure 10.16 Schematic diagram showing the integration of a polarization modulated birefringence apparatus within a laser Doppler velocimeter. This shows the side view. L: light source (a diode laser was used); PSG: rotating half-wave plate design; LS: lens; FC: flow cell (flow is into the plane of the figure); CP: circular polarizer; D: detector; 2D-T: two dimensional translation stage; 3D-T: three dimensional translation stage; LDVP: laser Doppler velocimeter probe.

The sample fluid was a 5% by weight of polyisobutylene (Vistanex-L120 from Exxon Chemical Co.) dissolved in tetradecane. This is a viscoelastic material with a thinning shear viscosity and possessing normal stresses. The zero shear viscosity at 25 °C was $\eta = 0.98$ Pa-s. The mean Maxwell relaxation time was measured to be $\lambda = 0.043$ s. This was determined by fitting a four-mode Maxwell model to the complex viscosity of the solution subjected to small-amplitude oscillatory simple shear flow. The fluid was pumped at a constant volumetric flow rate through the channel and this established an average upstream velocity, U. From these parameters, the Deborah number is defined as $De = \lambda U/R$, where $R = 2$ mm is the radius of the cylinder. The length of the cylinder was 64 mm. Using the polarization modulated birefringence experiment, the birefringence, $\Delta n'$, and the orientation angle, θ, are found. From these measurements, the shear stress, $\tau_{xy} = \frac{1}{2C}\Delta n' \sin 2\theta$, and the first normal stress difference, $N_1 = \frac{1}{C}\Delta n' \cos 2\theta$, can be calculated. Here C is the stress optic coefficient. These stresses can then be non-dimensionalized with $(3\eta U/R)$. The stress optical coefficient for this solution was $C = 1.87 \times 10^{-9} \text{Pa}^{-1}$, as reported by Quinzani *et al.* [192].

The experiments were run by pumping the fluid through the channel at a specified flow rate. Once a steady state flow had been achieved, velocity and birefringence measurements were carried out at various locations in the flow field by translating the flow channel relative to the optical train. The translation of the channel was in the direction perpendicular to the mean flow direction and this was repeated for a number of locations along the axis of the channel.

The plots in Figure 10.17 show the axial velocity profiles for various positions along the channel for a Deborah number, $De = 1.87$. The vertical, dashed lines indicate the locations of each scan of the channel. The initial profile, which is furthest upstream of the cylinder, shows an expected, parabolic shape. As the cylinder is approached, the velocity field becomes distorted, with a larger portion of the flow being diverted toward the lower region where the gap between the cylinder and the boundary of the channel is the largest.

As the flow accelerates into the gaps around the cylinder, it possesses a greater relative amount of extension. Ultimately, at distances far downstream from the cylinder, the flow is expected to relax back toward a parabolic profile. In these plots, the symbols represent the measured velocities and the solid curves are the results of a finite element, numerical simulation. The constitutive equation used was a four constant, Phan-Thien-Tanner model[193], which was adjusted to fit steady, simple shear flow shear and first normal stress difference measurements. The fit to the velocity data is very satisfactory.

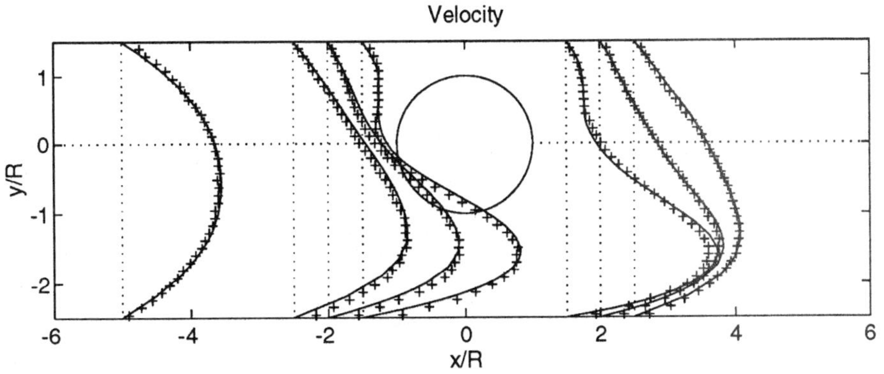

Figure 10.17 Axial velocity profiles for a Deborah number of $De = 1.87$. The vertical, dashed lines indicate the axial positions at which the channel was translated.

The shear and normal stress profiles are shown in Figures 10.18 and 10.19 for the same condition as Figure 10.17. Far upstream, the shear stress profile is approximately linear as a function of position, and the normal stress difference is nearly quadratic in position. These profiles are expected in this region since the flow is predominately a shearing motion. However, as the fluid approaches the cylinder, both profiles become highly nonsymmetric as the lower portion of the flow accelerates into the larger gap between the cylinder and the lower wall. The solid curves, representing the numerical simulation of the flow, are shown to be in excellent agreement with the measured profiles. The results of this work indicate that the Phan Thien-Tanner constitutive model used in the simulation accurately reproduces the nonlinear fluid dynamics of this complex flow, even at the relatively high Deborah number of $De = 1.87$.

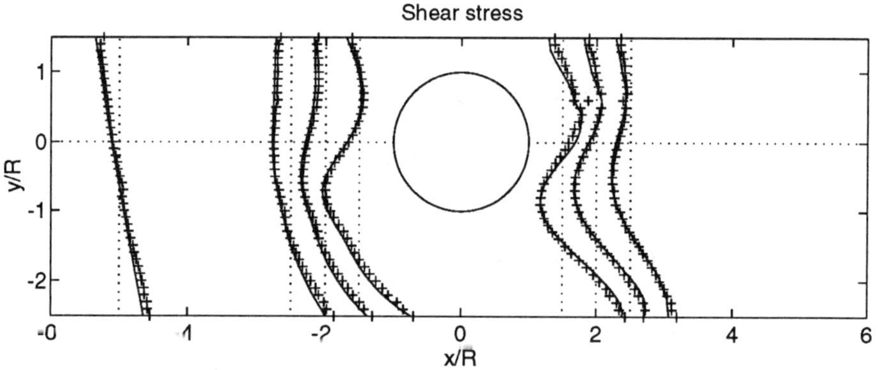

Figure 10.18 Shear stress profiles for the conditions indicated in Figure 10.18.

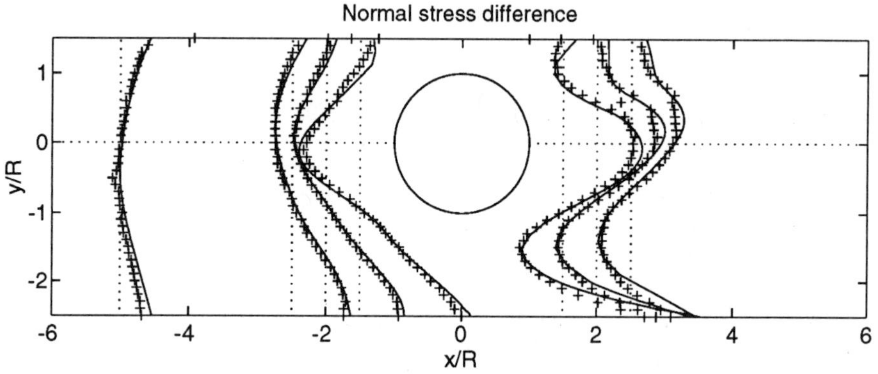

Figure 10.19 Normal stress profiles for the conditions indicated in Figure 10.19.

Appendix I

List of Jones and Mueller Matrices

The following list of Jones and Mueller matrices has been compiled for most optical elements encountered in optical instruments where polarization effects must be taken into account. In writing these matrices, the following notation has been used:

$\alpha' = 2\pi n' d/\lambda$; $\alpha'' = 2\pi n'' d/\lambda$, where $n = n' - in''$ is the isotropic refractive index, d is the sample thickness;

$\delta' = 2\pi\Delta n' d/\lambda$: retardation of a sample, where $\Delta n'$ is the birefringence;

$\delta'' = 2\pi\Delta n'' d/\lambda$: extinction of a sample, where $\Delta n''$ is the dichroism;

θ: the orientation of an optical element;

$s_\theta = \sin\theta$; $c_\theta = \cos\theta$; $S_\theta = \sinh\theta$; $C_\theta = \cosh\theta$;

$\mathbf{I}$ is the 2x2 unit matrix and $\mathbf{I}_M$ is the 4x4 unit matrix.

I.1 Isotropic Retarder

$$\mathbf{J}_1(\alpha') = e^{i\alpha'}\mathbf{I};\ \mathbf{M}_1(\alpha') = \mathbf{I}_M. \tag{I.1}$$

I.2 Isotropic Attenuator

$$\mathbf{J}_2(\alpha') = e^{-\alpha''}\mathbf{I};\ \mathbf{M}_2(\alpha') = e^{-2\alpha''}\mathbf{I}_M. \tag{I.2}$$

I.3 Birefringent Medium (0° orientation)

$$\mathbf{J}_3(\delta') = \begin{bmatrix} e^{i\frac{\delta'}{2}} & 0 \\ 0 & e^{-i\frac{\delta'}{2}} \end{bmatrix}; \ \mathbf{M}_3(\delta') = \begin{bmatrix} 1 & 0 & 0 & 0 \\ 0 & 1 & 0 & 0 \\ 0 & 0 & \cos\delta' & \sin\delta' \\ 0 & 0 & -\sin\delta' & \cos\delta' \end{bmatrix}. \tag{I.3}$$

I.4 Dichroic Medium (0° orientation)

$$\mathbf{J}_4 = \begin{bmatrix} e^{-\frac{\delta''}{2}} & 0 \\ 0 & e^{\frac{\delta''}{2}} \end{bmatrix}; \ \mathbf{M}_4 = \begin{bmatrix} C_{\delta''} & -S_{\delta''} & 0 & 0 \\ -S_{\delta''} & C_{\delta''} & 0 & 0 \\ 0 & 0 & 1 & 0 \\ 0 & 0 & 0 & 1 \end{bmatrix}. \tag{I.4}$$

I.5 Quarter-Wave Plate (0° orientation)

$$\mathbf{J}_5 = \begin{bmatrix} e^{i\frac{\pi}{4}} & 0 \\ 0 & e^{-i\frac{\pi}{4}} \end{bmatrix}; \ \mathbf{M}_5 = \begin{bmatrix} 1 & 0 & 0 & 0 \\ 0 & 1 & 0 & 0 \\ 0 & 0 & 0 & 1 \\ 0 & 0 & -1 & 0 \end{bmatrix}. \tag{I.5}$$

I.6 Half-Wave Plate (0° orientation)

$$\mathbf{J}_6 = \begin{bmatrix} i & 0 \\ 0 & -i \end{bmatrix}; \ \mathbf{M}_6 = \begin{bmatrix} 1 & 0 & 0 & 0 \\ 0 & 1 & 0 & 0 \\ 0 & 0 & -1 & 0 \\ 0 & 0 & 0 & -1 \end{bmatrix}. \tag{I.6}$$

I.7 Rotation Matrices

$$\mathbf{R}(\theta) = \begin{bmatrix} c_\theta & s_\theta \\ -s_\theta & c_\theta \end{bmatrix}; \ \mathbf{R}_M(\theta) = \begin{bmatrix} 1 & 0 & 0 & 0 \\ 0 & c_{2\theta} & s_{2\theta} & 0 \\ 0 & -s_{2\theta} & c_{2\theta} & 0 \\ 0 & 0 & 0 & 1 \end{bmatrix}. \tag{I.7}$$

I.8 Coaxial Birefringent/Dichroic Element (0° orientation)

$$\mathbf{J}_8(\delta', \delta'') = \begin{bmatrix} e^{(i\delta' - \delta'')/2} & 0 \\ 0 & e^{-(i\delta' - \delta'')/2} \end{bmatrix}; \quad \mathbf{M}_8(\delta', \delta'') = \begin{bmatrix} C_{\delta''} & -S_{\delta''} & 0 & 0 \\ -S_{\delta''} & C_{\delta''} & 0 & 0 \\ 0 & 0 & c_{\delta'} & s_{\delta'} \\ 0 & 0 & -s_{\delta'} & c_{\delta'} \end{bmatrix}. \quad (I.8)$$

I.9 Birefringent Element Oriented at θ

$$\mathbf{J}_9(\delta', \theta) = \begin{bmatrix} (c_{\delta'/2} + ic_{2\theta}s_{\delta'/2}) & is_{2\theta}s_{\delta'/2} \\ is_{2\theta}s_{\delta'/2} & (c_{\delta'/2} - ic_{2\theta}s_{\delta'/2}) \end{bmatrix}.$$

$$\mathbf{M}_9(\delta', \theta) = \begin{bmatrix} 1 & 0 & 0 & 0 \\ 0 & (c_{2\theta}^2 + s_{2\theta}^2 c_{\delta'}) & s_{2\theta}c_{2\theta}(1 - c_{\delta'}) & -s_{2\theta}s_{\delta'} \\ 0 & s_{2\theta}c_{2\theta}(1 - c_{\delta'}) & (s_{2\theta}^2 + c_{2\theta}^2 c_{\delta'}) & c_{2\theta}s_{\delta'} \\ 0 & s_{2\theta}s_{\delta'} & -c_{2\theta}s_{\delta'} & c_{\delta'} \end{bmatrix}. \quad (I.9)$$

I.10 Dichroic Element Oriented at θ

$$\mathbf{J}_{10}(\delta'', \theta) = \begin{bmatrix} (C_{\delta''/2} - c_{2\theta}S_{\delta''/2}) & -s_{2\theta}S_{\delta'/2} \\ -s_{2\theta}S_{\delta''/2} & (C_{\delta''/2} + c_{2\theta}S_{\delta''/2}) \end{bmatrix}.$$

$$\mathbf{M}_{10}(\delta'', \theta) = \begin{bmatrix} C_{\delta''} & -c_{2\theta}S_{\delta''} & -s_{2\theta}S_{\delta''} & 0 \\ -c_{2\theta}S_{\delta''} & (c_{2\theta}^2 C_{\delta''} + s_{2\theta}^2) & s_{2\theta}c_{2\theta}(C_{\delta''} - 1) & 0 \\ -s_{2\theta}S_{\delta''} & s_{2\theta}c_{2\theta}(C_{\delta''} - 1) & (s_{2\theta}^2 C_{\delta''} + c_{2\theta}^2) & 0 \\ 0 & 0 & 0 & 1 \end{bmatrix}. \quad (I.10)$$

I.11 Ideal Polarizer Oriented at θ

$$\mathbf{J}_{11}(\theta) = \begin{bmatrix} c_\theta^2 & s_\theta c_\theta \\ s_\theta c_\theta & s_\theta^2 \end{bmatrix}; \quad \mathbf{M}_{11}(\theta) = \frac{1}{2}\begin{bmatrix} 1 & c_{2\theta} & s_{2\theta} & 0 \\ c_{2\theta} & c_{2\theta}^2 & s_{2\theta}c_{2\theta} & 0 \\ s_{2\theta} & s_{2\theta}c_{2\theta} & s_{2\theta}^2 & 0 \\ 0 & 0 & 0 & 0 \end{bmatrix}. \tag{I.11}$$

I.12 Quarter-Wave Plate Oriented at θ

$$\mathbf{J}_{12}(\theta) = \frac{1}{\sqrt{2}}\begin{bmatrix} 1+ic_{2\theta} & is_{2\theta} \\ is_{2\theta} & 1-ic_{2\theta} \end{bmatrix}; \quad \mathbf{M}_{12}(\theta) = \begin{bmatrix} 1 & 0 & 0 & 0 \\ 0 & c_{2\theta}^2 & s_{2\theta}c_{2\theta} & -s_{2\theta} \\ 0 & s_{2\theta}c_{2\theta} & s_{2\theta}^2 & c_{2\theta} \\ 0 & s_{2\theta} & -c_{2\theta} & 0 \end{bmatrix}. \tag{I.12}$$

I.13 Half Wave Plate Oriented at θ

$$\mathbf{J}_{13}(\theta) = \begin{bmatrix} c_{2\theta} & s_{2\theta} \\ s_{2\theta} & -c_{2\theta} \end{bmatrix}; \quad \mathbf{M}_{13}(\theta) = \begin{bmatrix} 1 & 0 & 0 & 0 \\ 0 & c_{4\theta} & s_{4\theta} & 0 \\ 0 & s_{4\theta} & -c_{4\theta} & 0 \\ 0 & 0 & 0 & -1 \end{bmatrix}. \tag{I.13}$$

I.14 Coaxial Birefringent/Dichroic Element Oriented at θ

$$\mathbf{J}_{14}(\delta', \delta'', \theta) = \begin{bmatrix} (C_{(i\delta'-\delta'')/2} + c_{2\theta}S_{(i\delta'-\delta'')/2}) & s_{2\theta}S_{(i\delta'-\delta'')/2} \\ s_{2\theta}S_{(i\delta'-\delta'')/2} & (C_{(i\delta'-\delta'')/2} - c_{2\theta}S_{(i\delta'-\delta'')/2}) \end{bmatrix}$$

$$\mathbf{M}_{14}(\delta', \delta'', \theta) = \begin{bmatrix} C_{\delta''} & (-c_{2\theta}S_{\delta''}) & (-s_{2\theta}S_{\delta''}) & 0 \\ -c_{2\theta}S_{\delta''} & (c_{2\theta}^2 C_{\delta''} + s_{2\theta}^2 c_{\delta'}) & s_{2\theta}c_{2\theta}(C_{\delta''} - c_{\delta'}) & -s_{2\theta}s_{\delta'} \\ -s_{2\theta}S_{\delta''} & s_{2\theta}c_{2\theta}(C_{\delta''} - c_{\delta'}) & (s_{2\theta}^2 C_{\delta''} + c_{2\theta}^2 c_{\delta'}) & c_{2\theta}s_{\delta'} \\ 0 & s_{2\theta}s_{\delta'} & -c_{2\theta}s_{\delta'} & c_{\delta'} \end{bmatrix}. \tag{I.14}$$

I.15 Circular Dichroic Element

$$\mathbf{J}_{16}(\delta''_{circ}) = \begin{bmatrix} C_{\delta''_{circ}/2} & -iS_{\delta''_{circ}/2} \\ iS_{\delta''_{circ}/2} & C_{\delta''_{circ}/2} \end{bmatrix};\ \mathbf{M}_{15}(\delta''_{circ}) = \begin{bmatrix} C_{\delta''_{circ}} & 0 & 0 & S_{\delta''_{circ}} \\ 0 & 1 & 0 & 0 \\ 0 & 0 & 1 & 0 \\ S_{\delta''_{circ}} & 0 & 0 & C_{\delta''_{circ}} \end{bmatrix}. \tag{I.15}$$

where $\delta''_{circ} = (2\pi\Delta n''_{circ}d)/\lambda$ and $\Delta n''_{circ} = n''_l - n''_r$ is the circular dichroism defined in equation (2.33).

I.16 Circular Birefringent Element

$$\mathbf{J}_{16}(\delta'_{circ}) = \mathbf{R}(\delta'_{circ});\ \mathbf{M}_{16}(\delta'_{circ}) = \mathbf{R}_M(\delta'_{circ}), \tag{I.16}$$

where $\delta'_{circ} = (2\pi\Delta n'_{circ}d)/\lambda$ and $\Delta n'_{circ} = n'_l - n'_r$ is the circular birefringence defined in equation (2.32).

I.17 Circular Polarizers

$$\mathbf{J}_{17,\,left} = \frac{1}{2}\begin{bmatrix} 1 & -i \\ i & 1 \end{bmatrix};\ \mathbf{M}_{18,\,left} = \frac{1}{2}\begin{bmatrix} 1 & 0 & 0 & 1 \\ 0 & 0 & 0 & 0 \\ 0 & 0 & 0 & 0 \\ 1 & 0 & 0 & 1 \end{bmatrix};$$

$$\mathbf{J}_{17,\,right} = \frac{1}{2}\begin{bmatrix} 1 & i \\ -i & 1 \end{bmatrix};\ \mathbf{M}_{18,\,right} = \frac{1}{2}\begin{bmatrix} 1 & 0 & 0 & -1 \\ 0 & 0 & 0 & 0 \\ 0 & 0 & 0 & 0 \\ -1 & 0 & 0 & 1 \end{bmatrix}; \tag{I.17}$$

where the subscripts, *left* and *right*, refer to left and right circularly polarized light.

I.18 Noncoaxial Birefringent/Dichroic Element Containing Circular Birefringence and Circular Dichroism

This composite material contains both linear and circular birefringence and dichroism. The orientation of the linear birefringence is at θ' and the orientation of the linear dichroism is at θ''.

$$\mathbf{J}_{18}\,(\delta', \delta'', \delta'_{circ}, \delta''_{circ}, \theta', \theta'') = \begin{bmatrix} X - Y & -(Z + W) \\ -(Z - W) & X + Y \end{bmatrix}$$

$$\mathbf{M}_{18}\,(\delta', \delta'', \delta'_{circ}, \delta''_{circ}, \theta', \theta'') =$$

$$\begin{bmatrix} |X|^2 + |Y|^2 + |Z|^2 + |W|^2 & 2Re\,(X^*Y + ZW^*) & 2Re\,(X^*Z + YW^*) & 2Im\,(YZ^* + X^*W \\ 2Re\,(X^*Y + Z^*W) & |X|^2 + |Y|^2 - |Z|^2 - |W|^2 & 2Re\,(YZ^* - XW^*) & 2Im\,(X^*Z - Y^*W \\ 2Re\,(X^*Z + Y^*W) & 2Re\,(YZ^* + XW^*) & |X|^2 - |Y|^2 + |Z|^2 - |W|^2 & 2Im\,(XY^* + ZW^* \\ 2Im\,(Y^*Z - X^*W) & 2Im\,(XZ^* + YW^*) & 2Im\,(X^*Y - Z^*W) & |X|^2 - |Y|^2 - |Z|^2 + | \end{bmatrix}$$

(I.18)

where

$$X = C_Q; \qquad Y = \frac{1}{2}\,[\delta'' c_{2\theta''} - i\delta' c_{2\theta'}]\,\frac{S_Q}{Q};$$

$$Z = \frac{1}{2}\,[\delta'' s_{2\theta''} - i\delta' s_{2\theta'}]\,\frac{S_Q}{Q}; \qquad W = \frac{1}{2}\,[i\delta''_{circ} - \delta'_{circ}]\,\frac{S_Q}{Q};$$

$$Q = q + i\bar{q}; \qquad q = \frac{1}{2\sqrt{2}}\sqrt{(4|Q|^2 + a)}\,;$$

I.19 Linear and Circular Birefringence/Dichroic Element in the Limit of Small Anisotropies

This element is identical to that described in the above example except that the anisotropies are restricted to the following limits: $(|\delta'|, |\delta''|, |\delta'_{circ}|, |\delta''_{circ}|) \ll 1$.

$$\mathbf{J}_{19}(\delta', \delta'', \delta'_{circ}, \delta''_{circ}, \theta', \theta'') =$$

$$\begin{bmatrix} 1 + \frac{1}{2}(i\delta' c_{2\theta'} - \delta'' c_{2\theta''}) & -\frac{1}{2}[\delta'' s_{2\theta''} - i\delta' s_{2\theta'} - (\delta'_{circ} - i\delta''_{circ})] \\ -\frac{1}{2}(\delta'' s_{2\theta''} - i\delta' s_{2\theta'} + (\delta'_{circ} - i\delta''_{circ})) & 1 - \frac{1}{2}(i\delta' c_{2\theta'} - \delta'' c_{2\theta''}) \end{bmatrix};$$

$$\mathbf{M}_{19}(\delta', \delta'', \delta'_{circ}, \delta''_{circ}, \theta', \theta'') = \begin{bmatrix} 1 & -\delta'' c_{2\theta''} & -\delta'' s_{2\theta''} & \delta''_{circ} \\ -\delta'' c_{2\theta''} & 1 & \delta'_{circ} & -\delta' s_{2\theta'} \\ -\delta'' s_{2\theta''} & -\delta'_{circ} & 1 & \delta' c_{2\theta'} \\ \delta''_{circ} & \delta' s_{2\theta'} & -\delta' c_{2\theta'} & 1 \end{bmatrix}.$$

(I.19)

Appendix II

Nomenclature

This appendix identifies the symbols used throughout this monograph. Where appropriate, each description is followed by the equation numbers where the symbol is introduced.

$\mathbf{A}$	Jones vector, (1.50).
$\tilde{\mathbf{A}}$	Amplitude of the Jones vector, (2.12).
$\tilde{\mathbf{A}}_i$	Abeles' electric vector used in the analysis of stratified films, (3.7).
A	Area; strength of polarization modulation, (8.22).
A_{ij}	Rouse matrix, (7.55); total absorption tensor of a macromolecule, (5.7).
$\mathbf{a}$	Unit vector defining the basis set for a polarization state, (1.54), (1.55); segmental absorption tensor, (5.4).
a	Length of a polymer segment, Figure 7.3.
$\mathbf{B}$	Magnetic induction field, (1.1).
C	Stress optical coefficient, (7.123), Jeffery orbit constant, (7.108).
C_θ	Shorthand for $\cosh\theta$.
$\mathbf{C}(\mathbf{k})$	Fourier transform of the dielectric tensor autocorrelation function, (4.89).

c	Speed of light in vacuum, (1.16).
$c(\mathbf{x})$	Concentration, (4.48).
c_θ	Shorthand for $\cos\theta$
$\mathbf{D}$	Electric displacement vector, (1.3).
$\boldsymbol{D}$	Rate of strain tensor, (7.61).
D_t	Translational diffusion coefficient, (6.15).
D_N	Determinant of $\mathbf{N}$, (2.47).
D_r^0	Rotational diffusivity in dilute solution, (7.61).
$\bar{D}_r$	Rotational diffusivity in semi-dilute solution, (7.80).
D_R	Diffusion coefficient for reptation, (7.66).
$D(\theta, \varphi)$	Fraunhofer diffraction function, (4.61).
d	Sample thickness.
d_{ij}	Matrix used in the calculation of the Mueller matrix, (2.4).
$\mathbf{E}$	Electric field vector, (1.1).
$\mathbf{F}$	Transformation matrix for linear and circular basis sets, (1.57); strain tensor, (7.41).
$\boldsymbol{F}$	Force.
F	Free energy functional, (7.94).
$F_1(\mathbf{q}, t)$, $F_2(\mathbf{q}, t)$	Heterodyne and homodyne autocorrelation functions, (6.8), (6.9).
$f(\mathbf{r}, N, t)$	Probability distribution for the transient network model, (7.64).
$\mathbf{f}^{\mathbf{a}}$ and $\mathbf{f}^{e}$	Direction cosines of the absorption and emission axes, (5.53).
$G(\mathbf{x}, \mathbf{x}')$, $\mathbf{G}(k;\mathbf{x}, \mathbf{x}')$	Greens function, (1.34).
$G_{sj}(\mathbf{x_j})$	Van-Hove self space-time correlation function, (6.14).
$\boldsymbol{G}$	Velocity gradient tensor, (7.40).

$G'(\omega)$	Storage modulus.
$\mathbf{H}$	Magnetic field vector, (1.3).
$H(\mathbf{r}, N)$	Entanglement creation function, (7.64).
h	Planck's constant, (5.2).
$\mathbf{I}$	Unit matrix or tensor.
I	Light intensity, (1.53).
I_0	Incident light intensity.
$\mathbf{I}_M$	4x4 unit matrix.
$\mathbf{J}$	Jones matrix, (2.1).
$J_m(A)$	Bessel function of order m, (8.25)
J	Current density, (1.1).
j_i	Vector used in the calculation of the Mueller matrix, (2.4).
$\mathbf{j}$	Mass flux, (7.93).
$\mathbf{K}$	Surface current density, (1.69).
$\mathbf{K}_{i, i+1}$	2x2 transformation matrix for interface $i, i+1$, (3.8).
$K(\mathbf{R}), K_B$	Entropic spring constant, (7.41), spring constant for a Rouse segment, (7.55).
$k, \mathbf{k}$	Wave number, wave vector, (1.19).
$\hat{\mathbf{k}} = \mathbf{k}/\lvert\mathbf{k}\rvert$	Unit vector in the direction of the wave vector, (1.20).
$\mathbf{L}_i$	2x2 transformation matrix for a film i, (3.8).
L	Length scale of a light scattering volume, (6.10); length of a rigid rod molecule, Figure 7.6.
L_1, L_2	Particle shape factors, (7.30).
$L^{-1}(r)$	Inverse Langevin function, (7.17).
l	Thickness or position measure.
$\mathbf{M}$	Mueller matrix, (2.2).

$\mathbf{M}_{i,j;k,l}$	Transition dipole moment, (5.1)
M	Molecular weight.
M_c, M_e	Critical molecular weight, entanglement molecular weight, (7.65).
m_i	Vector used in the calculation of the Mueller matrix, (2.4).
$\mathbf{N}$	Differential propagation Jones matrix, (2.40).
N_K	Number of Kuhn segments making up a polymer chain, Figure 7.3.
N_1	First normal stress difference, (7.125).
$\mathbf{n}$	unit vector.
n, $\boldsymbol{n}$	Refractive index scalar, refractive index tensor, (1.15).
n', $\boldsymbol{n}'$	Real part of the refractive index, (2.12), (1.24).
n'', $\boldsymbol{n}''$	Imaginary part of the refractive index, (2.12), (2.20).
$\boldsymbol{n}_p$	Particle contribution to the refractive index tensor, (4.70).
$\Delta n'$	Linear birefringence, (2.16).
$\Delta n''$	Linear dichroism, (2.22).
$\Delta n'_{circ}$	Circular birefringence, (2.32).
$\Delta n''_{circ}$	Circular dichroism, (2.33).
N	Number concentration.
Q_N	Discriminant of $\mathbf{N}$, (2.47).
$\mathbf{q} = \mathbf{k}_s - \mathbf{k}_i$	Scattering vector, (4.32).
$\mathbf{P}(z, \Delta z)$	Jones matrix for a differential element, (2.37).
$\mathbf{P}(\mathbf{r}_e, \mathbf{r}_n)$	Dipole function, (5.2).
$\boldsymbol{P}$	Polarization, (7.1).
PSA	Polarization state analyzer, (Table 8.2).
PSG	Polarization state generator, (Table 8.2).

$\mathrm{P}_2(\mathbf{r})$	Pair distribution function, (7.118).
$P_2(\langle \cos^2\theta \rangle)$	Second order Legendre polynomial, (5.11).
p, $\mathbf{p}$	Scalar dipole, vector dipole, (4.6).
Q_k	Amplitude of the k th vibrational mode, (5.26).
$\mathbf{R}$	2x2 rotation matrix, (1.30), end-to-end vector of a polymer, Figure 7.3.
$\mathbf{R}_M$	4x4 rotation matrix, (2.9).
R	Dichroic ratio, (5.13).
R_{ij}	Reflection coefficient for incident and refected light polarized in the i and j directions, respectively, (1.75), (1.77), (3.2).
$R(\theta, \varphi)$	Rayleigh form factor, (4.34).
$\mathbf{S}$	Stokes vector, (1.59).
$\mathbf{S}(\theta, \varphi)$	Scattering Jones matrix, (4.65).
S_{ij}	Saupe orientation tensor, (5.12).
$S(\mathbf{q})$	Structure factor, (4.53).
$S(\langle \cos^2\theta \rangle)$	Herman's orientation function, (5.11).
S	Entropy, (7.42).
S_θ	Shorthand for $\sinh\theta$.
s_θ	Shorthand for $\sin\theta$.
s_{ij}	Matrix used in the calculation of the Mueller matrix, (2.4).
T	Time duration, temperature.
T_N	Trace of $\mathbf{N}$, (2.47).
$\mathbf{T}$	Three-dimensional rotation tensor, (5.5).
$\mathbf{T}_{ik}$	Electric field interaction tensor, (7.34).

T_{ij}	Transmission coefficient for incident and transmitted light polarized in the i and j directions, respectively, (1.74), (1.76), (3.4).
t	Time.
$\mathbf{U}$	Average velocity, (6.21).
$\mathbf{u}$	Unit vector.
V	Volume, (4.29).
$\mathbf{v}$	Velocity field vector.
v	Speed of light in a material, (1.16).
$\boldsymbol{W} = (\boldsymbol{G} - \boldsymbol{G}^{+})/2$	Vorticity tensor, (7.105).
W	Free energy, (7.42).
$\mathbf{Y}_i$	Normal mode vector, (7.58).
α	Orientation angle or scalar polarizability.
α_{ij}, $\underline{\alpha}$	Polarizability tensor, (4.10).
α'_{ij}	Derived polarizability tensor, (5.27).
$\bar{\alpha}$	Average polarizability.
β	Inverse Langevin function, (7.17).
$\beta(\mathbf{r}, N)$	Entanglement destruction function, (7.64).
χ	Orientation angle, (7.13).
γ	Rotary power, (1.25).
$\dot{\gamma}$	Velocity gradient, (6.26).
$\delta(\mathbf{x})$	Dirac delta function.
δ_{ij}	Kronecker delta function (unit tensor).
δ'	Retardation of light, (2.17).
δ''	Extinction of light, (2.23).
Δ	Ellipsometric angle, (3.6).
ε_0	Permittivity of free space, (1.1).

ε, ε_{ij}	Electric permittivity or dielectric tensor, (1.4).
ε_f	Form contribution to the dielectric tensor, (4.91).
$\varepsilon^{(2)}_{ijk}$, $\varepsilon^{(3)}_{ijkl}$	Nonlinear permittivities, (1.5).
ζ_{ij}, ζ'_{ij}	Optical rotation tensors, (1.4).
ζ	Friction factor, (7.39).
η	Contour integration parameter in the calculation of the Green's function, (1.42).
η_s	Solvent viscosity.
θ	Orientation angle, or polar spherical coordinate.
$\vartheta(\mathbf{r}, t)$	Brownian noise function, (7.96).
Θ_i, Θ_f	Coefficients for intrinsic and form birefringence, (7.36).
κ_{ij}	impermittivity tensor, (2.64).
κ	Measure of the Faraday effect, (8.23); Jeffery orbit constant, (7.108)
λ	Wavelength of light, (1.18).
λ_i	Rouse matrix eigenvalue, (7.58).
Λ	Evanescent wavelength scale, (1.83).
μ	Particle aspect ratio, (7.106).
μ_{ij}	Magnetic permittivity tensor, (1.4).
μ_0	Magnetic permittivity of free space, (1.1).
ν_{ij}	impermeability tensor, (2.64).
ν	Beer's law coefficient, (2.14).
ν	Number density of molecules or particles, (7.3).
ν	Frequency of light.
ξ	Correlation length, (4.52).
Ξ	Mobility, (7.93).

ρ	Density (charge or material), (1.2), (4.48).
ρ_i	Fractional end to end vector of a polymer chain, (7.27)
ρ_r, ρ_t	Ratios of reflection and transmission coefficients, (3.6).
ϕ	Angle of incidence, (1.72).
ϕ_B	Brewster angle, (1.78).
ϕ_{TIR}	Angle of incidence for total internal reflection, (1.79).
φ	Azimuthal spherical coordinate.
$\Phi(\mathbf{x})$	Potential function, (4.19).
$\sigma(\mathbf{x})$	Surface charge density, (1.69).
σ_{ij}	Conductivity tensor, (1.4).
ς	friction factor, (7.39).
τ	Relaxation time.
$\underline{\tau}$	Stress tensor, (7.119).
τ_{xy}	Shear stress, (7.125).
ψ	Ellipsometric angle, (3.6).
$\psi(\mathbf{r}, t)$	Order parameter, (7.91).
Ψ	Probability distribution function.
$\Psi_{i,j}(\mathbf{r}_e, \mathbf{r}_n)$	Schroedinger wave function, (5.1).
ω	Angular frequency of light, (1.14).
Ω	Solid angle, (4.92); angle between the electric vector and the transition dipole moment, (5.3); angular velocity of rotational modulation (section 8.3.1).

References

1. J. D. Jackson, *Classical electrodynamics*, 2nd ed., Wiley, New York (1975).
2. M. Born and E. Wolf, *Principles of optics: electromagnetic theory of propagation interference and diffraction of light*, 6th ed., Pergamon Press, New York (1980).
3. J. A. Kong, *Electromagnetic Wave Theory,* 2nd Edition, John Wiley & Sons, New York (1990).
4. A. L. Fetter and J. D. Walecka, *Theoretical mechanics of particles and continua,* McGraw-Hill, NY (1980).
5. R.M.A. Azzam and N.M. Bashara, *Ellipsometry and Polarized Light*, Elsevier Sci. Publ., New York (1977).
6. R. C. Jones, "A new calculus for the treatment of optical systems I. description and discussion of the calculus," J. Opt. Soc. Am., **31**, 488 (1941).
7. R. C. Jones, "A new calculus for the treatment of optical systems. VII. Properties of the n-matrices," J. Opt. Soc. Am., **48**, 671 (1948).
8. B. J. Berne and R. Pecora, *Dynamic light scattering: with applications to chemistry, biology and physics*, John Wiley, New York (1990).
9. H. C. van de Hulst, *Light scattering by small particles*, John Wiley and Sons, New York (1957).
10. G. H. Meeten, "Conservative dichroism in the Rayleigh-Gans-Debye approximation," J. Coll. Inter. Sci., **84**, 235 (1981).
11. N. Saito and Y. Ikeda, "The light scattering by non-spherical particles in solutions," J. Phys. Soc. Japan, **6**, 305 (1951).
12. P. L. Frattini, and G. G. Fuller, "Conservative dichroism of a sheared suspension in the Rayleigh-Gans light scattering approximation," J. Coll. Inter. Sci., **119**, 335 (1987).

13. A. Onuki and M. Doi, "Flow birefringence and dichroism of polymers. I. General theory and application to the dilute case," J. Chem. Phys., **85**, 1190 (1986).

14. G. I. Zahalak and S. P. Sutera, "Fraunhofer diffraction pattern of an oriented monodisperse system of prolate ellipsoids," J. Colloid Inter. Sci., **82**, 423 (1981).

15. A. J. Salem and G. G. Fuller, "Small angle light scattering as a probe of flow-induced particle orientation," J. Coll. Inter. Sci., **108**, 149 (1985).

16. G. H. Meeten, "The birefringence of colloidal dispersions in the Rayleigh and anomalous diffraction approximations," J. Coll. Int. Sci., **73**, 38 (1980).

17. I. M. Ward (Ed.), *Structure and Properties of Oriented Polymers.*, Applied Science Publishers, London (1975).

18. K. Sondergaard and J. Lyngaae-Jorgensen, eds., *Rheo-physics of multiphase polymer systems: characterization by rheo-optical techniques*, Technomic Publ. Co., Lancaster, PA (1994).

19. J. L. Koenig, *Spectroscopy of polymers*, ACS, Washington (1975).

20. C. V. Raman and K. S. Krishnon, "A new type of secondary radiation," Nature, **121**, 501 (1928).

21. D. A. Long, *Raman Spectroscopy*, McGraw-Hill, New York (1977).

22. L. A. Archer, G. G. Fuller, and L. Nunnelley, "Dynamics of polymer liquids using polarization modulated laser Raman scattering," Polymer, **33**, 3574 (1992).

23. Lakowicz, J. R., *Principles of fluorescence spectroscopy*, Plenum Press, New York, 1983.

24. P. Lapersonne, J.-F. Tassin, P. Sergot, and L. Monnerie, "Fluorescence polarization characterization of biaxial orientation," Polymer, **30**, 1558 (1989).

25. P. S. Hauge, "Recent developments in instrumentation in ellipsometry," Surf. Sci., **96**, 108 (1980).

26. K. Osaki, N. Bessho, T. Kojimoto, and M. Kurata, "Flow birefringence of polymer solutions in time-dependent field," J. Rheol., **23**, 457 (1979).

27. K. Hongladarom, W. R. Burghardt, S. G. Baek, S. Cementwala, and J. J. Magda, "Molecular alignment of polymer liquid crystals in shear flows. I. Spectroscopic birefringence technique, steady-state orientation, and normal stress behavior in poly(benzyl glutamate) solutions," Macromolecules, **26**, 772 (1993).

28. A.W. Chow and G.G. Fuller, "Response of moderately concentrated xanthan gum solutions to time-dependent flows using two-color birefringence," J. Rheol., **28**, 23 (1984).

29. H. Janeschitz-Kreigl, *Polymer melt rheology and flow birefringence*, Springer/Verlag (1983).

30. J.C. Kemp, "Piezo-optical birefringence modulators: new use for a long-known effect," J. Opt. Soc. Am., **59**, 950 (1969).

31. S. N. Jasperson and S. E. Schnatterly, "An improved method for high reflectivity ellipsometry based on a new polarization modulation technique," Rev. Sci. Instr., **40**, 761 (1969).

32. C. T. O'Konski, *Molecular electro-optics*, Vols. 1 & 2, M. Dekker, New York (1977).

33. B. Koeman and H. Janeschitz-Kriegl, "Analysis of slightly elliptically polarized light," J. Phys. E: Sci. Instrum., **12**, 625 (1979).

34. H. J. Hofrichter and J. A. Schellman, "The optical properties of oriented biopolymers," Jerusalem Symposium on Quantum Chemistry and Biochemistry, **5**, 787 (1973); J. Schellman and H. P. Jensen, "Optical spectroscopy of oriented molecules," Chem. Rev., **87**, 1359 (1987).

35. P. L. Frattini and G. G. Fuller, "The dynamics of dilute colloidal suspensions subject to time-dependent flow fields by conservative dichroism," J. Coll. Int. Sci., **100**, 506 (1984).

36. B. Drevillon, J. Perrin, R. Marbot, A. Violet and J. L. Dalby, "Fast polarization modulated ellipsometer using a microprocessor system for digital fourier analysis," Rev. Sci. Inst., **53**, 969 (1982).

37. S. J. Johnson, P. L. Frattini and G. G. Fuller, "Simultaneous dichroism and birefringence measurements of dilute colloidal suspensions in transient shear flow," J. Colloid Interface Sci., **104**, 440 (1985); S. J. Johnson and G. G. Fuller, "Flowing colloidal suspensions in non-Newtonian suspending fluids: decoupling the composite birefringence," Rheol. Acta, **25**, 405 (1986).

38. P. L. Frattini and G. G. Fuller, "A note on phase-modulated flow birefringence: a promising rheo-optical method," J. Rheol., **28** 61, (1984).

39. G. G. Fuller, "Optical rheometry," Ann. Rev. Fluid Mech. **22**, 387 (1990).

40. S. J. Johnson and G. G. Fuller, "The optical anisotropy of sheared hematite suspensions," J. Coll. Int. Sci., **124**, 441 (1988).

41. R. M. A. Azzam, "Mueller matrix measurement using the four detector photopolarimeter," Optics Lett., **11**, (1986).

42. R. C. Thompson, J. R. Bottiger and E. S. Fry, "Measurement of polarized light interactions via the Mueller matrix," Appl. Opt., **19**, 1323 (1980).

43. B. Chu, *Dynamic light scattering*, John Wiley and Sons, New York, (1987).

44. G. G. Fuller, J. M. Rallison, R. L. Schmidt, L. G. Leal, "The measurement of velocity gradients in laminar flow by homodyne light-scattering spectroscopy," J. Fluid Mech., **100**, 555 (1980).

45. J. J. Wang, D. Yavich, and L. G. Leal, "Time-resolved measurement of velocity gradient tensor in linear flow by photon correlation spectroscopy," Phys. Fluids A, submitted (1995).

46. F. R. Cottrell, E. W. Merrill, and K. A. Smith, "Conformation of polyisobutylene in dilute solution subjected to a hydrodynamic shear field," J. Polym. Sci., Part A, **27**, 1415 (1969).

47. R. S. Stein and M. Srinivasarao, "Fifty years in light scattering: A perspective," J. Polym. Sci.: Part B: Polym. Phys., **31**, 2003 (1993).

48. J. van Egmond and G.G. Fuller, "Time dependent small angle light scattering of shear-induced concentration fluctuations in polymer solutions," J. of Chem. Physics, **96**, 7742 (1992).

49. R. S. Stein, R. J. Taor, and M. B. Long, "A two-dimensional position sensitive detector for small angle light scattering," J. Polym. Sci., Polym. Phys. Ed., **20**, 2041 (1982).

50. A. J. Salem and G. G. Fuller, "Small angle light scattering as a probe of flow-induced particle orientation," J. Colloid Int. Sci., **108**, 149 (1985).

51. W. Shurcliff, *Polarized light*, van Nostrand, Princeton, NJ (1964).

52. E. H. Land and C. D. West, "Dichroism and dichroic polarizers," Coll. Chem., **46**, 160 (1946); E. H. Land, "Some aspects of the development of sheet polarizers," J. Opt. Soc. Am., **41**, 957 (1951).

53. Pancharatnam, "Achromatic combination of birefringent plates," Proc. Indian Acad. Sci., **A41**, 137 (1955).

54. A. M. Title, "Improvement of birefringent filters. II: Achromatic waveplates," Applied Optics, **14**, 229 (1975).

55. F. Durst, A. Melling, and J. H. Whitelaw, *Principles and practice of laser-Doppler anemometry*, Academic Press, New York (1976).

56. L. R. G. Treloar, *The physics of rubber elasticity*, 2nd edition, Oxford University Press, Oxford (1958).

57. W. Kuhn and F. Grun, "Beziehungen zwischen elastischen Konstanten und Dehnungsdoppelbrechung hochelastischer Stoffe," Kolloidzschr., **101**, 248 (1942).

58. A. Peterlin and H. A. Stuart, "The determination of the size and shape, as well as the electrical, optical and magnetic anisotropy of submicroscopic particles with the aid of artificial double regraction and inner viscosity," Z. Phys., **112**, 129 (1939).

59. M. Copic, "Streaming birefringence of polymer solutions: anisotropy of internal field," J. Chem. Phys. **26**, 1382 (1957).

60. J. van Egmond and G. G. Fuller, "Concentration fluctuation enhancement in polymer solutions by extensional flow," Macromolecules, **26**, 7182 (1993).

61. D. Wirtz, D. E. Werner, and G. G. Fuller, "Structure and optical anisotropies of critical polymer solutions in electric fields," J. Chem. Phys., **101**, 1679 (1994).

62. R. B. Bird, R. C. Armstrong, and O. Hassager, *Dynamics of polymeric liquids: Vol. II. Kinetic theory*, 2nd Ed., John Wiley and Sons, New York (1987).

63. M. Doi and S. F. Edwards, *The theory of polymer dynamics*, Oxford University Press, Oxford (1986).

64. R. G. Larson, *Constitutive equations for polymer melts and solutions*, Butterworths, Boston (1988).

65. H. Yamakawa, *Modern theory of polymer solutions*, Harper's Chemistry Series, Harper and Row, New York (1971).

66. C. Reinhold and A. Peterlin, "Hydrodynamics of linear macromolecules. I. Finite segment length," J. Chem. Phys., **44**, 4333 (1966).

67. R. Cerf, "La macromolecule en chaine dans un champ hydrodynamique. Theorie generale. Proprietes dynamo-optiques," J. Polym. Sci., **23**, 125 (1957).

68. H. C. Booij and P. H. van Wiechen, "Effect of internal viscosity on the deformation of a linear macromolecule in a sheared solution," J. Chem. Phys., **52**, 5056 (1968).

69. G. G. Fuller and L. G. Leal, "The effects of conformation-dependent friction and internal viscosity on the dynamics of the nonlinear dumbbell model for a dilute polymer solution," J. Non-Newt. Fluid Mech. **8**, 271 (1981).

70. P. G. de Gennes, " Coil-stretch transition of dilute flexible polymers under ultra-high velocity gradients," J. Chem. Phys., **60**, 5030 (1974).

71. E. J. Hinch, "Mechanical models of dilute polymer solutions in strong flows," Phys. Fluids, **20**, S22 (1977).

72. G. G. Fuller and L. G. Leal, "Flow birefringence of dilute polymer solutions in two-dimensional flows," Rheol. Acta **19**, 580 (1980).

73. P. E. Rouse, "A theory of the linear viscoelastic properties of dilute solutions of coiling polymers," J. Chem. Phys., **21**, 1272 (1953).

74. B. H. Zimm, "Dynamics of polymer molecules in dilute solution: viscoelasticity, flowbirefringence and dielectric loss," **24**, 269 (1956).

75. M. Yamamoto, "The viscoelastic properties of network structure. I. General formalism," J. Phys. Soc. Japan, **11**, 413 (1956).

76. A. S. Lodge, "A network theory of flow birefringence and stress in concentrated polymer solutions," Trans. Faraday Soc., **52**, 120 (1956).

77. G. G. Fuller and L.G. Leal, "Network models of concentrated polymer solutions derived from the Yamamoto network theory," J. Polym. Sci.: Phys. Ed., **19**, 531 (1981).

78. J. E. Mark, A. Eisenberg, W. M. Graessley, L. Mandelkern, J. L. Koenig, *Physical Properties of Polymers*, ACS, Washington (1984).

79. P.G. de Gennes, "Reptation of a polymer chain in the presence of fixed obstacles," J. Chem. Phys., **55**, 572 (1971).

80. M. Doi and S. F. Edwards, "Dynamics of concentrated polymer systems. Part I. Brownian motion in the equilibrium state," J. Chem. Soc. Faraday Trans. II, **74**, 1789 (1978); M. Doi and S. F. Edwards, "Dynamics of concentrated polymer systems. Part 2. Molecular motion under flow," J. Chem. Soc. Faraday Trans.II, **74**, 1802 (1978); M. Doi and S. F. Edwards, "Dynamics of concentrated polymer systems. Part 3. The constitutive equation," J. Chem. Soc. Faraday Trans. II, **74**, 1818 (1978); M. Doi and S. F. Edwards, "Dynamics of concentrated polymer systems. Part 4. Rheological properties," J. Chem. Soc. Faraday Trans. II, **75**, 38 (1979).

81. W. Philippoff, "Studies of flow birefringence of polystyrene solutions," Proc. 4th Int. Congr. Rheol., **2**, 343, (1965).

82. W.H. Talbott and J.D. Goddard, "Streaming birefringence in extensional flow of polymer solutions," Rheol. Acta, **18**, 507 (1979).

83. C. A. Cathey and G. G. Fuller, "The optical and mechanical response of flexible polymer solutions to extensional flow," J. Non-Newt. Fluid Mech., **34**, 63 (1990).

84. J. L. S. Wales, *The Application of Flow Birefringence to Rheological Studies of Polymer Melts*, Delft University Press, Rotterdam (1976).

85. M. Fukuda, G. L. Wilkes, and R. S. Stein, J. Polym. Sci., Polym. Phys. Ed., **9**, 1417 (1971).

86. H. Giesekus, "Elasto-vikose Flussigkeiten, fur die in stationaren Schichtstromungen samtliche normalspannungskomponenten verschieden GROb sind," *Rheol. Acta*, **2**, 50 (1962).

87. G. Marrucci and N. Grizzuti, "Predicted effect of polydispersity on rodlike polymer behaviour in concentrated solutions," J. Non-Newt. Fluid Mech., **14**, 103 (1984).

88. M. Doi and S. F. Edwards, "Dynamics of rod-like macromolecules in concentrated solution. Part I.," J. Chem. Soc. Faraday Trans. II, **74**, 560 (1978).

89. C. Rangel-Nafaile, A. Metzner, K. Wissbrun, "Analysis of stress-induced phase separations in polymer solutions," Macromolecules, **17**, 1187 (1984).

90. R. G. Larson, "Flow-induced mixing, demixing, and phase transitions in polymeric fluids," Rheol. Acta, **31**, 497 (1992).

91. N. Goldenfeld, *Lectures on phase transitions and the renormalization group*, Frontiers in Physics, Vol. 85, Addison Wesley, Reading, MA (1992).

92. E. Helfand and G. H. Fredrickson, "Large fluctuations in polymer solutions under shear," Phys. Rev. Lett., **62**, 2468 (1989).

93. A. Onuki, "Elastic effects in the phase transition of polymer solutions under shear flow," Phys. Rev. Lett., **62**, 2472 (1989).

94. S. T. Milner, "Hydrodynamics of semidilute polymer solutions," Phys. Rev. Lett., **66**, 1477 (1991).

95. Jefferey, "The motion of ellipsoidal particles immersed in a viscous fluid," Proc. R. Soc. London, A, **102**, 161 (1922).

96. E. J. Hinch and L. G. Leal, "The effect of Brownian motion on the rheological properties of a suspension of non-spherical particles," J. Fluid Mech., **52**, 683 (1972).

97. A. Okagawa and S. G. Mason, "The kinetics of flowing dispersions. VII. Oscillatory behavior of rods and discs in shear flow," J. Coll. Int. Sci., **45** 330 (1973)

98. W. B. Russel, D. A. Saville, and W. R. Schowalter, *Colloidal Dispersions*, Cambridge University Press, Cambridge (1991).

99. G. K. Batchelor and J. T. Green, "The determination of the bulk stress in a suspension of spherical particles to order c^2." J. Fluid Mech., **56**, 401 (1972).

100. W. R. Smith and D. Henderson, "Analytical representation of the Percus-Yevick hard sphere radial distribution function," Mol. Phys., **19**, 411 (1970).

101. W. Philippoff, "Flow Birefringence and Stress," J. Appl. Phys., **27**, 984, (1956).

102. J. A. van Aken and H. Janeschitz-Kriegl, Rheol. Acta, "Simultaneous measurement of transient stress and flow birefringence in one sided compression (biaxial extension) of a polymer melt," Rheol. Acta, **20**, 419 (1981).

103. W. H. Talbott and J. D. Goddard, "Streaming birefringence in extensional flow of polymer solutions," Rheol. Acta, **18**, 505 (1979).

104. F. H. Gortemaker, M. G. Hanssen, B. de Cindio, H. M. Laun, and H. Janeschitz-Kriegl, "Flow birefringence of polymer melts: application to the investigation of time-dependent rheological properties," Rheol. Acta, **15**, 256 (1976).

105. J. Miller and J. Schrag, "Oscillatory flow birefringence of low molecular weight polystyrene solutions. High frequency behavior," Macromolecules, **8**, 361 (1975); T. P. Lodge, J. Miller, and J. Schrag, "Infinite-dilution oscillatory flow birefrin-

gence properties of polystyrene and poly(methylstyrene) solutions," J. Poly. Sci. Poly. Phys. Ed., **20**, 1409 (1982).

106. G. V. Vinogradov, A. I. Isayev, D. A. Mustafaev, and Y. Y. Podolsky, "Polarization-optical investigation of polymers in fluid and high elastic states under oscillatory deformation," J. Appl. Polym. Sci., **22**, 665 (1978).

107. J. A. Zawada, *Component contributions to the dynamics of miscible polymer blends*, Ph.D. Thesis, Stanford University (1993).

108. T. Inoue, H. Hayashihara, H. Okamoto, and K. Osaki, "Birefringence of amorphous polymers. I. Dynamic measurements on polystyrene," Macromolecules, **24**, 5670 (1991): T. Inoue, H. Okamoto, and K. Osaki, "Birefringence of amorphous polymers, I. Dynamic measurement and relaxation measurement," J. Polym. Sci. Part B. Polym. Phys., **30**, 409 (1992).

109. D. W. Mead and R. G. Larson, "Rheooptical study of isotropic solutions of stiff polymers," Macromolecules, **23**,2524 (1990).

110. B. E. Zebrowski and G. G. Fuller, "Rheooptical studies of concentrated polystyrene solutions subjected to transient simple shear flow," J. Polym. Sci., Polym. Phys. Ed., **23**, 575 (1985).

111. D. S. Pearson, A. D. Kiss, L. J. Fetters, and M. Doi, "Flow-induced birefringence of concentrated polyisoprene solutions," J. Rheol., **33**, 517 (1989); D. S. Pearson, E. Herbolzheimer, N. Grizzuti, and G. Marrucci, "Transient behavior of entangled polymers at high shear rates," J. Polym. Sci., Part B, **29**, 1589 (1991).

112. R. G. Larson, S. A. Khan, and V. R. Raju, "Relaxation of stress and birefringence in polymers of high molecular weight," J. Rheol., **32**, 145 (1988).

113. A.W. Chow, and G.G. Fuller, "The rheo-optical response of rod-like chains subject to transient shear flow. Part I: Model calculations on the effects of polydispersity," Macromolecules **18**, 786 (1985); A.W. Chow, G.G. Fuller, D.G. Wallace and J.A. Madri, "The rheo-optical response of rod-like chains subject to transient shear flow. Part II. Two-color flow birefringence measurements," Macromolecules **18** , 793 (1985); A.W. Chow, G.G. Fuller, D.G. Wallace and J.A. Madri, "The rheo-optical response of rod-like shortened collagen protein to transient shear flow," Macromolecules, **18**, 805 (1985).

114. G. Murrucci and N. Grizzuti, "The effect of polydispersity on rotational diffusivity and shear viscosity of rodlike polymer in concentrated solutions," J. Polym. Sci., Polym. Lett. Ed., **21**, 83 (1983).

115. A. J. McHugh, M. E. Mackay, and B. Khomami, "Measurement of birefringence by the method of isoclinics," J. Rheol., **31**, 619 (1987); errata **32**, 813 (1988).

116. N. Checker, M. R. Mackley, and D. W. Mead, "On the flow of molten polymers into, within and out of ducts," Philos. Trans. R. Soc. London, Ser. A, **308**, 451 (1983).

117. J.S. Lee and G.G. Fuller, "The spatial development of transient couette flow and shear wave propagation in polymeric liquids by flow birefringence," J. of Non Newt. Fluid Mech., **26**, 57 (1987).

118. J. S. Lee and G. G. Fuller, "Shear wave propagation in polymer solutions following a step increase of shear rate," J. Non-Newt. Fluid Mech., **39**, 1 (1991).

119. D. Rajagopalan, J. A. Byars, R. C. Armstrong, R. A. Brown, J. S. Lee, and G. G. Fuller, "Comparison of numerical simulations and birefringence measurements in viscoelastic flow between eccentric rotating cylinders," J. Rheology, **36**, 1349 (1992).

120. D. D. Joseph, O. Riccius, and M. Arney, "Shear-wave speeds and elastic moduli for different liquids. Part 2. Experiments," J. Fluid Mech., **171**, 309 (1986).

121. S. R. Galante and P. L. Frattini, "Spatially resolved birefringence studies of planar entry flow," J. Non-Newt. Fluid Mech., **47**, 289 (1993).

122. M. M. Denn, "Issues in viscoelastic fluid mechanics," Annu. Rev. Fluid Mech., **22**, 13 (1990).

123. D. G. Kiriakidis, H. J. Park, E. Mitsoulis, B. Bergnes, and J.-F. Agassant, "A study of stress distribution in contraction flows of an LLDPE melt," J. Non-Newt. Fluid Mech., **47**, 339 (1993).

124. D. G. Baird, M. D. Read, and J. N. Reddy, "Comparison of flow birefringence data with a numerical simulation of the hole pressure," J. Rheol., **32**, 621 (1988).

125. L. Monnerie, "Segmental orientation and chain relaxation of polymer by spectroscopic techniques: A molecular approach to polymer viscoelasticity," in *Developments in Oriented Polymers - 2*, I. M. Ward, ed., page 199, Elsevier Applied Science, London and New York (1987).

126. J. F. Tassin and L. Monnerie, "A fluorescence polarization study of matrix molecular weight on the relaxation of a labeled molecule in a stretched polymer melt," J. Polym. Sci. Polym. Phys. Ed., **21**, 1981 (1983).

127. Y. Zhao, B. Jasse, and L. Monnerie, "Orientation and relaxation in uniaxially stretched poly(methyl methacrylate)-poly(ethylene oxide) blends," Polymer, **30**, 1643 (1989).

128. B. Amram, L. Bokobza, P. Sergot, and L. Monnerie, "Effect of temperature on intermolecular orientational correlations between chain segments in strained polyisoprene: A Fourier transform infrared dichroism investigation," Macromolecules, **23**, 1212 (1990).

129. J.-P. Jarry and L. Monnerie, "Effects of a nematic-like interaction in rubber elasticity theory," Macromolecules, **12**, 316 (1979).

130. W. W. Merrill, M. Tirrell, J.-F. Tassin and L. Monnerie, "Diffusion and relaxation in oriented polymer media," Macromolecules, **22**, 896 (1989).

131. B. Deloche and E. T. Samulski, "Short-range nematic-like orientational order in strained elastomers: A deuterium magnetic resonance study," Macromolecules, **14**, 575 (1981).

132. J.A. Kornfield, G.G. Fuller, and D.S. Pearson, "Infrared dichroism measurements of molecular relaxation in binary blend melt rheology," Macromolecules, **22**, 1334 (1989).

133. M. Doi, D. Pearson, J. Kornfield, and G. Fuller, "Effect of nematic interaction in the orientational relaxation of polymer melts," Macromolecules, **22**, 1488 (1989).

134. J. A. Kornfield, G. G. Fuller, and D. S. Pearson, "Third normal stress difference and component relaxation spectra for bidisperse melts under oscillatory shear," Macromolecules, **24**, 5429 (1991).

135. C. M. Ylitalo, G. G. Fuller, V. Abetz, and R. Stadler, and D.S. Pearson, "Relaxation dynamics of selected polymer chain segments and comparison with theoretical models," Rheol. Acta, **29**, 543 (1990).

136. C. M. Ylitalo, J. A. Kornfield, G. G. Fuller, and D. S. Pearson, "Molecular weight dependence of component dynamics in bidisperse melt rheology," Macromolecules, **24**, 749 (1991).

137. C. Ylitalo and G.G. Fuller, "Temperature effects on the magnitude of orientational coupling interactions in polymer melts," Macromolecules, **24**, 5736 (1991).

138. C. Ylitalo, J.A. Zawada, G.G. Fuller, and V. Abetz, "Oligomers as molecular probes of orientational coupling interactions in polymer melts and networks," Polymer, **33**, 2949 (1992).

139. J. Zawada, C. Ylitalo, G. Fuller, R. Colby, and T. Long, "Component relaxation dynamics in a miscible polymer blend: Poly(ethylene oxide)/poly(methyl methacrylate)," Macromolecules, **25**, 2896 (1992).

140. Noda, I., "Two dimensional infrared spectroscopy," J. Am. Chem. Soc., **111**, 8116 (1989).

141. Noda, I., "Two-dimensional infrared spectroscopy: Theory and application," Applied Spectroscopy, **44**, 550 (1990).

142. Noda, I., "Generalizd two-dimensional correlation method applicable to infrared, Raman, and other types of spectroscopy," Applied Spectroscopy, **47**, 1329 (1993).

143. Huang, K., Ph.D. Thesis, Department of Chemical Engineering, Stanford University, 1995.

144. A. Keller and J. Odell, "The extensibility of macromolecules in solution; a new focus for macromolecular science," Coll. Polym. Sci., **263**, 181 (1985).

145. D. P. Pope and A. Keller, "Alignment of macromolecules in solution by elongational flow; a study of the effect of pure shear in a four roll mill," Coll. Polym. Sci., **255**, 633 (1977).

146. C.J. Farrell, A. Keller, M.J. Miles, and D.P. Pope, "Conformational relaxation time in polymer solutions by elongational flow experiments: 1. Determination of extensional relaxation time and its molecular weight dependence," Polymer, **21**, 1292 (1980).

147. W. L. Olbricht, J. M. Rallison, and L. G. Leal, "Strong flow criteria based on microstructure deformation," J. Non-Newt. Fluid Mech., **10**, 291 (1982).

148. M. J. Menosveta and D. A. Hoagland, "Light scattering from dilute poly(styrene) solutions in uniaxial extensional flow," Macromolecules, **24**, 3427 (1991).

149. P. N. Dunlap and L. G. Leal, "Dilute polystyrene solutions in extensional flow: birefringence and flow modification," J. Non-Newt. Fluid Mech., **23**, 5 (1987). E. Geffroy and L. G. Leal, "Flow birefringence studies of a concentrated polystyrene solution in a two-roll mill. 1. Steady flow and start-up of steady flow," J. Polym. Sci., Polym. Phys., **30**, 1329 (1992).

150. A. Keller, A. J. Muller, and J. A. Odell, "Entanglements in semi-dilute solutions as revealed by elongational flow studies," Prog. Coll. Polym. Sci., **75**, 179 (1987).

151. O. G. Harlen, E. J. Hinch, and J. M. Rallison, "Birefringent pipes: The steady flow of a dilute polymer solutions near a stagnation point," J. Non-Newt. Fluid Mech., **44**, 229 (1992).

152. G. ver Strate and W. Philippoff, "Phase separation in flowing polymer solutions," Polym. Lett. Ed., **12**, 267 (1974).

153. T. Hashimoto and T. Kume, "'Butterfly' light scattering pattern in shear-enhanced concentration fluctuations in polymer solutions and anomaly at higher shear rates," J. Phys. Soc. Japan, **61**, 1839 (1992).

154. H. Yanase, P. Moldenaers, V. Abetz, J. van Egmond, G. G. Fuller, and J. Mewis, "Structure and dynamics of a polymer solution subject to flow-induced phase separation," Rheol. Acta, **30**, 89 (1991).

155. X.-L. Wu, D. J. Pine, and P. K. Dixon, "Enhanced concentration fluctuations in polymer solutions under shear flow," Phys. Rev. Lett., **66**, 2408 (1991).

156. F. Boue, J. Bastide, M. Buzier, A. Lapp, J. Herz and T.A. Vilgis, "Strain induced large fluctuations during stress relaxation in polymer melts observed by small-angle neutron scattering. 'Lozenges', 'butterflies', and related theory," Coll. Polym. Sci., **269**, 195 (1991).

157. D. Wirtz, K. Berend and G.G. Fuller, "Electric field induced structure in polymer solutions near the critical point," Macromolecules, **25**, 7234-7246 (1992); D. Wirtz and G. G. Fuller, "Phase transitions induced by electric fields in near-critical polymer solutions," Phys. Rev. Lett., **4**, 2236 (1993).

158. P. Navard, "Rheo-optical and rheo-X-ray investigations of liquid crystalline polymers," in refenence 18.

159. G. Kiss and R. S. Porter, "Rheology of concentrated solutions of poly (γ-benzyl-glutamate)," J. Polym. Sci., Polym. Symp., **65**, 193 (1978).

160. P. Navard, "Formation of band textures in hydroxypropylcellulose liquid crystals," J. Polym. Sci., Phys. Ed., **24**, 435 (1986).

161. P. Moldenaers, "Rheological behavior of lyoropic polymeric liquid crystals," Ph.D. thesis, Katholiecke Universiteit Leuven (1987).

162. R. G. Larson and D. W. Mead, "Time and shear-rate scaling laws for liquid crystal polymers," J. Rheol., **33**, 1251 (1989).

163. F. M. Leslie, "Theory of flow phenomena in liquid crystals," Adv. Liq. Cryst., **4**, 1 (1979).

164. P. G. de Gennes, *The Physics of Liquid Crystals*, Oxford University Press, New York (1974).

165. R. G. Larson and M. Doi, "Mesoscopic domain theory for textured liquid crystalline polymers," J. Rheol., **35**, 539 (1991).

166. M. Doi, "Rheological properties of rodlike polymers in isotropic and liquid crystalline phases," Ferroelectrics, **30**, 247 (1980).

167. M. Marrucci and P. L. Maffettone, "Description of the liquid-crystalline phase of rodlike polymers at high shear rates," Macromolecules, **22**, 4076 (1989).

168. R. G. Larson, "Arrested tumbling in shearing flows of liquid crystal polymers," Macromolecules, **23**, 3983 (1990).

169. T. Asada, H. Muramatsu, R. Watanabe, and S. Onogi, "Rheo-optical studies of racemic poly (γ-benzyl glutamate) liquid crystals," Macromolecules, **13**, 867 (1980).

170. K. Hongladarom and W. R. Burghardt, "Molecular alignment of polymer liquid crystals in shear flows. 2. Transient flow behavior in poly(benzyl glutamate) solutions," Macromolecules, **26**, 785 (1993).

171. P. Moldenaers, G. G. Fuller and J. Mewis, "Mechanical and optical rheometry of polymer liquid-crystal domain structure," Macromolecules, **22**, 960 (1989).

172. W. R. Burghardt and G. G. Fuller, "Role of director tumbling in the rheology of polymer liquid crystal solutions," Macromolecules, **24**, 2546 (1991).

173. B. Ernst and P. Navard, "Shear flow of liquid-crystalline polymer solutions as investigated by small-angle light scattering techniques," Macromolecules, **23**, 1370 (1990).

174. L. Nunnelley and G.G. Fuller, "Optical measurements of particle orientation in magnetic media," J. Appl. Phys., ***63***, 1687 (1988).

175. M. M. Fromjmovic, A. Okagawa, and S. G. Mason, "Rheo-optical transients in erythrocyte suspensions," Biochem. and Biophys. Research Comm., **62**, 17 (1975).

176. A. Okagawa and S. G. Mason, "Kinetics of flowing dispersions. X. Oscillations in optical properties of streaming suspensions of spheroids," Can. J. Chem., **55**, 4243 (1977).

177. P. L. Frattini and G. G. Fuller, "Rheo-optical studies of the effect of weak Brownian rotations in sheared suspensions," J. Fluid Mech., **168**, 119 (1986).

178. S. J. Johnson, A. J. Salem, and G. G. Fuller, "Dynamics of colloidal particles in sheared, non-Newtonian fluids," J. Non. Newt. Fluid Mech., **34**, 89 (1990).

179. B. J. Ackerson and N. A. Clark, "Shear-induced melting," Phys. Rev. Lett., **46**, 123 (1981).

180. N.J. Wagner, G.G. Fuller, and W.B. Russel, "The dichroism and birefringence of a hard-sphere suspension under shear," J. Chem. Phys., **89**, 1580 (1988).

181. P. D'Haene, J. Mewis, and G.G. Fuller, "Scattering dichroism measurements of flow-induced structure of a shear thickening suspension," J. of Col. and Interf. Sci., **156**, 350 (1993).

182. J. J. L. Higdon, "The kinematics of the four roll mill," Phys. Fluids A, **5**, 1 (1993).

183. J. O. Park and G. C. Berry, "Moderately concentrated solutions of polystyrene. III. Viscoelastic properties at thc Flory theta temperature", Macromolecules, **22**, 3022 (1989).

184. U. Seidel, R. Stadler, and G. G. Fuller, "Relaxation of Bidisperse Temporary Networks," Macromolecules, **27**, 2066 (1994).

185. A. N. Semonov, Sov. Phys. JETP, **61**, 732 (1985).

186. K. Almdal, J. H. Rosedale, F. S. Bates, G. D. Wignall, and G. H. Fredrickson, Gaussian-to stretch-coil transition in block copolymer melts," Phys. Rev. Lett., **65**, 1112 (1990).

187. T. P. Lodge and G. H. Fredrickson, "Optical anisotropies of tethered chains," Macromolecules, **25**, 5643 (1992).

188. M. J. Folkes and A. Keller, "The birefringence and mechanical properties of a 'single crystal' from a tri-block copolymer," Polymer, **12**, 222 (1971); M. J. Folkes and A. Keller, "Optical and swelling properties of macroscopic 'single crystals' of an S-B-S copolymer. I. Samples possessing a lamellar morphology," *J. Polym. Sci., Polym. Phys. Ed.*, **14**, 833 (1976).

189. L. A. Archer and G. G. Fuller, "Segment orientation in a quiescent block copolymer melts studied by Raman scattering," Macromolecules, **27**, 4359, 1994.

190. T. Inoue, H. Okamoto, and K. Osaki, "Birefringence of amorphous polymers. I. Dynamic measurements of polystyrene," Macromolecules, **24**, 5670 (1991).

191. J. P. W. Baaijens, *Evaluation of Constitutive Equations for Polymer Melts and Solutions in Complex Flows*, Eindhoven University of Technology, Department of Mechanical Engineering, Eindhoven, The Netherlands (1994).

192. L. M. Quinzani, R. C. Armstrong, and R. A. Brown, Birefringence and laser-Doppler velocimetry studies of viscoelastic flow through a planar contraction," J. Non-Newt. Fluid Mech., **52**, 1 (1994).

193. N. Phan Thien and R. I. Tanner, "A new constitutive equation derived from network theory," J. Non-Newt. Fluid Mech., **2**, 353 (1977).

Authors Cited

A

B

C

D

E

F

G

H

I

J

K

L

M

S

T

V

W

Y

Z

Index

S

T

V

W